be re____ d on or before
____ate ____ ____low.

Electronic Systems

M. W. BRIMICOMBE

M.A., D.Phil.

Nelson

Thomas Nelson and Sons Ltd
Nelson House Mayfield Road
Walton-on-Thames Surrey
KT12 5PL UK

Thomas Nelson Australia
102 Dodds Street
South Melbourne
Victoria 3205 Australia

Nelson Canada
1120 Birchmount Road
Scarborough Ontario
M1K 5G4 Canada

© M. W. Brimicombe 1985

First published by Thomas Nelson and Sons Ltd 1985

I(T)P Thomas Nelson is an International
Thomson Publishing Company

I(T)P is used under licence

ISBN 0-17-448067-9
NPN 15 14 13 12

Printed in China

Contents

Introduction

This book aims to teach electronics to a beginner, giving a thorough introduction to the principles and practices of modern electronics. No previous knowledge or experience of electronics is assumed, although it would probably help if you have studied O-level physics and maths. The contents of the book have been chosen so that they cover the requirements of the A-level electronics syllabus of the Cambridge Examinations Board and a large part of the AEB's electronics systems A-level syllabus. It discusses much of the material which is offered to university and polytechnic students during the first year of an electronics degree course and it will be helpful to students who are studying electronics at colleges of further education. Furthermore, the book will be of value to students of physics and engineering who need to know about electronics for their degree courses.

The emphasis, throughout the book, is on a systems approach to the subject. This means that we have concentrated on describing the function and use of a device, and have completely ignored the physics behind its operation. So the book treats all of the electronic components as black boxes, and shows how they can be assembled to make useful systems. We have been forced to adopt a systems approach because electronic technology is changing very rapidly. There is no point in learning about what makes an electronic component work if that knowledge is soon going to be redundant! You will find that if you know what a device does and how its behaviour is modified when it is linked to other devices, that is all you need to know in order to design a functional electronic system. Furthermore, although the manufacture of a device may change, its function will not. By concentrating on what happens when systems are linked together, your knowledge of electronics will remain useful to you over the years.

You cannot learn about electronics by simply reading about it. So each chapter is punctuated in several places with batches of problems. These have been designed to help you increase your understanding of the preceding section, so you ought to have a go at all of them. Answers (a bit sketchy at times) are provided at the back of the book. At the end of each chapter is a set of revision questions. These act as a summary; you should be able to answer all of them with ease once you have mastered the contents of that chapter.

Vitally important, are the experiments at the end of each section of a chapter. An experiment may invite you to assemble a circuit which has been discussed in the preceding section so that you can verify its behaviour. Or you may be asked to measure some properties of a component or system. You may even be told to assemble a circuit of your own design to perform a particular function. Since electronics is meaningless unless you have a feel for its practical aspects, it is crucial that you assemble as many circuits as you can! Appendix B contains details of all the integrated circuits and other components that you will need. Appendix C describes how to assemble circuits on solderless breadboard and explains how to go about fault finding.

The microprocessor chapter has no experiments, but you should have no difficulty in adapting our program suggestions for use with any microprocessor system that you can get access to. Details of a system which is similar to the one described in Chapter Eleven are given in Appendix B.

We have written the book on the assumption that when you read a particular chapter you are familiar with the contents of all the preceding ones. Nevertheless, the last three chapters could be tackled in any order, and the chapters on audio systems and FETs could be left until later. All of the calculus has been banished to Appendix A, but some trigonometry has been left in the main section of the book. Throughout, we have tried to emphasise that the calculations needed for electronics need not be precise.

1
Logic systems

What does the title of this chapter mean? Why have we chosen it to be the first one in the book?

This chapter is typical of most of the others in the book. It starts off by introducing a very simple example of a **system** which obeys the rules of formal logic. Then it shows how a number of these elementary systems can be put together to make a new system which appears to be more than just the sum of its parts. Furthermore, it shows how to connect a number of the new systems to make an even more complex system. This approach illustrates the few central concepts which lie behind electronic circuit design. It should help you to appreciate that any complex system can easily be understood if you break it up into a number of inter-related simple systems.

By the time you have worked through this chapter you will know, for example, how complex memory systems which can store thousands of bits of information can be assembled from devices called multiplexers and demultiplexers. You will know how to make a complex multiplexer out of a number of simple ones, and how to build a simple multiplexer out of NAND gates. The NAND gate is the simplest useful logic system. Any other logic system can be built out of NAND gates, so it is effectively a basic component of such a system.

This chapter will not tell you how to break up a NAND gate into its constituent parts. We are going to adopt a **systems approach** and treat the NAND gate as a **black box.** (The interior of that black box will be discussed in the chapter which deals with transistors.) We are going to have to introduce you to the basic ideas of current electricity before you can make much headway with logic systems, and leave the complex behaviour of electronic components until you are familiar with those ideas.

All of the concepts which you will meet, perhaps for the first time, in this chapter will reappear in subsequent chapters. There they will be built upon, elaborated and used in different contexts. You will become aware of the existence of a small number of central ideas in electronics which can be used to design, or understand, any complex electronic system.

SWITCH/RESISTOR LOGIC SYSTEMS

The simplest logic systems are made out of switches and resistors. Before we try to explain what a logic system is, we are going to go over some basic electricity. Then you will be able to appreciate how switches and resistors can be used to construct circuits which obey the rules of logic.

ELECTRICITY

Look at the **circuit diagram** of figure 1.1. Circuit diagrams have two functions in electronics. Firstly they are a series of coded instructions which tell
you how to assemble the circuit. Secondly they allow you to work out how
the system will behave if it is assembled. So before we go any further, it is
worthwhile ensuring that you are quite clear about what figure 1.1 shows.

The two horizontal lines in the diagram are the **supply rails** of the circuit.
They are held at different **voltages** by a **power supply** that is not shown in
the diagram. That power supply holds the top rail at +5 V (plus five volts)
relative to the bottom rail at 0 V (sometimes called **ground** or **earth**). All of
the other lines drawn in the diagram represent wires that can carry electric
current. Each wire is at a definite voltage, and every bit of the same wire
will be at the same voltage. Between the two supply rails lie two components, a **push switch** and a **resistor**. Finally, there is an output terminal
which is labelled Q. It can be connected to a piece of test equipment, such
as a voltmeter (not shown) to find out what its voltage is relative to ground.

In order to sort out what happens to Q as the switch is pressed and
released, you need to understand something about voltage and current.
You need to have a **model** which allows you to work out the behaviour of
an electrical system. A full understanding of the concepts of voltage,
current and resistance belongs to the realm of physics. Electronics requires
a set of rules and is not too interested in the laws of nature behind those
rules! So we are not going to define the words current and voltage in
precise, unambiguous terms. Instead, we shall offer useful pictures for you
to keep in your mind's eye when the words voltage and current are used.

The voltage of a wire is a property conferred upon it by the power
supply, rather like the way in which its temperature is fixed by the central
heating supply of the room. It helps to think of the circuit diagram built out
of hollow pipes and stood up with the 0 V supply rail on the floor. The
voltage of a part of the circuit is then represented by its height above the
ground.

The current which flows through a wire is a measure of the rate at which
charge flows through it. We are going to imagine that charge is a fluid
which can flow through certain materials (**conductors**) and not through
others (**insulators**). Think of charge as water completely filling the hollow
tubes of our model. The water will flow down the tubes from the high parts
to the low parts of the model. It should be obvious that the power supply
has to ensure that charge is pushed into the top supply rail as fast as it
flows out of the bottom supply rail if the flow of charge through the system
is to be maintained.

Resistance

Now let's try to sort out how the system of figure 1.1 behaves when the
switch is pressed. The push switch simply connects the two wires on either
side of it when it is pressed. Figure 1.2 shows that Q will be directly connected to the 0 V supply rail when the switch is pressed. So Q will be at
0 V.

What happens to the rest of the circuit when the switch is pressed? Look
at figure 1.3. The resistor has a voltage of +5 V at its top end and a
voltage of 0 V at the other end. This means that an electric current has to
flow through it. The direction in which the charge flows is shown with
arrows drawn on the wires. So the current (i.e. charge per second) flows out

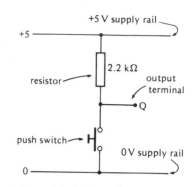

Figure 1.1 A circuit diagram

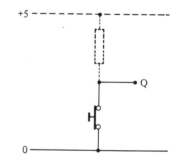

Figure 1.2 A circuit with the
switch closed

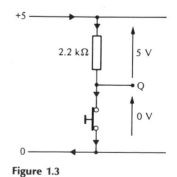

Figure 1.3

of the top supply rail, through the resistor, through the switch and into the bottom supply rail. The amount of current flowing out of the top supply rail is always precisely equal to the amount of current flowing into the bottom supply rail. You never gain or lose any charge as it flows down through the circuit.

We can calculate the size of the current with a formula. It could be called the **resistance formula**, but it is common practice in electronics to call it **Ohm's Law**. Strictly speaking, Ohm's Law says something completely different; it states that the resistance of a chunk of metal is independent of the current flowing through it, provided that it does not heat up. Of course, for this to make any sense you need to be able to calculate the resistance of something. You do that with the following formula;

$$R = \frac{V}{I}$$

where R is the resistance of the object,
$\quad\quad V$ is the voltage between its ends, and
$\quad\quad I$ is the current which flows through it.

If the voltage is measured in volts and the current in amps, then the resistance will be measured in ohms. These units are not very convenient for electronics. We will therefore use the following units throughout the rest of this book; V in **volts**, I in **milliamps** and R in **kilohms**. (One milliamp (mA) is 10^{-3} amps and one kilohm (kΩ) is 10^3 ohms.)

Now let's use the formula to calculate how much current flows through the resistor of figure 1.3. The voltage drop across the resistor is 5 V, so $V = 5$. We are told that the resistance of the resistor is 2.2 kilohms, so $R = 2.2$.

$$R = \frac{V}{I}$$

$$\text{Therefore} \quad 2.2 = \frac{5}{I}$$

$$\text{Therefore} \quad I = \frac{5}{2.2} = 2.3$$

So a current of 2.3 milliamps (or mA) flows through the resistor.

Pull-up resistors

Before we leave the circuit of figure 1.1 we need to work out what happens to Q when the switch is not being pressed. When the switch is open, no current can flow through it. It is an insulator. So an open switch means that no current flows into the bottom supply rail. Therefore the current which flows through the 2.2 kilohm resistor is zero i.e. $I = 0$. Using Ohm's Law to calculate the voltage drop across the resistor;

$$R = \frac{V}{I}$$

$$\text{Therefore} \quad 2.2 = \frac{V}{0}$$

$$\text{Therefore} \quad V = 2.2 \times 0 = 0$$

So there is no voltage drop across the resistor. Q must be at the same voltage as the top supply rail. Look at figure 1.4. The resistor pulls Q up to +5 V when the switch is open. Resistors which have this function in a circuit are called **pull-up resistors**; in the absence of any current flow through them they tie a point in a circuit to the top supply rail. You can probably work out what a pull-down resistor does.

BLACK BOXES

Our analysis of the circuit of figure 1.1 has been fairly lengthy. It illustrates how the behaviour of a system can be worked out if you understand what each of its components does and the laws of electric current. Once we have decided how the output states of the system depend on its input states we can replace the whole system, in our minds eye, with a **black box**. Such a black box is shown in figure 1.5. The word black indicates that we do not see the insides of the device, only the terminals which go in and out of it. The system has one input (A) and one output (Q). A represents the push switch and Q represents the output wire. The overall behaviour of the system is as follows; when A is pressed Q is at 0 V and when A is not pressed Q is at +5 V.

Note that the diagram does not represent an electrical circuit directly. The arrows on the input and output lines show the direction in which information travels through the system, not the flow of electric current.

The systems approach

Why bother with black boxes? Electronic circuit design is primarily concerned with the function of a system, rather than the individual components which are contained within it. By banishing the components into a black box we can turn our full attention to the behaviour of the system in its entirety. In fact, once we know what sort of input signals the system will notice (the state of a switch), the type of signals it feeds out (the voltage of a wire) and how the output is related to the input, the contents need not concern us. Indeed, it is quite common in electronics for several circuits to look exactly the same once they are reduced to a black box. For example, figure 1.6 shows three circuits which behave, in electronic terms, just like the one shown in figure 1.1.

It should be obvious that the use of black boxes means that our understanding of the electronic behaviour of a system becomes independent of the technology used to actually build the system. This is a very important freedom as the technology of electronics is changing very rapidly and is likely to carry on changing in the foreseeable future. If we concentrate on the function of a system and do not worry too much about its technology, the skills that we acquire in electronics are not going to be out of date

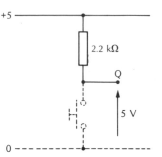

Figure 1.4 A circuit with the switch open

Figure 1.5 A black box

Figure 1.6 Three circuits with the same function

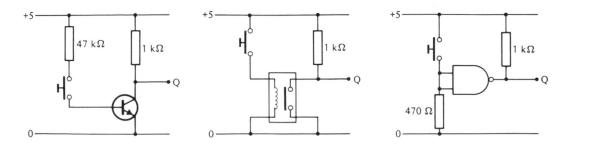

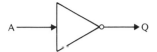

Figure 1.7 A NOT gate

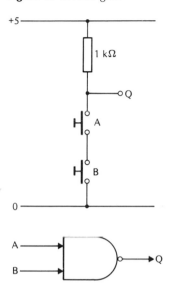

Figure 1.8 A switch/resistor NAND gate

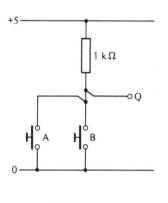

Figure 1.9 A switch/resistor NOR gate

when the next advance in semiconductor manufacture makes the currently available devices obsolete.

LOGIC SYSTEMS

A logic system is one whose outputs can only have one of two **states**. So the black box of figure 1.5 must be a logic system as Q is either at +5 V or 0 V. The two states of a logic system are given special names. An output of +5 V is called **logical 1**; 0 V is called **logical 0**. The names of the two states are often abbreviated to **1** or **0**. Although the outputs of a logic system are always at 1 or 0, not all systems have +5 V for 1 and 0 V for 0.

There are various standard logic systems. The simplest one of all is called a **NOT gate**; its black box symbol is shown in figure 1.7. The switch/resistor network which we have been considering behaves like a NOT gate; its output is 1 when the switch is not pressed.

Another standard logic system – perhaps the most important – is the **NAND gate**. A switch/resistor version of a NAND gate, together with its symbol, is shown in figure 1.8. If you study the circuit diagram you will see that current can only flow through the resistor when both switches are pressed. So Q is pulled up to logical 1 by the resistor unless A and B are pressed. In black box terms, a NAND gate will only feed out a logical 0 when both of its inputs are signalled.

Finally, figure 1.9 shows what a switch/resistor **NOR gate** looks like. A bit of thought should convince you that a black box NOR gate only feeds out a logical 1 when neither of its inputs are signalled.

Linking logic systems

Switch/resistor logic gates have mechanical inputs and electrical outputs. They are very useful when human beings need to feed information into electronic systems. Fingers appear to be perfectly adapted to pressing switches, and banks of buttons in the form of calculator keypads or computer keyboards, are very good at interfacing human beings to complex electronic devices. Indeed, you will be using the switch/resistor NOT gate a lot as you work your way through the experiments described in this book.

Switch/resistor logic does, however, suffer from one serious drawback. The output of one unit cannot be directly fed into the inputs of similar units. This means that we cannot link the gates together to make larger systems. Instead, a large system has to be designed in one fell swoop. This lack of flexibility means that switch/resistor logic gates can only perform simple functions. If you want a complex logic system you are going to have to use gates which have the same type of input and output.

For example, look at the logic system shown in figure 1.10. Don't worry about what the whole thing does (Q is 1 when A and B are the same!). A

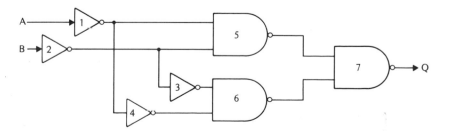

Figure 1.10 A logic system

and B could be mechanical inputs, so that gates 1 and 2 could be switch/resistor NOT gates. All the other gates, however, have to be purely electrical as they have to feed their outputs into similar units. You can see that once you have worked out how to assemble an all electric logic gate, the sky is the limit as far as the complexity of a logic system is concerned.

PROBLEMS

1 Study the circuits shown in figure 1.11. State in words how the state of Q (1 or 0) depends on the state of the switches. Calculate how much current flows through the resistor when Q is 1.

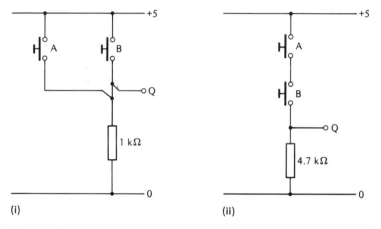

(i)   (ii)

Figure 1.11

2 A circuit has three push switches A, B and C. Its output, Q, is 1 only when A is pressed and either B or C is pressed. So Q is 1 when A and B are pressed, or when A and C are pressed; or when all three are pressed; otherwise it is 0. Draw the circuit diagram.

3 An electronic door lock has a bank of nine switches. In order to open the lock, three specific switches have to be simultaneously pressed; a logical 0 will open the lock mechanism. If any of the other six switches are pressed, an alarm is triggered; a logical 1 will trigger the alarm. Draw a circuit diagram of the system and explain how it works. Use only switches and resistors.

RELAY LOGIC SYSTEMS

There are many ways in which you can assemble a purely electrical logic system. The first way in which it was done used **relays**. These are electrically controlled switches and are used extensively in telephone networks. Although logic systems based on relays are more or less obsolete (modern integrated circuits function much better), they are fairly easy to understand. So after we have discussed some of the properties of logic systems in general, we shall illustrate how they can be built out of relays.

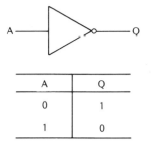

A	Q
0	1
1	0

Figure 1.12 The truth table for a NOT gate

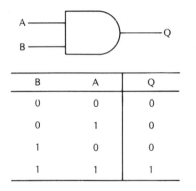

B	A	Q
0	0	0
0	1	0
1	0	0
1	1	1

Figure 1.13 The truth table for an AND gate

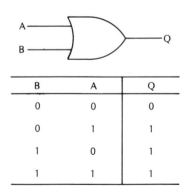

B	A	Q
0	0	0
0	1	1
1	0	1
1	1	1

Figure 1.14 The truth table for an OR gate

BASIC FUNCTIONS

Any logic system can be built out of just three basic components. They are called gates. More precisely, the **NOT gate**, the **AND gate** and the **OR gate**. They can be strung together because their inputs can be fed from the outputs of other gates.

NOT gates

A NOT gate has one input (A) and one output (Q). The output always has the opposite state to the input. So if A is logical 0, then Q will be logical 1. On the other hand, if A is logical 0 then Q will be logical 1. Q equals NOT A.

The behaviour of a logic gate is best summarised in a **truth table**. The truth table of a NOT gate is shown in figure 1.12. Each row of the table represents one of the possible input states with its consequent output states. In the case of a NOT gate, there is only one input (with two possible states) so the truth table has only two rows.

AND gates

Figure 1.13 illustrates the symbol of an AND gate, and gives its truth table. As you can see, the output of the gate (Q) is only a logical 1 when both of its inputs (A and B) are logical 1. The behaviour of the gate is conveniently encapsulated in this statement; Q equals A AND B. In other words, Q is 1 when both A is 1 and B is 1, otherwise Q is 0.

OR gates

The output of an OR gate is a logical 1 when either, or both, of its inputs are at logical 1. Otherwise the output is logical 0. The truth table and symbol are shown in figure 1.14. The gate shown in that diagram can be described by this statement; Q equals A OR B.

Solving a problem

Let us suppose that we want to design a logic system which will feed out a logical 1 when its two inputs have the same state. How do we do it with NOT, AND and OR gates?

Start off by labelling the inputs and outputs. We will call the inputs A and B, and the output Q. Then write down how the output state depends on the input states. Q is 1 when A is 1 and when B is 1. Q is also 1 when A is 0 and when B is 0. Otherwise Q is 0.

The next step is tricky. Using our NOT, AND and OR jargon, we can describe the behaviour of the system with this statement;

Q equals (A AND B) OR ((NOT A) AND (NOT B))

The brackets show the order in which the operations have to be done. So you have to NOT A and NOT B before you AND them. The NOT/AND/OR statement is an abbreviation of our description of how Q depends on A and B; note how NOT A is used when A is 0.

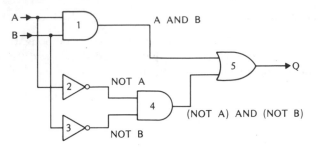

Figure 1.15

Finally, you draw a circuit diagram like the one shown in figure 1.15. Gate 1 generates (A AND B). Gates 2, 3 and 4 generate ((NOT A) AND (NOT B)). These two signals are combined by gate 5, an OR gate, to generate Q.

Analysing a system

This method of writing down a NOT/AND/OR statement can also be used to analyse the behaviour of a logic system. That is, you can use it to work out the truth table of a logic system.

For example, consider the circuit shown in figure 1.16. The output of the OR gate is A OR B. This is fed into one input of an AND gate and NOT B is fed into the other. So Q equals (A OR B) AND (NOT B). Now we expand this jargon into standard English. Q is 1 when A is 1 or when B is 1, provided that B is also 0. This can then be used to fill in the truth table. If you look at figure 1.16, you will see that Q is 1 when A is 1 and B is 0. A little thought should convince you that this matches our English description of the circuit's behaviour.

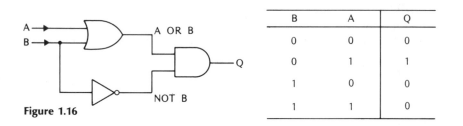

Figure 1.16

B	A	Q
0	0	0
0	1	1
1	0	0
1	1	0

Later on in this chapter you will be introduced to better ways of designing and analysing logic systems. Those methods involve the use of **Boolean algebra**, a branch of mathematics specifically designed to simplify the whole business of combining logic gates together. The NOT/AND/OR method gets very cumbersome when it is applied to complex logic systems. Nevertheless, it is a useful starting point in understanding how logic gates behave when they are connected to each other.

RELAY GATES

We are now going to show you how NOT, AND and OR gates can be built out of relays. You would never use such gates in practice because integrated circuits are a much neater alternative, but a brief study of how they work will help you to understand how logic gates can be built out of simple components.

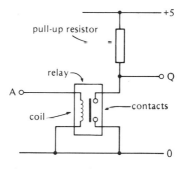

Figure 1.17 A relay NOT gate

Relay NOT gates

Look at figure 1.17. The circuit contains a relay. This is a **switch** which is controlled by an **electromagnet**. The switch is normally open, as drawn in its circuit symbol. If enough current flows through the **coil** of the electromagnet the switch is pulled closed. So if the input A is held at logical 0 by some device not shown in the diagram, the switch will be open. The resistor will therefore pull Q up to logical 1. If A is now raised to logical 1 i.e. $+5$ V, one end of the coil will be at a higher voltage than the other. So current will flow through it, the switch will close and Q will be held down at 0 V i.e. Q will be logical 0.

So the state of Q is the opposite of the state of A. The circuit has the characteristic behaviour of a NOT gate.

Relay AND and OR gates

Figures 1.18 and 1.19 illustrate how two relays and a pull-down resistor can be arranged to make systems which behave like AND and OR gates. Study each one carefully and convince yourself that they do work.

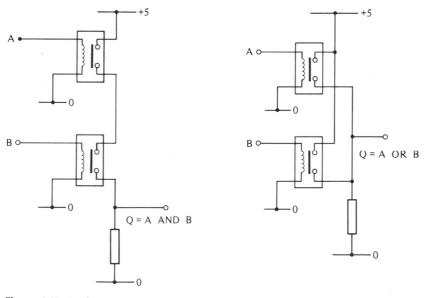

Figure 1.18 A relay AND gate

Figure 1.19 A relay OR gate

PROBLEMS

1 Draw the circuit diagram of a two input NAND gate built out of a pair of relays. The truth table of a NAND gate is shown below;

A	B	Q
0	0	1
0	1	1
1	0	1
1	1	0

Figure 1.8 may give you a few clues.

2 The output of a two input NOR gate is only 1 when both of its inputs are 0. Draw the truth table of a NOR gate. Draw the circuit diagram of one built from two relays and a pull-up resistor.

3 Write down a NOT/AND/OR statement for a NOR gate. Draw a circuit diagram to show how one can be built out of a couple of NOT gates and an AND gate.

4 Write down a NOT/AND/OR statement for a NAND gate. Use that statement to draw the circuit diagram of a NAND gate using NOT, AND and OR gates; you will need four NOT gates, three AND ones and two OR ones.

5 A system has two inputs (called A and B) and a single output (called Q). Q equals NOT (A AND B). Draw up a truth table for the system. Draw a circuit diagram of it, using NOT and AND gates.

EXPERIMENT
Building some logic systems with relays

Start off by assembling the circuit shown in figure 1.17. Connect a voltmeter between Q and the 0 V rail. Use a 100 ohm pull-up resistor. Verify that Q is +5 V when A is 0 V and that Q is 0 V when A is +5 V.

Design a relay logic system which has the truth table shown on the right. You will only need to use one relay and a 100 ohm pull-down resistor. Try out your design and see if it works. Then draw its circuit diagram and explain carefully how it works.

Finally, see if you can design the following logic system with two relays and a single pull-up resistor.

A	B	Q
0	0	0
0	1	1
1	0	1
1	1	0

C	B	A	Q
0	0	0	1
0	0	1	1
0	1	0	0
0	1	1	1
1	0	0	1
1	0	1	0
1	1	0	1
1	1	1	1

The system has three inputs (A, B and C) and a single output (Q). When you have established that your design works, draw its circuit diagram.

TTL LOGIC GATES

By now you should be aware that a logic system which is built out of a number of logic gates can have properties that are different from those of the individual gates. The whole system is more than simply the sum of its parts. By using enough basic logic gates (i.e. the NOT, AND and OR type) and linking them up the right way you can assemble as complex a logic system as you like. These three gates allow you to build systems with as many inputs and outputs as you like and which will obey any truth table that you care to think up. The most complex logic systems that exist today, including computers, are just vast assemblies of enormous numbers of simple logic gates.

In theory, there is no limit to the complexity of a logic system manufactured from basic logic gates. By the time you have worked your way through to the end of this chapter you will know how the rules of Boolean algebra and the Russian doll principle make the design and analysis of large logic systems relatively easy. In practice, there is a limit to the size of a relay logic system. Relay gates take up a lot of room; each gate takes up the same space as a box of matches. Then each gate has a finite lifetime. A relay has moving parts which are liable to fail after a million switchings. Furthermore, relay logic systems are very slow. Even the fastest reed relays have reaction times of a few milliseconds (there are 1000 milliseconds in 1 second). Finally, relays are expensive to make.

Modern integrated circuit technology uses **transistors** to build logic gates. These transistor gates beat logic gates hands down. They take up the space of a grain of sand, have an indefinite lifetime with reaction times that are a million times shorter than those of relays, and they can be mass produced cheaply.

THE NAND GATE

Although you can build any logic system out of NOT, AND and OR gates, in practice you only use one type of gate. This simplifies the wiring of a system and makes more efficient use of an integrated circuit. You can choose to use either a NOR gate or a NAND gate as the basic logic unit but the industry standard is the NAND gate. Its symbol and truth table are shown in figure 1.20. Notice that the output of a NAND gate is logical 0 only when all of its inputs are at logical 1.

NAND gates can be bought in groups of four, for a few pence each. The 7400 integrated circuit contains four independent NAND gates which have been etched onto a single slice of silicon and encased within a black plastic package. Fourteen metal legs (or **pins**) emerge from that package to allow electrical connections to be made to the gates within it. Two of the pins are for the +5 V and 0 V supply rails, the other twelve provide inputs and outputs for the gates. If you look at figure 1.21 you can see where each pin appears to go within the package. The diagram does not show what the interior of the package looks like. All four gates are on a small piece of silicon in the centre of the package, and each gate is made from a number of inter-connected transistors. The technology used to manufacture the 7400 IC (integrated circuit) is called **TTL** (transistor-transistor logic) and has been around for a number of years. We shall be discussing other IC technologies in a later section.

TTL input and output

Before you can start exploring the properties of TTL NAND gates you have to know how to get signals in and out of them. In particular, you need to know how to connect switches to their inputs so that you can feed in 1's

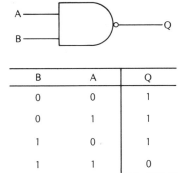

B	A	Q
0	0	1
0	1	1
1	0	1
1	1	0

Figure 1.20 The truth table for a NAND gate

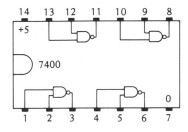

Figure 1.21 The pinout of a 7400 TTL IC

and 0's. Furthermore, you have to have some method of knowing what the state of an output pin is.

We shall deal with inputs first. As you might expect, an input held at +5 V will be read in as a logical 1 and one held at 0 V will be read in as logical 0. But TTL inputs have an extra property which is very useful. They **float up** to logical 1 if they are not pulled down to logical 0. A floating input rises to about +1.4 V; any input held below +1 V is read in as a logical 0.

Figure 1.22 shows how to use push switches to feed 1's and 0's into a TTL input. The circuit at the top has a pull-up resistor which hauls the input up to +5 V when the switch is open. Since virtually no current flows into a TTL input when it is at logical 1, the value of the pull-up resistor is not critical. Indeed, you can dispense with it altogether and let the input float up to logical 1 on its own! The circuit at the bottom uses a pull-down resistor, so that a logical 1 is fed into the input when the switch is pressed. The value of the pull-down resistor is fairly critical as a current of 1 mA flows out of a TTL input when it is hauled down to logical 0. That current has to flow through the pull-down resistor to the 0 V supply rail. We can use Ohm's Law to work out the consequences of this. Since 470 Ω is 0.47 kΩ;

$$R = \frac{V}{I}$$

therefore $0.47 = \dfrac{V}{1}$

therefore $V = 0.47\,V.$

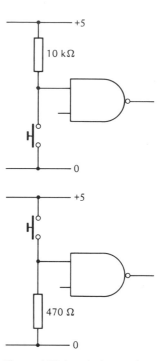

Figure 1.22 Interfacing push switches to TTL inputs

So the input gets pushed up to +0.5 V by the current flowing through the pull-down resistor. As the input has to be below +1.0 V for it to be read in as a logical 0, it should be fairly obvious that the pull-down resistor must not be bigger than 1 kilohm.

LEDs

The most convenient way of monitoring the state of the output of a logic system is to use a **light-emitting diode** or **LED**. This is an electronic component which gives out light when current flows the correct way through it; its circuit symbol is shown in figure 1.23. That diagram shows how an LED can be hooked up to a TTL output. Note that you have to put a resistor in series with the LED. The size of the resistor dictates how much current will flow through the LED. Since you are going to be using LEDs a lot as voltage-to-light converters, we shall spend a few moments discussing their properties.

Current will only flow one way through an LED, from **anode** to **cathode** (marked A and C respectively in figure 1.23). The voltage drop across the diode is virtually independent of the current which flows through it; for red LEDs it is 2.0 V. The more current you pump through the LED the brighter it will glow, but if you exceed its maximum current rating it will be damaged. The LED can also be damaged if you try to make the current flow across it the wrong way, from cathode to anode.

Returning to the circuit of figure 1.23, we shall now run over a calculation of the current which will flow through the LED. When the TTL output is at logical 1 the current flowing through the LED must be zero i.e. the LED will not glow. Once the output goes down to logical 0 current will flow

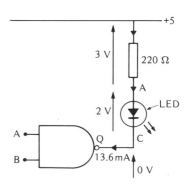
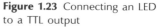

Figure 1.23 Connecting an LED to a TTL output

through the LED and it will light up. Q will be held at 0 V by the circuitry within the gate, so it will pull the cathode down; the anode will be pulled up by the 220 ohm resistor. As the voltage drop across the LED will be 2 V, the voltage drop across the resistor will be $5 - 2 = 3$ V. Using Ohm's Law, we can calculate the current which flows through the resistor;

$$R = \frac{V}{I}$$

therefore $\quad 0.22 = \frac{3}{I}$

therefore $\quad I = \frac{3}{0.22} = 13.6 \text{ mA}.$

So the current which flows through the LED into the TTL output must be 13.6 mA; that current will be delivered to the lower supply rail via the 0 V supply line of the gate.

TTL outputs are designed to be compatible with TTL inputs. Since you do not have to push much current into a TTL input when it is being held at logical 1, TTL outputs cannot deliver much current at logical 1. Using the standard electronic jargon, we say that TTL outputs are not very good at **sourcing** current. If you pull more than a couple of milliamps out of a TTL output at logical 1, its voltage starts to drop from its usual +4 V (if you want +5 V as logical 1, use a pull-up resistor). On the other hand, TTL outputs are very good at **sinking** current. That is, a TTL output at logical 0 can swallow a current of up to 16 mA.

The asymmetry of the current-handling capability of TTL outputs means that you are stuck with using a glowing LED to indicate logical 0. For some reason, most people seem to assume that a lit LED ought to indicate a logical 1. You can connect an LED directly between a TTL output and the 0 V supply rail if you want it to light up whenever the output tries to go up to logical 1. The output will only get up to +2 V, not a very good logical 1 and the LED won't be very bright either.

EXPERIMENT

Investigating the properties of a red LED

Assemble the circuit shown in figure 1.24, making sure that you get the LED round the right way. Use a voltmeter to measure the voltage across the LED. Replace the resistor with each of the following values in turn, and note the voltage across the LED; 4.7 kΩ, 2.2 kΩ, 1.0 kΩ, 470 Ω, 220 Ω and 100 Ω. Use the known supply voltage and the voltage across the LED to calculate the current which flows through it. For example, suppose that for the 1.0 kΩ resistor the LED has 1.8 V across it. Then the voltage drop across the resistor will be $5 - 1.8 = 3.2$ V. Finally, using Ohm's Law, we can calculate the current;

$$R = \frac{V}{I}$$

therefore $\quad 1 = \frac{3.2}{I}$

therefore $\quad I = \frac{3.2}{1} = 3.2 \text{ mA}.$

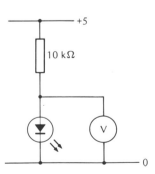

Figure 1.24

Use your results to draw a graph to show how the current through an LED depends on the voltage across it. Draw the current axis vertical.

You may care to repeat the experiment with a green LED.

COMBINING NAND GATES

Earlier on we stated that NAND gates were all you needed to build any logic system. We are now going to try to convince you of the truth of this, by showing you how NOT, AND and OR gates can be synthesised from NAND gates.

The NOT gate

Look at figure 1.25. The inputs of the gate are joined together, so it really has only two input states. A = B = 1 or A = B = 0. A glance at the truth table of a NAND gate (figure 1.20) should convince you that Q will always have the opposite state to that of A if A = B. So the circuit of figure 1.25 is a NOT gate.

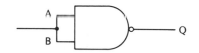

Figure 1.25 A NOT gate

The AND gate

An AND gate can be made out of two NAND gates as shown in figure 1.26. Its operation is best explained with the help of a truth table.

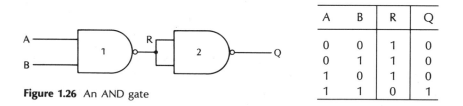

Figure 1.26 An AND gate

A	B	R	Q
0	0	1	0
0	1	1	0
1	0	1	0
1	1	0	1

R is the output of gate 1. It therefore has the standard truth table for a NAND gate i.e. only 0 when A and B are 1. Gate 2 has been connected as a NOT gate, so Q equals NOT R. So the overall behaviour is that of a system which only feeds out 1 when both its inputs are 1. Q equals A AND B.

The OR gate

This one is shown in figure 1.27. The truth table of the system is shown below. Work through it and convince yourself that it is correct.

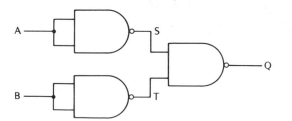

Figure 1.27 An OR gate

A	B	S	T	Q
0	0	1	1	0
0	1	1	0	1
1	0	0	1	1
1	1	0	0	1

EXPERIMENT

*Finding the truth tables of a number of systems assembled
from the four NAND gates on a 7400 IC*

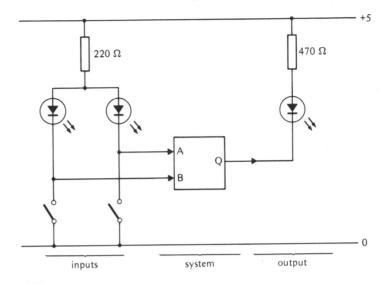

Figure 1.28

Figure 1.28 shows the circuit diagram of the switches and LEDs used to set
up the input states and monitor the output state. On the left, a pair of
switches are used to generate A and B. A pair of LEDs give a visual indica-
tion of the state of those inputs. On the right, a single LED displays the
state of Q. For all three LEDs, a logical 0 will make them glow. Set up the
circuit on your breadboard, with a 7400 IC sitting between the input and
output circuitry.

The various systems that you are going to explore are shown in figure
1.29. Set up each one in turn. Feed in inputs corresponding to each of the
four lines of the truth table and note the output. Use your results to write
down a truth table for each system and to describe its behaviour with a
NOT/AND/OR statement.

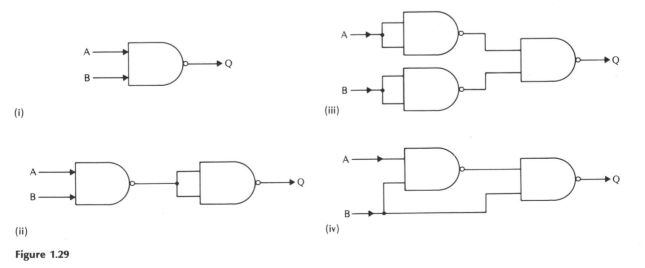

(i)

(ii)

(iii)

(iv)

Figure 1.29

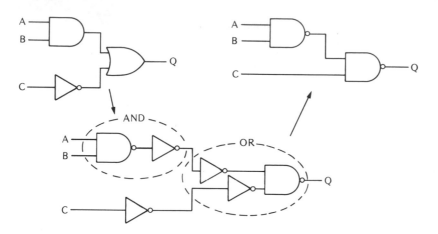

Figure 1.30 Working out the equivalent of a system in NAND gates

BUILDING WITH NAND GATES

We are going to end this section of the chapter with an example of the use of NAND gates to assemble a logic system. The design technique will become obsolete when you have made the acquaintance of Boolean algebra in the next section, but it is a useful standby if you ever get confused or lost.

Here is the task. We want to make a system which has one output (Q) and three inputs (A, B and C), is made out of interconnected NAND gates and obeys this statement;

Q equals (A AND B) OR (NOT C).

The first step is to draw out a suitable logic system using NOT, AND and OR gates; this is shown in figure 1.30 on the left. Then you replace each gate with its NAND gate equivalent. Finally you look for adjacent NOT gates and eliminate them from the circuit (NOT(NOT X) equals X?); this gives the final circuit, shown on the right of figure 1.30.

This method is not very elegant, can be very demanding in terms of paper and does not always lead to a very efficient use of NAND gates. The methods of Boolean algebra can indicate a way of implementing a given circuit using fewer gates.

PROBLEMS

1 Design a logic system built out of NAND gates which obeys this statement;

Q equals A OR B OR C OR D

2 Design a logic system built out of NAND gates which obeys this statement;

Q equals (A AND B) OR ((NOT A) AND (NOT B))

3 Work out the truth tables for the systems shown in figure 1.31.

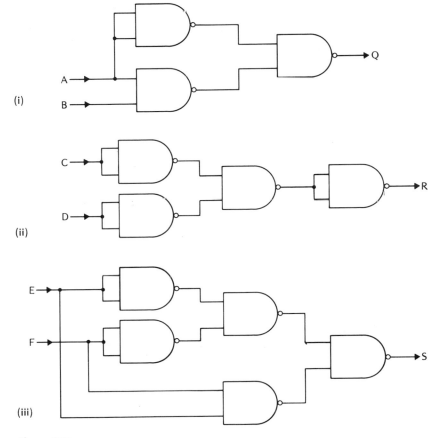

(i)

(ii)

(iii)

Figure 1.31

4 The circuit of figure 1.32 shows a pull-down resistor holding a pair of TTL inputs down to logical 0.

a) If each input sources 1 mA into the resistor and has to be held below +1 V to be read in as a logical 0, what is the maximum value of R that will allow the LED to go off when the switch is not pressed?

b) How much current flows through the LED when the switch is closed? Assume a voltage drop of 2 V across the LED.

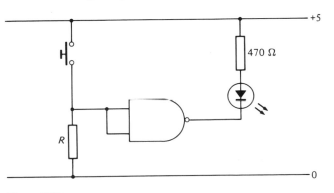

Figure 1.32

BOOLEAN ALGEBRA

Boolean algebra is a branch of mathematics which considerably simplifies the business of working out the truth table of a logic system. It can also be used backwards; starting with a truth table, it can be used to work out how basic logic units have to be linked to make up a system. The algebra is so useful that you will be applying it extensively as you work through the rest of this book.

We are going to state the properties of Boolean algebra without trying to justify them. The rules of the algebra give us systems which have precisely the truth tables they were designed to have. If we adopt the precept that "the proof of the pudding is in the eating", then the rules of Boolean algebra need no theoretical justification for the designer of practical logic systems!

Variables

Capital letters, such as A, B and Q, represent the variables of Boolean algebra. Each variable can have one of two values. They can be logical 1 or logical 0. The statement $A = 1$ means that the variable A has the value of logical 1. Similarly, if B is a logical 0, then $B = 0$.

If $A = Q$ then A and Q have the same value; they are either both 1 or both 0.

There are three basic operations which can be performed on variables, inversion or the NOT operation, the AND operation and the OR operation.

Inversion

The statement that $\bar{A} = 1$ means that $A = 0$. The bar over the top of the variable changes its value, or inverts it. It is often called the NOT operation, and $\bar{B}$ is usually read as "not B".

It should be obvious that $\bar{\bar{A}} = A$, i.e. double inversion, leaves the value of a variable untouched.

The AND operation

This one can be defined by three statements.

$$A \cdot 1 = A$$
$$A \cdot 0 = 0$$
$$A \cdot B = B \cdot A$$

So if $A \cdot B = 1$ then both A and B are 1. On the other hand, if $A \cdot B = 0$ then either $A = 0$ or $B = 0$, or both. $A \cdot B$ is read as "A and B".

It follows from the definitions of the NOT and AND operations that $A \cdot \bar{A} = 0$ regardless of whether $A = 1$ or $A = 0$. Similarly, $A \cdot A = A$.

The OR operation

Like the AND operation, this needs three statements for its definition.

$$A + 1 = 1$$
$$A + 0 = A$$
$$A + B = B + A$$

So if $A + B = 0$, then both A and B must be 0. If $A + B = 1$, then either, or both, of A and B are 1.

A little thought should convince you that $A + \bar{A} = 1$, regardless of the value of A. Similarly, $A + A = A$.

$A + B$ is read as "A or B", not as "A plus B".

Brackets

Boolean algebra uses brackets to define the order in which operations must be performed. Ordinary arithmetic uses brackets in exactly the same way. The following examples should make the use of brackets clear.

$$A.(B + C) = A.B + A.C$$
$$\bar{P}.(\bar{Q} + S) = \bar{P}.\bar{Q} + \bar{P}.S$$
$$\overline{P.Q} + \bar{P}.S \neq \bar{P}.(\bar{Q} + S)$$
$$W.X + Y \neq W.(X + Y)$$

SIMPLIFYING

Boolean algebra is very useful for making sure that you do not use more gates than you have to when you design a logic system. By simplifying a Boolean expression, you reduce its complexity. Let's look at an example.

Suppose that we start off with a fairly complex expression;

$$Q = A.\bar{B} + A.B$$

We then split up the expression with brackets;

$$Q = A.(\bar{B} + B)$$

But $\bar{B} + B = 1$ (have a look at the definition of the OR function). So $Q = A.1$. Finally, because of the way that the AND operation is defined, $Q = A$.

The basic definitions of Boolean algebra lead to three theorems which are very useful for simplifying Boolean expressions. These are the Redundancy theorem, the Race Hazard theorem and De Morgans theorem.

The Redundancy theorem

$$A + A.B = A$$

For example, suppose that we had the statement;

$$Q = A.B.C + \bar{A}.B + B.C$$

Compare the first and last terms on the right hand side of the expression. The first term is redundant with respect to the last one; that is $B.C + B.C.A = B.C$. Whenever $B.C.A = 1$, $B.C$ will also be 1 (think about it). So we might as well forget $B.C.A$ altogether;

$$Q = \bar{A}.B + B.C$$

Here is another example.

$$Q = (X + Y) . (X + Z)$$

Multiplying out the brackets;

$$Q = X . X + X . Z + Y . Z + Y . X$$

Now, you will remember that $X . X = X$. So the expression becomes;

$$Q = X + X . Z + X . Y + Y . Z$$

Now we use the Redundancy theorem (or RT) to eliminate the two middle terms of the right hand side. Both are redundant with respect to X. So our final, simplified, expression is;

$$Q = X + Y . Z$$

The Race Hazard theorem

$$A . B + \bar{A} . C = A . B + \bar{A} . C + B . C$$

The Race Hazard theorem (called RHT for short) allows you to make a Boolean expression more complex. If you study its statement above, you will see that it allows you to add **redundant terms** to an expression. Believe it or not, this can sometimes be the first step in the simplification of an expression! For example, suppose that you have the expression;

$$Q = A . B . C + \bar{A} . B$$

Apply the Race Hazard theorem;

$$Q = A . B . C + \bar{A} . B + B . C . B$$
$$\text{therefore} \quad Q = B . C . A + \bar{A} . B + B . C$$

Noticing that the first term is redundant with respect to the last one;

$$Q = \bar{A} . B + B . C$$

Our final expression is simpler than the original one. The technique of using the RHT followed by the RT is very useful. You will have had lots of practice at it by the time you get to the end of this chapter.

De Morgans theorem

$$\overline{(A + B)} = \bar{A} . \bar{B}$$
$$\bar{A} + \bar{B} = \overline{(A . B)}$$

The theorem states that if you wish to invert an expression, you replace AND with OR and each variable with its inverse. It is important to obey the brackets carefully; novices tend to fall into the trap of saying that $\overline{A . B} = \bar{A} . \bar{B}$!

Here is an example. If $Q = (A + \bar{B}) . C$, what is $\bar{Q}$?

$$\bar{Q} = \overline{(A + \bar{B}) . C}$$
$$= \overline{(A + \bar{B})} + \bar{C}$$
$$= \bar{A} . B + \bar{C}$$

So by applying De Morgans theorem (or DMT) twice, we have shown that $\bar{Q} = \bar{A} . B + \bar{C}$.

PROBLEMS

1 Use De Morgans theorem to find $\bar{P}$ if

    a) $P = A.(B + \bar{C})$

    b) $P = A + B.C$

    c) $P = \overline{A.B} + \bar{C}.D$

    d) $P = \overline{(A + B)}.(C + \bar{D})$

2 Show that

    a) $(U + V).(U + \bar{V}) = U$

    b) $\overline{(X + Y).(\bar{Y}.X)} = Y + \bar{X}$

    c) $\overline{(W.Z + W)} = \bar{W}$

3 Use the Race Hazard theorem followed by the Redundancy theorem to show that

    a) $A.C + \bar{A}.B.C = C.(A + B)$

    b) $A.B + \bar{A}.B.C + \bar{B} = A + \bar{B} + C$

    c) $(A + \bar{B}).(A + C.B) = A$

    d) $A.\bar{B}.\bar{C} + A.B.\bar{C} + \bar{A}.B.\bar{C} = A.\bar{C} + B.\bar{C}$

ANALYSING LOGIC SYSTEMS

We are now in a position to run through a few examples of logic system analysis using Boolean algebra. Given the circuit diagram of a logic system, you can use Boolean algebra to work out what its truth table is going to be. Each gate in the system can be represented by a Boolean expression. By stringing those expressions together and then simplifying them, you end up with an expression which describes the behaviour of the whole system.

Our first example is illustrated in figure 1.33. The system is a single NAND gate, with one of its inputs (B) held at logical 1. The NAND gate is represented by this chunk of algebra;

$$Q = \bar{A} + \bar{B}$$

This is fairly obvious if you think about the NOT/AND/OR expression for a NAND gate, namely Q equals (NOT A) OR (NOT B). (The output of a NAND gate is 1 when either, or both, of its inputs are 0.) If we apply DMT to the expression;

$$Q = \overline{A.B}$$

In other words, Q equals NOT (A AND B); hence the name NAND gate.

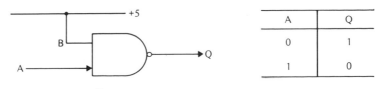

A	Q
0	1
1	0

Figure 1.33 $Q = \bar{A}$

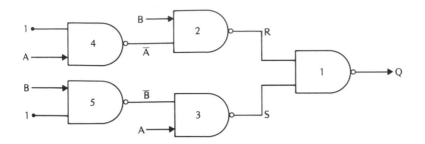

B	A	Q
0	0	0
0	1	1
1	0	1
1	1	0

Figure 1.34 $Q = B . \bar{A} + \bar{B} . A$

Returning to figure 1.33, B = 1. So;

$$Q = \bar{A} + \bar{1}$$
$$= \bar{A} + 0$$
$$= \bar{A}$$

The whole system is therefore represented by the expression $Q = \bar{A}$. It has a single input, so there are only two possible input states (A = 0 and A = 1). The truth table will therefore only have two lines in it. Since Q can only be 1 when A is 0, the truth table will be as shown in figure 1.33.

Our second example is more complicated. Look at figure 1.34. In order to get a truth table for this system we start off by writing down an expression for Q. From gate 1;

$$Q = \bar{R} + \bar{S}$$

We then write down expressions for R and S by writing down the expressions for gates 2 and 3;

$$R = \overline{B . \bar{A}}$$
$$S = \overline{\bar{B} . A}$$

The functions for gates 4 and 5 should be obvious. Now combine the two sets of expressions;

$$Q = \bar{R} + \bar{S}$$
$$= \overline{\overline{(B . \bar{A})}} + \overline{\overline{(\bar{B} . A)}}$$
$$= B . \bar{A} + \bar{B} . A$$

The expression for Q is now in standard form. It is a sum of products, in the form that you need in order to write out the truth table of the system. The products are $B . \bar{A}$ and $\bar{B} . A$. Each product tells you of a line in the truth table for which Q = 1. Consider the $\bar{B} . A$ product first. It will only be 1 when B is 0 and A is 1. Similarly, the other product, $B . \bar{A}$, is only 1 when B is 1 and A is 0. So the second and third lines of the truth table (shown in figure 1.34) have Q = 1; the other two lines have Q = 0.

The final example concerns the system shown in figure 1.35. It has one output (Q) and three inputs (A, B and C). We have introduced R, S and T as intermediate variables to help us get an expression linking Q with A, B and C. This time, we are going to analyse the system by moving from left to right across the circuit diagram. This is the same direction as the flow of information. By starting off at the left and simplifying the expressions as we go we shall avoid the ungainly many-inversion expression for Q that we obtained in our previous example.

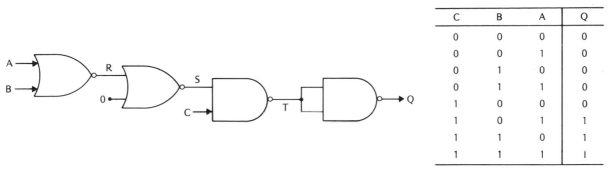

C	B	A	Q
0	0	0	0
0	0	1	0
0	1	0	0
0	1	1	0
1	0	0	0
1	0	1	1
1	1	0	1
1	1	1	I

Figure 1.35 $Q = A . C + B . C$

What Boolean expression do we use for a NOR gate? Well, the output of a NOR gate is 1 only when both of its inputs are 0. So R is (NOT A) AND (NOT B), namely $R = \bar{A} . \bar{B}$. If we apply DMT to this expression, the reason for the name NOR gate should be clear; $R = \overline{(A + B)}$, or in words, R equals NOT (A OR B).

The second NOR gate gives us the expression;

$$S = \bar{R} . \bar{0} = \bar{R} . 1 = \bar{R}$$

Combining the expressions for R and S and using DMT, we obtain;

$$S = \bar{R} = \overline{\bar{A} . \bar{B}} = A + B$$

Now for the first NAND gate;

$$T = \overline{S . C} = \overline{(A + B) . C}$$

The final NAND gate gives (remember that $T . T = T$);

$$Q = \overline{T . T} = \bar{T}$$

So

$$Q = \overline{\overline{(A + B) . C}} = (A + B) . C = A . C + B . C$$

This is the simplest expression for Q, but it is not in **standard form**. To be in standard form, each product in the expression should contain all three variables. So we have to expand our expression for Q before we can write out the truth table of the system. Here is how you do it;

$$Q = A . C + B . C$$
$$\text{therefore} \quad Q = A . 1 . C + 1 . B . C$$
$$\text{therefore} \quad Q = A . (B + \bar{B}) . C + (A + \bar{A}) . B . C$$
$$\text{therefore} \quad Q = A . B . C + A . \bar{B} . C + A . B . C + \bar{A} . B . C$$
$$\text{therefore} \quad Q = A . B . C + A . \bar{B} . C + \bar{A} . B . C$$

The final expression tells us that Q will be 1 under three circumstances. When $A = B = C = 1$, when $A = C = 1$ and $B = 0$, when $B = C = 1$ and $A = 0$. So the truth table must be as shown in figure 1.35.

PROBLEMS

1 Each of the expressions shown below represent systems with two inputs (A and B) and a single output (Q). Draw up their truth tables after writing down the expressions in standard form, if necessary.

    a) $Q = \bar{A}.\bar{B} + A.B$

    b) $Q = A + \bar{A}.\bar{B}$

    c) $Q = \overline{(A + B)} + \bar{A}$

    d) $Q = (\bar{B} + \bar{A}).(B + A)$

2 This question is concerned with the circuits shown in figure 1.29. Use the techniques of Boolean algebra to write down an expression which describes each system. Put the expression into standard form and hence draw up a truth table for each system.

3 Figure 1.36 shows a number of logic systems built out of NAND and NOR gates. Work out a Boolean expression to describe each one, and use them to write out their truth tables.

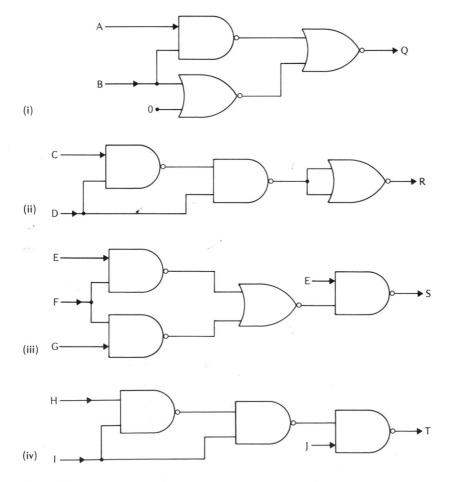

Figure 1.36

SOLVING LOGIC PROBLEMS

A logic problem is a task which requires you to make up a logic system to perform a particular function. It is in this area that Boolean algebra comes into its own. In the last section you used the algebra to analyse systems. You started off with a circuit diagram and worked out a truth table. In this section you will learn how to get from the truth table to a circuit diagram; i.e. you will learn how to use Boolean algebra to synthesise logic systems.

Our method of designing logic systems is quite straightforward and will always give a circuit that works. It may not give the design which uses the minimum number of basic gates, nor will it necessarily give the most elegant circuit. Nevertheless, it can, in principle, allow you to design systems with as many inputs as you like. Furthermore, it works equally well for NOR or NAND gates as the basic component of the final system.

We shall go over three examples of logic system synthesis in detail. Then you will be able to have a go at it yourself!

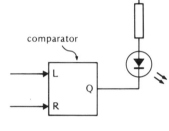

comparator

A COMPARATOR

Look at figure 1.37. We are going to design the contents of the black box called the comparator. The comparator looks at the state of its inputs (L and R) and makes the LED light up if they are the same. Since the LED will only glow when Q is 0, Q can only be 1 when $L \neq R$. Having established what the system is supposed to do, we can draw up its truth table (see figure 1.37).

The next step is to use the truth table to write down a Boolean expression which describes the system. Q is 1 for two lines of the table. $Q = 1$ when $L = 1$ and $R = 0$; $Q = 1$ when $L = 0$ and $R = 1$. So the Boolean expression is;

$$Q = \bar{L}.R + L.\bar{R}$$

L	R	Q
0	0	0
0	1	1
1	0	1
1	1	0

Figure 1.37 A comparator

Using NAND gates

If we want to assemble the system from NAND gates, the expression for Q must be a **sum of products**. This is because a NAND gate is described by a Boolean expression which is a sum of two terms. Look at figure 1.38; it shows a NAND gate being fed with $\bar{A}$ and $\bar{B}$. Its output will be given by $\overline{\bar{A}.\bar{B}}$. If we use DMT, an alternative expression for the NAND gate output is $A + B$.

As Q is a sum of two products, it can be generated by a two input NAND gate, provided that it has the inverse of the products fed into it. If you look at the circuit of figure 1.39, you will see that Q emerges from gate

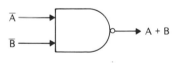

Figure 1.38

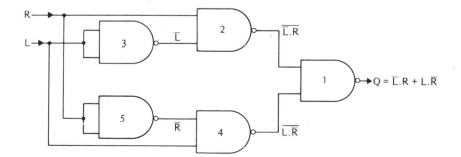

Figure 1.39 A comparator made out of NAND gates

1 because $\overline{L.R}$ and $\overline{L.R}$ are fed into it. Gate 2 generates $\overline{\overline{L}.R}$, so it has to have $\overline{L}$ and R fed into it. $\overline{L}$ is, in turn, generated by gate 3; its output is $\overline{L.L} = \overline{L}$. Similarly, gates 4 and 5 generate the other term $\overline{L.\overline{R}}$

The solution to the problem is now complete. It has been implemented with five dual input NAND gates. Since you can only get four such gates in a 7400 IC, the final circuit is going to consist of two ICs. In a sense, a solution to the problem which only involved the use of a single IC would be more elegant. In fact, the problem can be solved four times over with a single IC, the 7486. This contains four **EXCLUSIVE-OR gates** (usually short-ened to **EOR**); one of these gates is shown in figure 1.40, together with its truth table. If you compare the truth table with that of the system which we have just designed, you will see that $Q = L \oplus R$. ($\oplus$ is the symbol for the EXCLUSIVE-OR operation.)

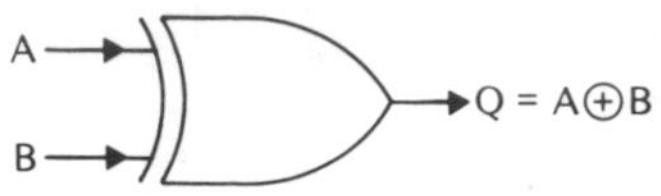

Although a single EOR gate will act as our complete system, the NAND gate solution is perfectly satisfactory. 7400 ICs are cheap and you can afford to keep lots of them in stock. Furthermore, because the NAND gate is a very widely used logic unit, you quickly become familiar with its posi-tion within the IC package and the business of interconnecting the pins of the IC packages becomes straightforward. On the other hand, the use of a 'special' IC (the 7486) means that you handle a relatively unfamiliar package. The special IC will cost you more than the 7400. Furthermore, you have to do some research in suppliers catalogues to find the special that solves your problem and in some cases there may not be one!

These drawbacks are insignificant when a system is to be manufactured on a large scale. Then the number of IC packages contributes substantially towards the cost of the system, and the use of a special IC may make economic sense. But on a small scale, where you are only considering the assembly of a few systems, the use of flexible, general purpose ICs like the 7400 has many advantages. No IC is perfect. You need experience to sort out some of the idiosyncracies of an integrated circuit. If you always use the same small family of ICs you will quickly learn all of their tricks, but if you use a special every time, then your experience may not help you very much if the IC misbehaves!

B	A	Q
0	0	0
0	1	1
1	0	1
1	1	0

Figure 1.40 An EOR gate

Using NOR gates

You do not have to solve a logic problem with NAND gates. Any logic system can also be built out of NOR gates, as well as out of NAND gates. Although NAND gates are usually used in the design of real systems, it is worth knowing how to use NOR gates as well. This is because NOR gates are particularly good at implementing Boolean expressions which are products of variables. For example, suppose that you want to generate $Q = A.B.C.D$. If you use DMT on this expression, you will see that

$$Q = \overline{(\overline{A} + \overline{B} + \overline{C} + \overline{D})}$$

So a four input NOR gate fed with $\overline{A}$, $\overline{B}$, $\overline{C}$ and $\overline{D}$ will generate Q.

We are now going to show how the system of figure 1.37 can be built out of NOR gates. The first step is to get an expression for $\overline{Q}$ from the truth table. $\overline{Q}$ is a sum of products, one product for each line of the truth table that has $Q = 0$ (that is $\overline{Q} = 1$).

$$\overline{Q} = \overline{L}.\overline{R} + L.R$$

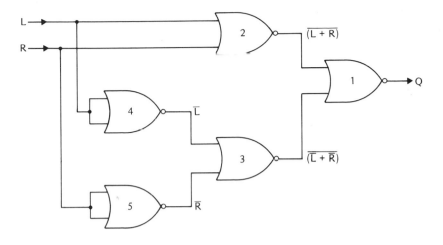

Figure 1.41 A comparator made from NOR gates

Next we apply DMT to get an expression for Q.

$$Q = \overline{(\overline{L}.\overline{R} + L.R)}$$
$$\text{therefore} \quad Q = (L + R).(\overline{L} + \overline{R})$$

The expression for Q is now a product of sums, just right for being generated by NOR gates. Figure 1.41 shows the circuit. The first gate that you draw is the one which generates Q i.e. gate 1. Then you draw gates 2 and 3 to generate $\overline{(L + R)}$ and $\overline{(\overline{L} + \overline{R})}$. Finally, you put in gates 4 and 5 to generate $\overline{L}$ and $\overline{R}$.

It is interesting to note that both of our solutions which used basic logic gates are very similar. They are equally economical in terms of gate numbers and wiring complexity. In general, any system which is built out of one sort of gate can be built out of the other sort with only a marginal difference in the number of gates used. Furthermore, a logic system can always be assembled in at least two different ways, and there is no way, from outside the system, in which you can tell what sort of gate was used to construct it.

A MACHINE CONTROL SYSTEM

Our second example is of a system which might be useful in industry. A piece of machinery is controlled by three press switches. The power supply is completely cut off if none of the switches are pressed; when this happens, a green light comes on, showing that it is safe to approach the machine and do things inside and around it. The power supply is switched on when any two of the switches are pressed. The switches are placed so that this requires the operator of the machine to have each hand on a switch. If his hands are pressing switches, they cannot be inside the machine at the same time! A red LED comes on when the machine is powered up.

Figure 1.42 (overleaf) shows how we are going to set up the switches and LEDs. The switches pull the inputs of the logic system down to logical 0 when they are pressed. The output labelled P controls the power supply of the machine. Since the green LED must come on when the power is off, we have chosen an **active-high** input for the machine power supply. That is,

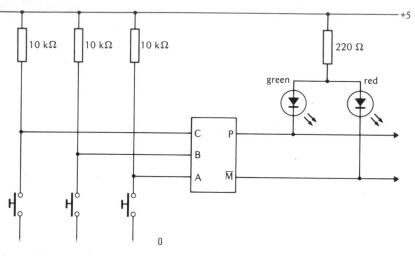

Figure 1.42

you have to feed a logical 1 into the power supply unit to switch on the power. (The signal from P could operate a relay.) The other output ($\overline{M}$) operates the machine. Since we want the red LED to come on when the machine is on, we have chosen an **active-low** input for it. So $\overline{M}$ has to be a logical 0 to switch on the machine.

Having specified the problem in detail, labelling and explaining the functions of all the inputs and outputs of the logic system, we can draw up a truth table for it.

C	B	A	P	$\overline{M}$
0	0	0	1	X
0	0	1	1	0
0	1	0	1	0
0	1	1	1	1
1	0	0	1	0
1	0	1	1	1
1	1	0	1	1
1	1	1	0	X

The system has three inputs, so it can have 2^3 independent input states. Its truth table must therefore have eight lines. If you study the truth table, you will notice that $\overline{M}$ has not been specified in the first and last lines. Instead, we have inserted an X, meaning "**don't care**". This is because the original problem does not state what $\overline{M}$ has to be when the power supply is cut off (last line) or that the operator may have more than two hands (first line). $\overline{M}$ will, of course, have a definite value in the first and last lines of the truth table when the final system has been designed. The X simply means that we can choose a 1 or a 0 without affecting the overall operation of the system. As you will find out later, this freedom will allow us to save a lot of hardware in our final design.

The P sub-system

The whole system can be considered as two separate systems. Each sub-system has a single output (P or $\overline{M}$), but they share the same input signals (A, B and C). So we can design each sub-system on its own.

The first step in designing the P sub-system is to write down an expression for P from the truth table. P would be a sum of seven products, one for each of the first seven lines of the table. It is more economical, and elegant, to write down the expression for $\bar{P}$ instead; P = 0 for only one line of the truth table.

$$\bar{P} = C.B.A$$

$$\text{therefore} \quad P = \overline{C.B.A}$$

So P can be generated by a single three input NAND gate. Although three input NAND gates are available, we have chosen to use a four input one instead. You can get two four input NAND gates on a 7420 IC; they are very useful when up to four variables have to be combined. Similarly, the 7430 IC contains a single eight input NAND gate. Both of these ICs are useful additions to the basic 7400 IC and are no more expensive in terms of pence per input. The terms **dual**, **quad** and **octal** are sometimes used to describe the type of NAND gate housed in the 7400, 7420 and 7430 ICs respectively.

If you look at figure 1.43 you will see how a quad NAND gate can be used to generate P. Notice that it is converted into a triple input gate by holding the unwanted input at logical 1. $(A.B.C.1 = A.B.C?)$

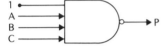

Figure 1.43 The P sub-system

The M̄ sub-system

If we ignore the two "don't care" conditions, the expression for $\bar{M}$ is;

$$\bar{M} = \bar{C}.B.A + C.\bar{B}.A + C.B.\bar{A}$$

This needs to be simplified. We could **brute force** a solution with four quad NAND gates and a trio of NOT gates, as shown in figure 1.44. Although this will work, it is not very elegant as we can build the same system with fewer gates. So let's try and simplify the expression for M̄.

Unfortunately, if we use the RHT on the expression for M̄, all of the extra products which we can insert are equal to 0. So we cannot subsequently use the RT to knock out some of the original products. For example;

$$\bar{M} = \bar{C}.B.A + C.\bar{B}.A + C.B.\bar{A}$$

$$\text{therefore} \quad \bar{M} = \bar{C}.B.A + C.\bar{B}.A + C.B.\bar{A} + B.A.\bar{A}$$

The last term, the product inserted with the aid of the RHT, is $B.A.\bar{A} = 0$. We are back where we started.

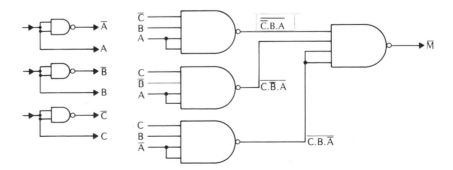

Figure 1.44 A brute force version of the M̄ sub-system

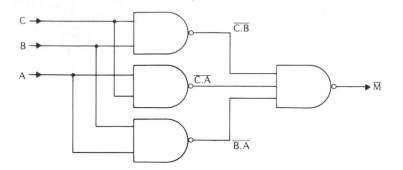

Figure 1.45 An elegant version of the $\bar{M}$ sub-system

However, we can use a "don't care" condition to insert an extra product. Let us choose $\bar{M} = 1$ when $P = 0$ i.e. that the X in the last line of the truth table is 1. Then the expression for $\bar{M}$ becomes;

$$\bar{M} = \bar{C}.B.A + C.\bar{B}.A + C.B.\bar{A} + C.B.A$$

Using the RHT followed by the RT;

$$\bar{M} = \bar{C}.B.A + C.\bar{B}.A + C.B.\bar{A} + C.B.A + C.B$$

therefore $\bar{M} = \bar{C}.B.A + C.\bar{B}.A + C.B$

The process can be repeated twice;

$$\bar{M} = \bar{C}.B.A + C.\bar{B}.A + C.B + C.A$$

therefore $\bar{M} = \bar{C}.B.A + C.B + C.A + B.A$

therefore $\bar{M} = C.B + C.A + B.A$

Our final expression is much simpler than the one that we started off with. Figure 1.45 shows how that expression can be realised with only four gates; it is much more elegant than the circuit of figure 1.44.

A KEYBOARD ENCODER

Our last example of logic system synthesis aims to caution you against getting too involved with Boolean algebra. The whole system is shown in figure 1.46. The four push switches are labelled 0, 1, 2 and 3; the logic system indicates which of the four switches, if any, is being pressed. The number of the switch which is being pressed is indicated, in **binary**, with a pair of logic signals (A and B). Before we discuss how to design a system of this sort, you need to know a bit about how electronic systems handle binary numbers.

A **two bit binary** number can be represented by BA. B and A can be 1 or 0, so there are four different two bit binary numbers. They are shown, with their decimal values, in the table below;

Binary	Decimal
00	0
01	1
10	2
11	3

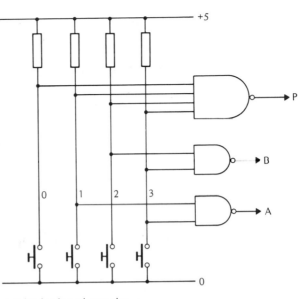

Figure 1.46 A 2 bit keyboard encoder

The decimal value of BA is given by 2B + A. If we have two logic signals called A and B, then they can be used to represent a binary number BA. So if B = 1 and A = 0, the two signals 'mean' the decimal number 2. Numbers larger than 3 need more than two bits. For example, numbers up to 15 can be represented by DCBA; the decimal value of DCBA is given by 8D + 4C + 2B + A.

The keyboard encoder has three outputs. Two of them (A and B) represent the number of the switch which is pressed, and the third (P) goes to 0 when none of the switches is pressed. The third output is quite important; it signals to other logic systems that B and A represent a number rather than nonsense. Strictly speaking, a four input system should have a sixteen line truth table (i.e. 2^4). However, since only one switch is supposed to be pressed at a time, most of those lines will be filled with X's. So we shall only write down the important lines of the truth table.

0	1	2	3	B	A	P
0	1	1	1	0	0	1
1	0	1	1	0	1	1
1	1	0	1	1	0	1
1	1	1	0	1	1	1
1	1	1	1	X	X	0

Generating P is straightforward. Using the last line of the truth table we can get an expression for $\bar{P}$;

$$\bar{P} = 0.1.2.3$$

$$\text{therefore} \quad P = \overline{0.1.2.3}$$

As you can see from figure 1.46, this is easily implemented with a quad NAND gate.

The business of generating B and A could be quite lengthy. Consider B.

$$B = 0.1.\bar{2}.3 + 0.1.2.\bar{3}$$

If you carefully select the missing lines of the truth table you can add extra products to this expression and reduce it to;

$$B = \bar{2} + \bar{3}$$

This expression should be intuitively obvious. After all, we want B to be 1 when switch 2 or switch 3 are pressed. A pressed switch pulls its input down to 0, so B is (NOT 2) OR (NOT 3). So there is little point in grinding through the algebra to simplify the original expression! Similarly, we can write down the simplified expression for A straight off;

$$A = \bar{1} + \bar{3}$$

Figure 1.46 shows how A and B can be generated with a pair of dual input NAND gates.

EXPERIMENT
Designing, assembling and evaluating the performance of three binary-to-decimal converters

Each system takes in a two bit binary number and displays its value with the help of some LEDs. Devices like this are very important; they allow the results of computations carried out by electronics to be understood by people.

Start off by assembling the circuitry shown in figure 1.47; put it at the left hand end of your breadboard. The switches can be used to generate all four values of the binary number BA, from 0 to 3. The two LEDs indicate the state of the switches; a glowing LED indicates a logical 0.

For each of the three devices described below, you will need to follow these steps:

1) Draw up a truth table to show how the outputs depend on the inputs.

2) Use the table to write down an expression for each of the outputs in terms of the inputs.

3) Simplify the expression.

4) Draw circuit diagrams to show how each output can be generated from the inputs with NAND gates. You can use quad and octal gates if you wish.

5) Assemble the complete system on a breadboard.

6) Evaluate its performance. That is, set up each of the four input states in turn and check that the LED display is correct.

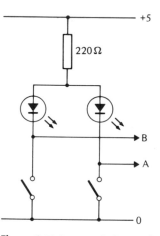

Figure 1.47 Input switches and indicators

A binary-to-barchart converter

This is the simplest of the three systems. Its outputs are shown in detail in figure 1.48. The three LEDs would be mounted in a row to form a coloured bar. The length of the bar indicates the size of the binary number BA. So no LEDs are lit when BA = 00, the bottom one is lit when BA = 01, the bottom and middle ones are lit when BA = 10 etc.

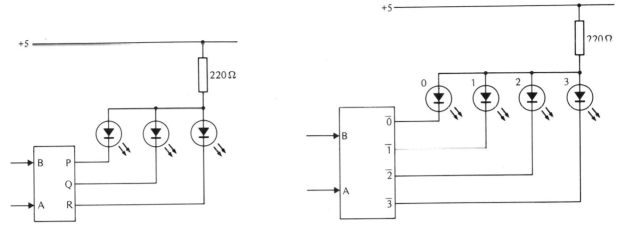

Figure 1.48 A binary-to-barchart converter

Figure 1.49 A binary-to-one-of-four converter

A binary-to-one-of-four converter

Have a look at figure 1.49. Each of the four LEDs has a number beside it. When an LED comes on, the number next to it tells you the value of BA. So if BA = 01, the LED marked "1" comes on; all the others stay off.

A binary-to-seven-segment converter

A **seven-segment display** is an array of LEDs arranged in a figure of eight. By lighting up the appropriate segments, any decimal number can be displayed directly. They are widely used in electronic calculators. The converter system has to decide which segments to sink current from in order to display the value of BA. The outline arrangement is shown in figure 1.50. Note that the common connection of the anodes means that the LEDs have to share one limiting resistor between them.

If you want to try your hand at designing a more complex system, have a go at a binary-to-barchart decoder which will cope with three bit binary numbers i.e. will display from 0 to 7.

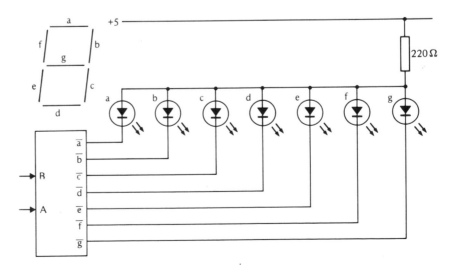

Figure 1.50 A binary-to-seven-segment converter

EXPERIMENT

Designing a keyboard encoder

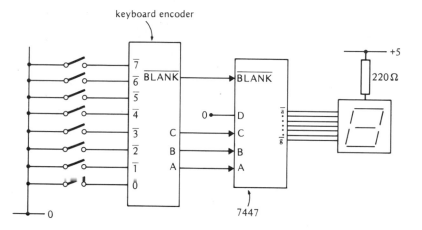

Figure 1.51 A 3 bit keyboard encoder and display

The system is shown in outline in figure 1.51. On the left is an array of eight switches labelled 0 to 7. A logic system (the keyboard encoder) looks at those switches and feeds a three bit binary number (CBA) into a 7447 IC. This IC converts that number into the seven signals which are needed to operate a seven segment display. It also has a pin labelled $\overline{\text{BLANK}}$ which can be used to turn off the display; it is an active low input. When one of the switches is closed, its value should be displayed by the seven segment LED; the display should be off when none of the switches are pressed.

Get the 7447 and LED display working at the far right hand end of your breadboard. Use a temporary switch to generate CBA and check that the display functions correctly. Then assemble the rest of the circuit; you should only need to use one 7430 and two 7420 ICs.

When you have got the system to work, draw its circuit diagram and explain how you designed it.

PROBLEMS

For each of the systems outlined below, draw up a truth table, get an expression for each output in terms of the inputs, simplify the expressions and draw circuit diagrams to show how they may be implemented with NAND gates.

1 The output of this system, $\overline{L}$, controls an LED. The inputs, R and S are generated by push switches (see figure 1.52). The LED must come on when either, but not both, of the switches are pressed.

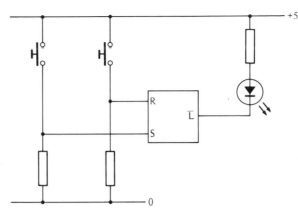

Figure 1.52

Figure 1.53

Figure 1.54

2 The system shown in figure 1.53 is a quad input AND gate. Its output is a logical 1 only when all of its inputs are logical 1.

3 A quad OR gate has an output which is only 0 when all of its inputs are 0. It is shown in figure 1.54.

4 A pair of astronauts sit in their capsule going through their checklists before takeoff. They start by putting a switch up; they have a switch each (A and B). When each has finished his checklist, he puts his switch down. A green light comes on when both checklists have been completed, a yellow light comes on when one checklist (either one) is complete and a red one comes on when neither checklist is complete.

5 Figure 1.55 shows a decimal-to-binary converter. The output represents, in binary, the number of switches which are being pressed. So if any two switches are pressed, BA = 10.

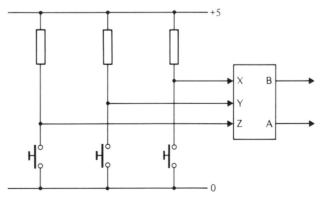

Figure 1.55

6 The keyboard encoder of figure 1.56 (overleaf) has ten inputs labelled $\bar{0}$ to $\bar{9}$. Four outputs represent the binary number DCBA and the fifth (P) only goes low when none of the switches are pressed. The value of DCBA corresponds to the number of the switch which is pressed. So if $\bar{6}$ is pulled low, DCBA = 0110.

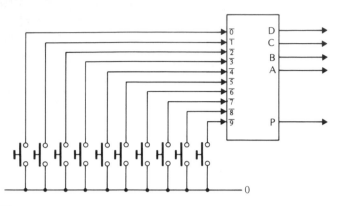

Figure 1.56

The final batch of problems have to be solved with NOR gates.

7 Show how a dual input NAND gate can be assembled from NOR gates.

8 A system has four LEDs as its output. One of the four glows to indicate the value of the binary number BA fed into the system. Each LED indicates one of the four possible values of BA.

9 The system outlined in figure 1.57 has three inputs (A, B and C) and two outputs (P and Q). P must go low if any one (and only one) of the inputs is pulled low. Q must only go low if more than one input is pulled low.

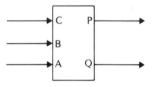

Figure 1.57

SCHMITT TRIGGERS

Now that you know how to build up complex logic systems from simple units like NAND gates, you need to realise their limitations. These stem from the technology used to manufacture the ICs which you have been using. There are other types of IC which can be used to assemble logic systems, based on different technologies.

Take a look at figure 1.58. It shows two circuit diagrams of a dual input NAND gate. The one marked TTL relies on **bipolar transistors**, and is the

Figure 1.58 Inside TTL and CMOS NAND gates

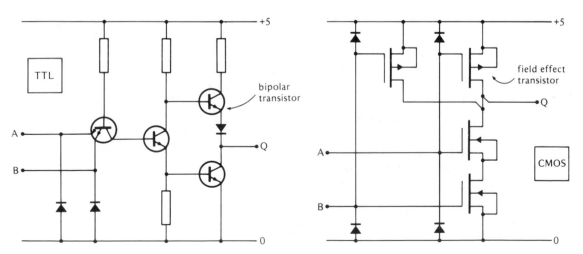

type of gate that you have been using up till now. The other circuit has precisely the same function (Q = A . B), but uses **field effect transistors**. Circuits like this are given the label **CMOS** (read as "sea moss"), and their input and output characteristics are quite different from those of TTL circuits. It is that difference which makes both logic families (TTL and CMOS) useful. You can do things with CMOS that you cannot do with TTL and vice versa.

Before we discuss CMOS characteristics we shall go over the TTL characteristics.

TTL characteristics

TTL gates are useful because they can sink up to 16 mA at logical 0. They can drive LEDs, small loudspeakers and relays, all of which need a fairly large current to operate. TTL outputs are designed to drive TTL inputs. Each TTL input at logical 0 has to have about 1 mA drawn from it, so a single TTL output can be connected to up to 16 inputs. In other words, the **fanout** of TTL is about sixteen.

TTL outputs are not very good at sourcing current. You can only pull a few mA out of a TTL output at logical 1. This does not matter when a TTL output feeds a TTL input, because the latter doesn't need much current pushed into it to haul it up to logical 1. Indeed, TTL inputs float up to logical 1 on their own.

Finally, if an input is held below +1.4 V it will think that it is being held at logical 0.

CMOS characteristics

CMOS inputs require virtually no current to flow in or out of them. This means that you can run as many CMOS inputs as you like off a single CMOS output. So CMOS gates have a very large fanout. On the other hand, a CMOS output can just about source 1 mA at logical 1 and sink 1 mA at logical 0. This does not matter when a CMOS output is driving a CMOS input, but it does matter when a CMOS output drives anything else. For example, CMOS outputs cannot make LEDs glow very brightly. If you want to drive LEDs with a CMOS system you will probably have to insert a TTL gate between the CMOS output and the LED. A CMOS output can just manage enough current to feed a single TTL input.

CMOS gates are not very fussy about the voltage of their power supply. Anything from 5 to 15 volts will do. When they are run off 5 V, an input which is held below +2.5 V will think that it is at logical 0.

NOISE IMMUNITY

In any logic system of sufficient complexity there will be signals that will stray into regions where they have no right to be. Signals flowing in one wire can stray into other adjacent wires. In particular, any sudden change of voltage of a wire can cause brief voltage changes in nearby wires. This is known as **crosstalk**. It can be reduced to a sensible level by having short leads between ICs and keeping wires apart from each other, but you have to accept it as a fact of life. So if you have a wire at logical 0, it will not necessarily be at 0 V. In a large system, where logic levels on other wires are rapidly changing, small spikes will appear on the wire so that its voltage wavers around 0 V in an irregular fashion. This is not serious if the voltage of the wire does not stray too far from 0 V. Large spikes can, however,

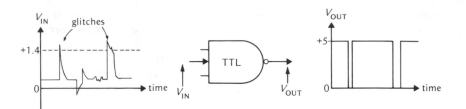

Figure 1.59 The effect of glitches on a logic signal

change a logical 0 into a logical 1 for a short space of time. This is illustrated in figure 1.59. V_{IN} is a noisy signal, supposedly at logical 0, which is fed into a TTL gate. The graph of V_{OUT} as a function of time shows that the gate interprets its input as logical 1 every time V_{IN} strays above $+1.4$ V. These **glitches** on the output of a logic gate can have very serious consequences for many of the systems described in later chapters of this book.

A noise spike only has to be about a volt high in order to upset a TTL input held at logical 0. Since TTL outputs handle relatively large currents which can be switched on and off in the space of a few nanoseconds (10^{-9}s), such noise spikes need not be rare in a complex TTL system. Furthermore, any noise spikes which get into the supply rails can upset the gates, especially as the power supply for TTL has to be between 4.75 V and 5.25 V for the gates to function properly. CMOS gates are much more tolerant of noise spikes. You can alter their supply voltage over a wide range without upsetting them. Then the spike has to be more than 2.5 V high before an input at logical 0 is read in as a logical 1. Finally, the currents involved in a CMOS system are relatively small and they are switched on and off relatively slowly. So there tends to be less crosstalk anyway.

You may wonder why TTL gates are used at all when CMOS ones appear to be more immune to noise. CMOS gates are not as rugged as TTL ones; you can be brutal with TTL inputs without destroying them, but if you pull a CMOS input too far above or below the supply rails you will destroy the chip. CMOS gates need to be carefully handled if they are to survive in a circuit. They are easily damaged by static electricity; you ought to earth yourself before handling them. Furthermore, if you manage to feed more than 50 mA into a CMOS input the gate will heat itself into oblivion!

POSITIVE FEEDBACK

Suppose that you have a CMOS gate which is run off a 5 V power supply. Any input between 0 and $+2.5$ V will be read in as a logical 0, and inputs held between $+2.5$ V and $+5$ V will be read in as logical 1. The 1/0 threshold is $+2.5$ V, halfway between the two supply rails. (This is the nominal value of the 1/0 threshold. It varies from one CMOS IC to another; all the gates on one chip will have the same threshold, but gates on different chips will have slightly different thresholds.) You may think that this is the best that we can do as far as noise immunity is concerned. After all, noise spikes have to be greater than 2.5 V high in order to change a logical 1 into a logical 0 and vice versa. Nevertheless, by using positive feedback we can increase the noise immunity even further.

Look at figure 1.60. The two CMOS NOT gates are represented by the expression A = Q i.e. the output follows the input. The graph shows how

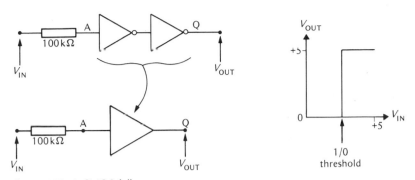

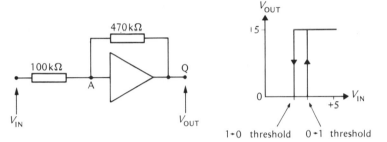

Figure 1.60 A CMOS follower

Figure 1.61 A Schmitt trigger

the output voltage of the system (V_{OUT}) depends on the input voltage (V_{IN}). (There is no significant voltage drop across the 100 kΩ resistor because a CMOS input draws very little current.)

Figure 1.61 shows how to apply positive feedback to the circuit. A 470 kΩ resistor connects Q to A. This dramatically alters the shape of the graph of V_{OUT} as a function of V_{IN}. The system now has two switching thresholds. V_{OUT} rises from 0 V to +5 V if V_{IN} goes above +3 V, but will subsequently dive back down to 0 V when V_{IN} goes below +2 V. By allowing the output of the system to feed a signal back to the input we have increased the range of voltages that are acceptable as logical 1 or logical 0.

We are now going to analyse the operation of this circuit (called a **Schmitt trigger**) in detail. Start off by assuming that Q is a logical 1 i.e. V_{OUT} = +5 V. We are going to calculate the value of V_{IN} which will just make Q change from 1 to 0. Since the 1/0 threshold of a CMOS gate is +2.5 V, Q will go from 1 to 0 when the A terminal goes below +2.5 V. So we need to find out the value of V_{IN} that will pull A down to +2.5 V when V_{OUT} is +5 V. The situation is shown in figure 1.62.

The 470 kΩ resistor has a voltage drop of $5 - 2.5 = 2.5$ V across it. The current which flows through it can be calculated with the help of Ohm's Law;

$$R = \frac{V}{I}$$

therefore $$470 = \frac{2.5}{I}$$

therefore $$I = \frac{2.5}{470}$$

therefore $$I = 0.005 \, \text{mA}$$

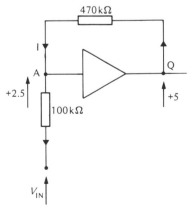

Figure 1.62

That current flows from Q towards A, then through the 100 kΩ resistor to the source of V_{IN}; no current flows into the CMOS input. Using Ohm's Law on the 100 kΩ resistor;

$$R = \frac{V}{I}$$

therefore $\quad 100 = \dfrac{2.5 - V_{IN}}{0.005}$

therefore $\quad 0.5 = 2.5 - V_{IN}$

therefore $\quad V_{IN} = 2.0\,V$

(Notice that we have not been very precise in our calculation. Electronics usually only requires an accuracy of plus or minus 10% in a calculation.)

So once V_{IN} goes below $+2.0\,V$, Q will fall from 1 to 0. The transition between the two logic states is going to be very fast. This is because of the **positive feedback**. Once V_{OUT} starts to drop from $+5\,V$, the voltage at A is pulled below the $+2.5\,V$ threshold by the 470 kΩ resistor. That drop in voltage at A ensures that Q will end up at logical 0. Once the circuit has been **triggered** by bringing V_{IN} below $+2.0\,V$, the feedback forces Q to change state rapidly.

When Q is 0, V_{IN} has to be raised up to $+3.0\,V$ in order to trigger the system again. Have a look at figure 1.63. V_{IN} has to pull A up to $+2.5\,V$ in order to get Q started on its journey from 0 to 1. Since we know the voltage drop across the 470 kΩ resistor we can calculate the current which flows through it;

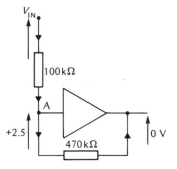

Figure 1.63

$$R = \frac{V}{I}$$

therefore $\quad 470 = \dfrac{2.5}{I}$

therefore $\quad I = \dfrac{2.5}{470} = 0.005$

That current flows through the 100 kΩ resistor as well, so we can use Ohm's Law again to find V_{IN};

$$R = \frac{V}{I}$$

therefore $\quad 100 = \dfrac{V_{IN} - 2.5}{0.005}$

therefore $\quad 0.5 = V_{IN} - 2.5$

therefore $\quad V_{IN} = 2.5 + 0.5 = 3.0\,V$

So if V_{IN} is raised above $+3.0\,V$, the system is triggered. V_{OUT} will rise from $0\,V$ to $+5\,V$, dragging the voltage at A up with it. Once the circuit has been triggered, Q effectively pulls itself up from logical 0 to logical 1.

Schmitt triggers are very good at cleaning up logic signals. They convert signals which rise slowly from 0 to 1 into very fast rising edges; this is a property which is indispensible for some of the devices that you will meet in the other chapters of this book.

EXPERIMENT

Measuring the switching thresholds of some Schmitt triggers

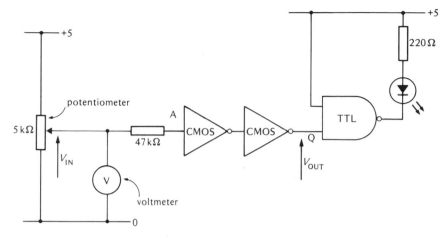

Figure 1.64

Assemble the circuit shown in figure 1.64. V_{IN} is a variable voltage which is generated by a $5\,k\Omega$ potentiometer; its value is monitored by a voltmeter. V_{OUT} is monitored by an LED, with a TTL NOT gate between Q and the LED. It should be clear that the LED will only glow when V_{OUT} is $+5\,V$ i.e. when Q is a logical 1.

Twiddle the potentiometer until you find the value of V_{IN} which causes Q to change state. Note the value of this threshold voltage.

Now add positive feedback by inserting a $100\,k\Omega$ resistor as shown in figure 1.65. Measure the two switching thresholds of the circuit i.e. the values of V_{IN} which make Q change state.

Use your measured value of the 1/0 threshold of the system without feedback and the values of the two resistors to work out what the two thresholds of the Schmitt trigger ought to be.

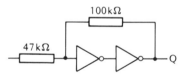

Figure 1.65

PROGRAMMABLE LOGIC SYSTEMS

By now you are probably quite competent at designing simple logic systems. By simple, we mean systems that have only a few outputs and inputs. At first glance it would appear that more complex systems are going to be more difficult to design and harder to understand. They will have larger truth tables, you will have to simplify longer Boolean expressions and they will have larger, and more complicated, circuit diagrams.

This is not necessarily true. It is possible to design simple logic units which can be slotted together to make as large and as complex a system as you like, with hardly any design effort at all. Furthermore, a particular type of complex circuit can be made to behave like **any** logic system. That system is known as a **programmable read only memory**, or **PROM**. It is the ultimate logic system; we are now going to show you how it is assembled and used.

RUSSIAN DOLLS

When you look at a Russian doll, it appears to be complete in itself. If you investigate it closely you will find that it can be opened. Inside, you will find another Russian doll, identical to the first one but a bit smaller. Similarly, when the second doll is opened a third doll will be discovered inside. Each Russian doll contains smaller versions of itself, except the smallest one. Starting off with the smallest doll (the one that does not open up) you can build as large a doll as you like by adding more and more identical layers.

There are some complex logic systems which exhibit similar behaviour when they are analysed. The complex system looks like an assembly of smaller versions of itself. This characteristic is especially useful when we have to build very large systems. For example, consider the simple AND gate shown in figure 1.66. Its output obeys the expression $Q = A.B$. A little Boolean algebra should convince you that the three AND gates shown in figure 1.67 make up a 4 input AND gate; i.e. a system whose output obeys the expression $Q = A.B.C.D$. So a complex AND gate can be synthesised out of a few simple AND gates.

Once you know that a system behaves like a Russian doll, you can easily make it as complex as you like. You can put five 4 input AND gates together to make a 16 input one (figure 1.68), without any difficulty. This

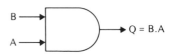

Figure 1.66 A two input AND gate

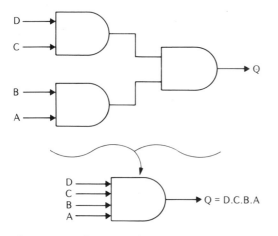

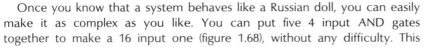

Figure 1.67 A four input AND gate

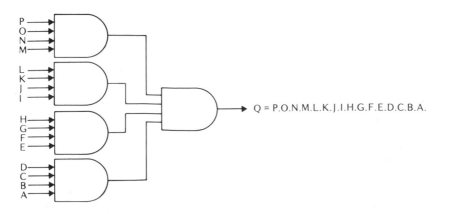

Figure 1.68 A sixteen input AND gate

ease of design effort is impressive because the system of figure 1.68 really represents 15 dual input AND gates. Furthermore, imagine that you are trying to design the system from scratch. There are sixteen inputs, requiring a 65536 line truth table! The worst part about designing a 256 input AND gate (the next Russian doll in the series) is the tedium of drawing out the seventeen 16 input AND gates, despite the fact that the circuit really consists of 255 dual input AND gates.

Multiplexers

A simple **multiplexer** is shown in figure 1.69. Two signals are fed into I_1 and I_0 from the left. The state of the address pin (A) decides which of those two signals will appear at Q. When $A = 1$, $Q = I_1$ and when $A = 0$, $Q = I_0$. In other words, the signal fed into the address pin selects the source of the signal fed out of Q.

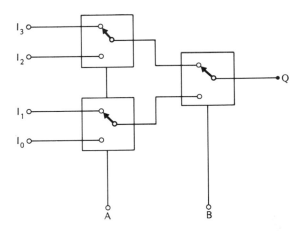

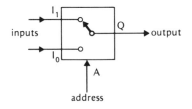

Figure 1.69 A 2 input multiplexer

Figure 1.70 A 4 input multiplexer

Multiplexers are examples of Russian doll systems. Big multiplexers can be made out of many smaller ones. For example, figure 1.70 shows how three 2 input multiplexers can be put together to make a 4 input one. That device needs to have two address pins, B and A, as it has to be able to select one out of four inputs to connect to the output. The binary number BA is the address of the input which is selected. So when BA = 10, $Q = I_2$.

The way is now clear to designing as complex a multiplexer as you like. A 16 input multiplexer can be assembled out of five 4 input ones, a 256 input one can be assembled out of seventeen 16 input ones etc. What is not so obvious is the fact that large multiplexers can be made to behave like any other logic system.

BLANK SYSTEMS

Take a look at figure 1.71. An eight input multiplexer has its inputs connected to logical 1 or logical 0, and it is those eight connections which fix the truth table of the whole system. If you treat the three address pins (C, B and A) as the inputs of the system and Q as its output, then its truth table

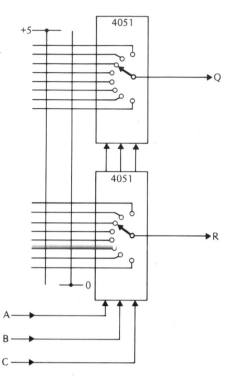

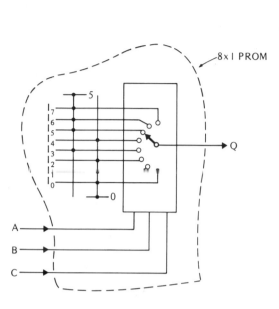

Figure 1.71 An eight input multiplexer

Figure 1.72 An 8 × 2 bit blank PROM

will be as shown below; convince yourself! (Think of the decimal value of the binary number CBA.)

C	B	A	Q
0	0	0	0
0	0	1	0
0	1	0	0
0	1	1	1
1	0	0	0
1	0	1	1
1	1	0	1
1	1	1	1

A little thought should convince you that by altering the connections of the eight inputs, you could make the system have **any** eight line truth table. Moreover, you don't need to shuffle any Boolean algebra in order to get the system to have a particular truth table; each line of the table gives you the address of an input and what its logic state needs to be. Multiplexers which are used this way make logic systems '**user-friendly**'; the multiplexers themselves are complex, but that complexity enables them to be easy to use.

You can get a single 8 input multiplexer on a 4051 CMOS IC. It is a very useful blank logic system to have in your arsenal of basic logic units. There are 2^8 (i.e. 256) different logic systems which have three inputs and one output, so a single 4051 IC is, in a sense, 256 different circuits. If you need to assemble a system which has more than one output, then you use more than one 4051. Figure 1.72 shows how to connect up two 8 input multiplexers to make a blank system which has three inputs and two outputs. That circuit can be programmed to store two eight line truth tables; it is therefore an 8 × 2 bit programmable read only memory (or PROM). The programming is done by connecting the sixteen multiplexer inputs to logical 1 or logical 0.

Multiplexers are Russian doll devices, so you can easily design PROMs that can store any number of truth tables of any length. So in principle you can use multiplexers to solve any logic problem! Furthermore, by using a PROM to solve a problem rather than an array of interconnected NAND gates you end up with a more flexible system; the PROM is easily reprogrammed whereas the array of NAND gates is not.

The use of multiplexers alone does not give the most elegant sort of PROM. You can get away with fewer logic gates if you use demultiplexers as well.

Demultiplexers

The simplest possible **demultiplexer** is shown in figure 1.73. It has one signal input, two signal outputs and an address pin. Logic signals fed into Q are routed through to either D_1 or D_0; their destination is fixed by the state of the address pin. So when $A = 0$, $D_0 = Q$; when $A = 1$, $D_1 = Q$. A demultiplexer is essentially a multiplexer which is being run backwards. Indeed, because signals can travel both ways through CMOS multiplexers, you can make an 8 output demultiplexer out of a 4051. (This trick doesn't work for TTL multiplexers; the CMOS ones are built around analogue switches, whereas the TTL ones are made out of logic gates.) It should be obvious that demultiplexers are Russian doll devices, so you can make them as big as you like without too much trouble.

Figure 1.74 shows a demultiplexer in action. CMOS devices have been

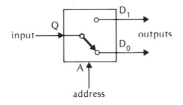

Figure 1.73 A 2 output demultiplexer

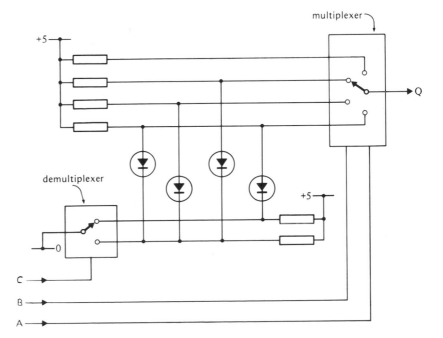

Figure 1.74 An 8 × 1 bit PROM

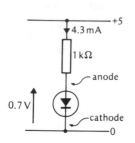

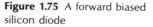

Figure 1.75 A forward biased silicon diode

Figure 1.76 A dual input AND gate

used to assemble the system, so it effectively contains four 2 input multi-plexers. The whole system has three inputs (C, B and A) and a single output Q; it is an 8 × 1 bit PROM. It has exactly the same function as the circuit shown in figure 1.71, a system which is built out of seven 2 input multi-plexers. The programming is done by connecting the outputs of the demul-tiplexer to the inputs of the multiplexer with **diodes**. Before we proceed to explain how the circuit works, you need to know a bit about the behaviour of diodes.

A diode is the electronic equivalent of the one-way valve. When it is connected as shown in figure 1.75 current will flow through it; its cathode is being pulled below its anode. When a **silicon diode** is **forward biased** in this way, the voltage drop across it is approximately 0.7 V. You **reverse bias** a diode by pulling its cathode above its anode. When this happens no current flows through the diode.

There are many ways in which this behaviour can be exploited in an electronic circuit. PROMs generally use diodes to make simple AND gates. If you look at figure 1.76, you will see how a couple of diodes and a pull-up resistor can be put together to make a dual input AND gate. Suppose that B is a logical 0, and A a logical 1. Diode D_1 will be forward biased (its cathode is being pulled below its anode), so Q will sit at +0.7 V. This means that diode D_2 will be reverse biased, as its anode will be 4.3 V below its cathode; no current will flow through it. Obviously, pulling either, or both, of A or B will pull Q almost down to 0 V. Only when both inputs are held at +5 V will the resistor pull Q up to logical 1.

Returning to the circuit of figure 1.74, we can now explain how it has been programmed. If you consult its truth table (the one for the system of figure 1.71), you will see that Q is 0 for four of the input states i.e. CBA = 000, 001, 010 and 100. C is a one bit address which feeds a logical 0 into one of the two outputs of the demultiplexer. BA is a two bit address which selects one of the four inputs of the multiplexer, feeding its state out to Q. If you want Q to be 0 for a particular address CBA then you have to hook a diode between the selected output of the demultiplexer and the selected input of the multiplexer.

Study the circuit of figure 1.74 and convince yourself that it has the truth table which we claim it has. Remember that the resistors normally pull the lines that they are connected to up to logical 1; a line is only pulled down to (almost) 0 if it is attached to the anode of a diode whose cathode is being held at logical 0.

EPROMs

Read only memories are a very convenient way of solving large and complex logic problems. Modern silicon chip technology can easily fit the circuitry needed to implement a 2048×8 bit PROM into a single IC package. Furthermore, because that package can be used to solve an enormous number of problems, it can be made in huge production runs which brings its price down. Of course, you can't get inside a PROM which has been made as an integrated circuit, so you can't program it by inserting diodes. Instead you purchase the PROM with a full complement of programming diodes built into the circuit. The PROM is programmed by selectively destroying (or "blowing") the diodes by pulsing an enormous current through them.

Once it has been programmed, the contents of a PROM cannot be changed. Nevertheless, if you use a PROM to solve a logic problem, you end up with a system which consists of only one IC. So by pulling out that IC and replacing it with a PROM which contains a different set of truth tables you can quickly and easily change the behaviour of the entire system.

EPROMs (erasable programmable read only memories) can be scrubbed clean by exposing them to ultra-violet light. A popular EPROM, the 2716, is shown in figure 1.77. It has ten address lines and eight outputs, so it can store eight truth tables, each of which is 2048 (i.e. 2^{10}) lines long. You program it as follows. For each address in turn, you feed into the outputs what they ought to be. Hold the V_{PP} pin at $+25\,V$, the $\overline{PGM}$ pin at $0\,V$ and pulse the $\overline{CE}$ pin down to logical 0 for 50 milliseconds. Once it has been programmed, you disconnect the $+25\,V$ supply, anchor $\overline{CE}$ at logical 0 and the IC behaves just like a 2048×8 bit PROM.

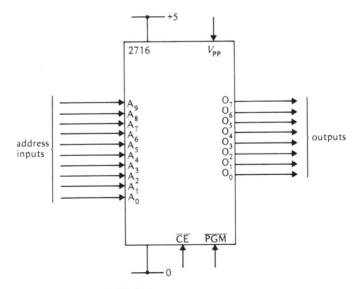

Figure 1.77 A 2716 EPROM

EXPERIMENT
Programming a 16 × 1 bit PROM

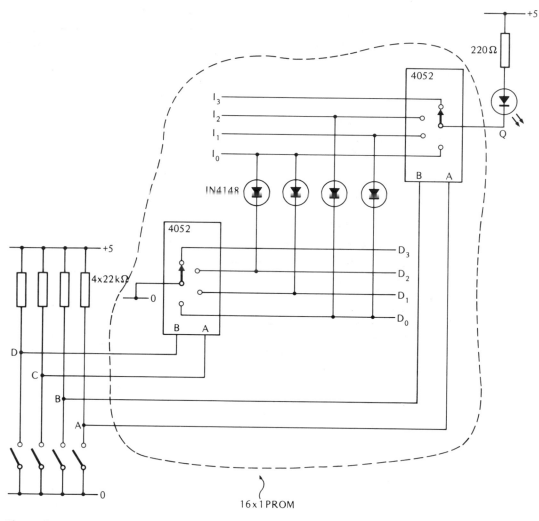

Figure 1.78

The blank PROM is shown in figure 1.78. Both the demultiplexer and the multiplexer can be found on a 4052 IC. You will have to use an IC for each. The four switches provide the four input signals for the system i.e. they generate the address DCBA. The LED monitors the state of the output, Q.

Assemble the system, with the four diodes inserted as shown in the circuit diagram. You should find that the LED only glows when any one (and only one) of the switches are open.

Reprogram the system so that the LED only comes on when only two of the switches are open. When you have got it to work, draw the circuit diagram of the system.

Finally, program the circuit so it obeys the Boolean expression $Q = A \cdot B + C \cdot D$. Draw the circuit diagram of the system.

PROBLEMS

1 Draw up the truth table of a 2 input multiplexer. Draw a circuit diagram to show how one can be built out of NAND gates.

2 Show how an 8 input multiplexer can be built out of 2 input multiplexers.

3 A TTL demultiplexer has the truth table shown below.

A	Q	D_0	D_1
0	0	0	1
0	1	1	1
1	0	1	0
1	1	1	1

Draw a circuit diagram to show how one can be made out of dual input NAND gates.

4 Show how an 8 output demultiplexer can be assembled from 2 input ones.

5 Draw a circuit diagram to show how you would use the two 4 input multiplexers on a 4052 IC to make a system with the following properties. It has two inputs and two outputs. One output is a logical 1 only when both inputs have the same state. The other is a logical 0 only when both inputs are 0.

REVISION QUESTIONS

1 A circuit contains three switches and a resistor. When one of the switches is pressed and either of the other two are pressed at the same time, the output of the circuit goes from logical 1 to logical 0. Draw a suitable circuit. Explain how it works.

2 Draw up truth tables for the following dual input logic gates; AND, OR, NOT, NAND, NOR and EOR. Draw the circuit symbol for each gate. If each gate has inputs labelled A and B, and an output labelled Q, write down a Boolean expression linking Q with A and B for each gate.

3 Explain how AND, OR and NOT gates may be assembled from NAND gates. Draw the circuit diagram of a system whose output is given by (A AND (NOT B)) OR ((NOT A) AND B) using

  a) any mixture of gates

  b) NAND gates only

  c) NOR gates only.

4 Draw a NOR gate based on two relays and a pull-up resistor. Explain how it works.

5 Discuss the electrical properties of TTL inputs and outputs (e.g. limits on logical 1 and logical 0, typical input and output currents etc.). Compare and contrast them with the equivalent properties of CMOS inputs and outputs.

6 Write down examples of the use of

   a) De Morgans theorem

   b) the Redundancy theorem

   c) the Race Hazard theorem.

7 Three push switches pull the inputs A, B and C of a logic system up to logical 1 when they are pressed. The output of the logic system controls a LED; a current of 10 mA must flow through it when two or three of the switches are pressed. Design a suitable logic system using

   a) NAND gates

   b) NOR gates.

Show, with reasons, the values of any resistors used in the circuit.

8 Draw up the truth table of the expression $Q = A + B . \bar{C}$.

9 Draw the circuit diagram of a system whose output is given by

$$\bar{Q} = \bar{A} . B . \bar{C} + A . \bar{B} . \bar{C} + \bar{A} . \bar{B} . C$$

using NAND and NOT gates only.

10 Design a keyboard encoder which will feed out a two bit binary number whose value corresponds to the number of the switch being pressed.

11 Draw the circuit diagram of a Schmitt trigger made out of two NOT gates and a pair of resistors. Describe the properties of the circuit.

12 Explain the meaning of the terms 'noise immunity', 'crosstalk', '1/0 threshold' and 'fanout'.

13 Describe the properties of a two input multiplexer. Show how seven of them can be connected to make an eight input multiplexer. Explain how the latter system can be made to look like any logic system which has three inputs and one output.

2

Timing systems

All of the systems which you met in the last chapter were static. Their outputs did not change until their inputs were changed. This chapter introduces you to dynamic systems i.e. systems whose outputs change with time even if their inputs do not change. Virtually all dynamic systems contain at least one **capacitor**. In a sense, the addition of a capacitor to a system allows it to detect the passing of time!

CAPACITORS

Figure 2.1 A capacitor

A capacitor can be constructed out of two flat sheets of metal which are placed next to each other without touching. The gap between the sheets has to be quite small, so a thin sheet of material that does not conduct current is used to separate them. The construction of the device is reflected in its circuit symbol (figure 2.1). The two metal sheets are known as the **plates** of the capacitor and the high resistance stuff between them is called the **dielectric**.

Capacitors come in a vast range of shapes, sizes and colours. The electrical size of a capacitor is measured in **farads**. A small capacitor would be of the order of a few picofarads, or pF ($1\,pF = 10^{-12}\,F$), whereas 10 000 microfarads, or μF ($1\,\mu F = 10^{-6}\,F$) would be reckoned a large capacitor. The physical size of a capacitor goes more or less hand in hand with its electrical size, as does the method of its construction. Small capacitors are made by depositing thin metal films on sheets of plastic or mica. The resulting sandwich is sometimes rolled up to make a cylinder. This type of capacitor is said to be **unpolarised**. Large capacitors (i.e. more than a microfarad) are polarised. Their construction means that they must always be operated with one plate at a higher voltage than the other.

STORING CHARGE

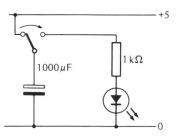

Figure 2.2

We will now look at what a capacitor can do. It is basically a rechargeable cell, a device which can store electrical energy. When a capacitor is connected across a pair of supply rails it charges up and looks like a battery. It stays charged up when it is disconnected from the supply and can be used to push current through a circuit that is subsequently attached to its plates.

Figure 2.2 illustrates this process. Initially the top plate of the capacitor is connected to the $+5\,V$ supply line, so the capacitor is charged up. When the switch changes over, the capacitor discharges through the LED which lights up for an instant. The capacitor shown in the diagram is a polarised one; the clear plate must always be at a higher voltage than the opaque one.

Every capacitor, whatever its size or construction, obeys this equation:

$$C = \frac{Q}{V}$$

where C is the capacitance of the capacitor,
 Q is the amount of charge stored on each plate and
 V is the voltage difference between the plates.

The standard units for C, Q and V are farads (F), coulombs (C) and volts (V). For electronics work it is convenient to adopt a different set of units, mainly because one farad represents an enormous capacitor. So for the rest of this book, unless otherwise indicated, we are going to assume that C is measured in microfarads, Q in microcoulombs and V in volts. (1 microfarad $(\mu F) = 10^{-6} F$.)

Charge

What is charge? Consider figure 2.3. The switch is closed, so there is a voltage difference of 5 V across the capacitor. Since we are told the size of the capacitor (1000 μF) we can calculate the amount of charge on each plate;

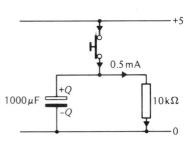

Figure 2.3 Charged up

$$C = \frac{Q}{V}$$

therefore $1000 = \dfrac{Q}{5}$

therefore $Q = 5 \times 1000$

therefore $Q = 5000\,\mu C$

So the top plate has $+5000\,\mu C$ of charge stored on it. The two plates always have equal amounts of charge on them, but with the opposite sign. The plate at the higher voltage always carries the positive charge. The bottom plate of the capacitor in figure 2.3 must therefore hold $-5000\,\mu C$ of charge.

You will recall from Chapter One that when a current flows through a resistor, charge is moved from one end of it to another. A current flow is a transfer of charge from high voltage to low voltage. For example, consider the resistor shown in figure 2.3. We can use Ohm's Law to calculate that a current of 0.5 mA flows through it, so charge is being pushed through the resistor at the rate of 0.5 millicoulombs per second.

We can state the link between current and charge more formally with this equation:

$$I = \frac{Q}{t}$$

where I is the current flowing through the system,
 Q is the amount of charge pushed through it and
 t the time taken for the charge to be transferred.

The standard units for I, Q and t are amps, coulombs and seconds respectively. As one second is rather a long time interval in electronic terms, we shall use a more convenient set of units throughout this book. Unless otherwise stated, we will use milliamps, microcoulombs and milliseconds.

Note that the only current which flows in figure 2.3 is the one through the resistor. The gap between the capacitor plates stops any charge being directly transferred from one to the other. So charge simply sits on the capacitor plates, whereas it continuously flows through the resistor. In practice, the dielectric will not have an infinite resistance, so there will be a small **leakage current** through it. (The leakage current of an unpolarised capacitor is usually minute, but that of a polarised capacitor can be of the order of a few microamps and is very temperature dependent.)

Discharging

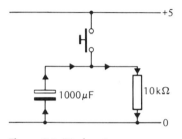

Figure 2.4 Discharging

Figure 2.4 illustrates what happens when the switch of figure 2.3 is opened. The capacitor acts like a battery and initially keeps the top end of the resistor 5 V above the ground end. This voltage drop across the resistor makes a current flow through it; you can use Ohm's Law to verify that it is going to be 0.5 mA. So positive charge will flow off the top plate, through the resistor into the ground supply rail. At the same time, positive charge will flow from the ground supply rail onto the bottom plate in order to keep the same amount of charge on both plates. (The positive charge which arrives from the ground supply rail will cancel out some of the negative charge already on the plate.)

This flow of charge through the resistor steadily depletes the store of charge on the capacitor plates. Eventually there will be no charge left on the plates and the capacitor will have been completely discharged.

How long will the capacitor be able to supply a current of 0.5 mA? We know the charge and the current so we should be able to calculate the time;

$$I = \frac{Q}{t}$$

$$\text{therefore} \quad 0.5 = \frac{5000}{t}$$

$$\text{therefore} \quad t = \frac{5000}{0.5}$$

$$\text{therefore} \quad t = 10\,000 \text{ ms}$$

So it looks as though the 5000 μC on the top plate takes 10 000 milliseconds, or 10 s, to flow through the resistor onto the bottom plate. This does, however, assume that the current is going to be a constant 0.5 mA until the capacitor is completely discharged. That assumption is not valid. As the capacitor runs out of charge the voltage across it will get progressively smaller ($C = Q/V$?). So as time passes, the amount of current flowing through the resistor will get smaller and smaller. The rate at which a capacitor is being discharged will depend on how much charge it has on its plates. It will lose charge rapidly when it is fully charged, but lose it slowly when nearly discharged.

Half-life

We shall now try to calculate the time taken for the voltage across the capacitor to get to half its original value. This is called the **half-life** of the

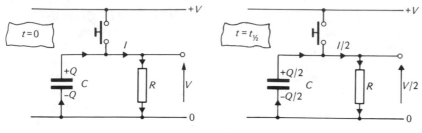

Figure 2.5

system, and will give us some idea of how quickly the capacitor will be discharged. We are going to use algebra in the calculation, so that we end up with a formula.

The symbols are defined in figure 2.5. The diagram on the left shows the situation when the switch has just been released i.e. when $t = 0$. The diagram on the right shows what has happened one half-life later, when $t = t_{1/2}$.

At $t = 0$ we can say that;

$$C = \frac{Q}{V} \tag{1}$$

$$R = \frac{V}{I} \tag{2}$$

When $t = t_{1/2}$, the voltage across the plates is $\frac{1}{2}V$ (by definition). Since C (which is constant) $= Q/V$, if you halve V you must also halve Q. So a positive charge of $\frac{1}{2}Q$ must have passed through the resistor by the time that $t = t_{1/2}$. The average current must equal $\frac{1}{2}Q/t_{1/2}$. If we assume that the current dropped steadily from I to $\frac{1}{2}I$, then the average value must also equal $\frac{3}{4}I$. (This is not going to be quite right as we know that the current will drop more rapidly at the start of the discharge than later on.)

So, combining our two expressions for the average current, we get;

$$\frac{Q}{2t_{1/2}} = \frac{3I}{4} \tag{3}$$

From equations (1) and (2);

$$Q = CV \tag{4}$$

$$I = \frac{V}{R} \tag{5}$$

Using equations (4) and (5) to eliminate Q and I from (3) we obtain;

$$\frac{CV}{2t_{1/2}} = \frac{3V}{4R}$$

If we cancel V from both sides and shuffle the remaining terms around we get our final expression;

$$t_{1/2} = 0.67RC$$

Notice that something interesting has happened. The formula for $t_{1/2}$ does not contain V, the original voltage across the capacitor. This means that the half-life of the system is independent of the voltage. Whatever value the voltage has at a particular instant, its value will be halved in the next 0.67RC milliseconds. If you think about it, this is a sensible result. If you start off with a large voltage there will be a correspondingly large current flow which will make the voltage drop quickly. On the other hand, a small initial voltage leads to a small current flow and a slow change of voltage with time.

Our expression for $t_{1/2}$ is slightly inaccurate. A more sophisticated calculation (see Appendix A for the details) gives the following relationship between the voltage and the time;

$$V = V_0 e^{-(t/RC)}$$

V is the voltage at time t, V_0 is the voltage when $t = 0$ and e is a number with very special properties (its value is about 2.71). If we shuffle this expression around we can get another, very useful, version of it;

$$t = RC \log_e (V_0/V)$$

You can easily verify for yourself that when V is half of V_0 the expression reduces to;

$$t = RC \log_e (2) = 0.69RC$$

So the half-life of the system should really be given by;

$$t_{1/2} = 0.69RC$$

We can use this expression to work out what happens to the voltage after the switch of figure 2.4 is released. Firstly, we calculate the value of $t_{1/2}$;

$$t_{1/2} = 0.69RC$$
$$= 0.69 \times 10 \times 1000$$
$$\text{therefore} \quad t_{1/2} = 6\,900 \text{ ms}$$
$$\text{therefore} \quad t_{1/2} = 6.9 \text{ s}$$

So the voltage across the resistor is halved for every 6.9 s that elapses. The initial value is $+5$ V. After 6.9 s it will be $+2.5$ V. After 13.8 s (i.e. two half-lives) it will be $+1.25$ V etc. The graph of figure 2.6 shows what happens; since the voltage is being continually halved every 6.9 s, it never gets to 0 V.

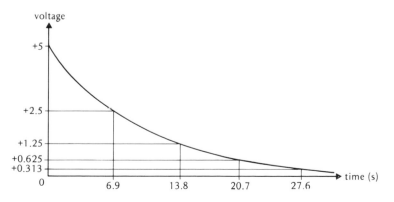

Figure 2.6 How the voltage across a discharging capacitor changes with time

Charging up

We are now going to look at what happens when a capacitor is charged up through a resistor. Take a look at figure 2.7. When the switch is closed, there will be no voltage drop across the capacitor. Both of its plates will be held at 0 V. A current of approximately 1 mA will flow through the resistor. Once the switch is released, that current flows onto the top plate of the capacitor instead of the bottom supply rail. So positive charge starts to accumulate on the top plate. At the same time, positive charge flows off the bottom plate in order to keep the amount of charge on both plates the same. So while a current of 1 mA flows onto the top plate, an identical current flows off the bottom plate. The current appears to flow right through the capacitor.

As the capacitor charges up, the voltage of the top plate steadily rises until it eventually reaches +5 V. No more current will flow after that, as there will be no voltage drop across the resistor. Of course, the current does not suddenly stop. It just gets steadily smaller as the capacitor charges up. If you look at figure 2.8 you will see how the voltage across the capacitor changes with time. Note that the voltage rises to half of its final value in 0.69RC milliseconds.

In Appendix A it is shown that for any RC network whose ends are kept at fixed voltages, the voltage across the resistor is given by this equation;

$$V = V_0 \, e^{-(t/RC)}$$

V_0 is the initial voltage across the resistor and V the voltage across it after t milliseconds. R and C are the values of the resistor and capacitor in our usual units.

Exactly the same formula applies to the charging up of a capacitor. If you go back over the arguments used in the last section you will recognise that V always represented the voltage across the resistor. Of course, once you know how the voltage across the resistor changes with time, it is simple to work out how the voltage across the capacitor changes. For example, for the system shown in figure 2.7, the voltage across the capacitor is always $5 - V$, where V is the voltage across the resistor. So if you want to work out how long it takes the capacitor in figure 2.7 to charge up to within 5% of its final voltage when the switch is released, you have to think about the voltage change across the resistor. At the instant that the switch is released the voltage across the resistor will be 5 V i.e. $V_0 = 5$. When the capacitor has acquired 95% of its ultimate charge, the voltage across it must be $0.95 \times 5 = 4.75$ V. So V (the voltage across the resistor) will be

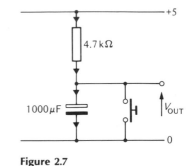

Figure 2.7

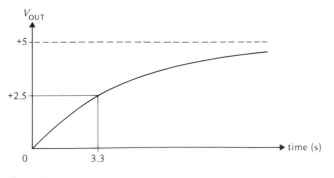

Figure 2.8

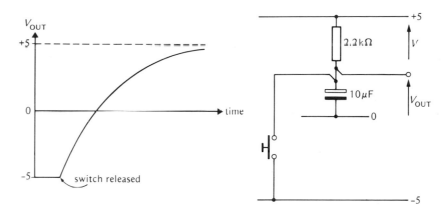

Figure 2.9

$5 - 4.75 = 0.25\,\text{V}$ at that instant. The product RC is $4.7 \times 1000 = 4700$, so we can substitute the numbers into the formula to find the elapsed time;

$$t = RC \log_e (V_0/V)$$
$$= 4700 \log_e (5/0.25)$$
$$\text{therefore} \quad t = 14\,080\,\text{ms}$$
$$\text{therefore} \quad t = 14\,\text{s}$$

Figure 2.9 shows the system which we shall use for our final two examples. The graph illustrates what happens to the voltage of the top plate of the capacitor when the switch is released. Convince yourself that the graph shows that the voltage across the resistor tends towards zero as time passes; this is a very useful rule of thumb which applies to any series RC network whose two ends are held at fixed voltages.

First of all, how long do we have to wait for V_{OUT} to get to $+2\,\text{V}$ when the switch is released? The initial value of V_{OUT} will be $-5\,\text{V}$. So the value of V_0 will be $10\,\text{V}$ (across the resistor, remember?). Once V_{OUT} has got to $+2\,\text{V}$, the value of V will be $+5 - 2 = 3\,\text{V}$.

$$t = RC \log_e (V_0/V)$$
$$= 2.2 \times 10 \times \log_e (10/3)$$
$$\text{therefore} \quad t = 26.5\,\text{ms}.$$

So V_{OUT} reaches $+2\,\text{V}$ at a time $26.5\,\text{ms}$ after the switch has been released.

Now let us work out the value of V_{OUT} at a time $50\,\text{ms}$ after the switch is opened. We have to calculate the voltage across the resistor first. The value of V_0 will still be $10\,\text{V}$ and RC will still be $22\,\text{ms}$.

$$V = V_0\,e^{-(t/RC)}$$
$$= 10 \times e^{-(50/22)}$$
$$\text{therefore} \quad V = 1.03\,\text{V}$$

Now if the top end of the resistor is held at $+5\,\text{V}$ and there is a $1\,\text{V}$ drop across it, the bottom end must be at $+4\,\text{V}$. So V_{OUT} is $+4\,\text{V}$ at a time $50\,\text{ms}$ after the switch has been opened.

EXPERIMENT

*Measuring the half-life of an RC network and comparing
its actual value with its theoretical one*

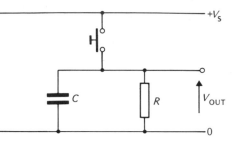

Figure 2.10 A circuit used for investigating the discharging of a capacitor

The circuit is indicated in figure 2.10. Make $R = 10\,k\Omega$ and $C = 1000\,\mu F$. The capacitor will be polarised; be careful to get it round the right way. Use a 5 V supply for V_S. Connect a voltmeter across the resistor so that it measures V_{OUT}.

Press the switch. When you release it, the value of V_{OUT} will slowly fall from $+5\,V$ to $0\,V$. When V_{OUT} gets to $+4\,V$ start a stopclock going and note its reading when V_{OUT} gets down to $+2\,V$, $+1\,V$, $\frac{1}{2}V$ and $\frac{1}{4}V$. Does 7 s elapse between each of these readings? (Remember that the value printed on the capacitor can is not going to be very accurate. Your value of the half-life may therefore be only approximately 7 s.)

Repeat the experiment twice, with $R = 4.7\,k\Omega$ and $2.2\,k\Omega$. Does the half-life change as you expect it to?

Go back to having $R = 10\,k\Omega$, but have two capacitors in parallel. Measure the half-life of the system as before, taking the average of your four readings. Then repeat the same experiment with the two capacitors in series with each other. You should now be able to work out the total capacitance of a pair of identical capacitors when they are placed a) in parallel and b) in series.

PROBLEMS

1 This question concerns the system of figure 2.10. The switch is pressed and then released. How long do we have to wait for V_{OUT} to get to $+3\,V$ if V_S, R and C are

a) $+5\,V$, $22\,k\Omega$ and $47\,\mu F$

b) $+10\,V$, $2.7\,k\Omega$ and $100\,nF$

c) $+15\,V$, $470\,k\Omega$ and $22\,nF$?
    $(1\,\mu F = 1000\,nF)$

2 For the system shown in figure 2.11, what is the value of V_{OUT} 100 ms after the switch is opened if V_S, R and C are

a) $10\,V$, $4.7\,k\Omega$ and $10\,\mu F$

b) $5\,V$, $220\,k\Omega$ and $1\,\mu F$

c) $15\,V$, $22\,k\Omega$ and $4.7\,\mu F$?

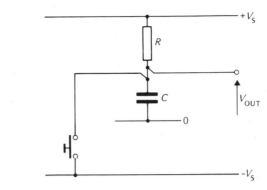

Figure 2.11

3 Go back to figure 2.10 for this one. Suppose that the system is being used to measure the size of a capacitor, by noting the value of V_{OUT} exactly 10 s after the switch is released. If V_s is $+5$ V and R is $22\,k\Omega$, what is the value of C if V_{OUT} gets to

a) $+1$ V

b) $+0.1$ V

c) $+4.5$ V

in the ten seconds?

4 Look at figure 2.11. Work out a formula for the time taken for V_{OUT} to get to 0 V after the switch has been released. Hence show that $V_{OUT} = 0$ V at a time $0.69RC$ ms after the switch is released.

5 The switch marked S in figure 2.12 is released for a long time. It is then pressed for exactly 14 s before being released again. Sketch a graph to show how the voltage at X varies with time. Calculate the time for which Q is a logical 0. You may assume that the NOT gates are CMOS, with 1/0 thresholds at $+2.5$ V.

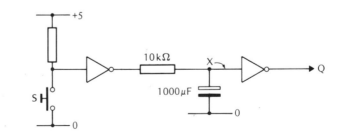

Figure 2.12

6 The system of the last question can be adapted to provide a fixed time delay of one second to a logic signal. You change the input signal and exactly one second later the output signal changes state. Draw a circuit diagram for the system, showing suitable component values. Explain how it works; some graphs might help.

RELAXATION OSCILLATORS

An *RC* network can be used to delay a change in the state of a logic signal. The length of the time delay depends on the size of the resistor and the capacitor, and can therefore be made as long or as short as you like. But if you connect the output of a logic system to its input via an *RC* network you end up with a device which can be used to measure time i.e. an **oscillator**.

SIGNAL DELAY

Consider the system shown in figure 2.13. The NOT gate is a CMOS one, so we can forget about any current which flows in or out of its input terminal. This means that we can use the ideas about the behaviour of the *RC* network that you met in the last section. Let's think about what happens when the state of A is suddenly changed.

Suppose that A is initially at logical 0. When it is suddenly raised to logical 1 (i.e. +5V), there will be a delay before the input of the NOT gate gets to its 1/0 threshold of +2.5 V. In fact, a little thought should convince you that Q will dive down to logical 0 at a time $0.7RC$ ms after A goes up to logical 1; the capacitor has to charge up to half its final voltage. Furthermore, provided that A stays at logical 1 for long enough ($3RC$ ms would do) there will be an identical delay between A going to 0 and Q going to 1. The whole system behaves like an inverter with a built-in time delay; its properties are summarised in the graph of figure 2.14.

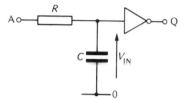

Figure 2.13 A primitive signal delay system

Glitches

You will have noticed in figure 2.14 that the output of our system does not change state cleanly. Q wobbles up and down a few times before settling

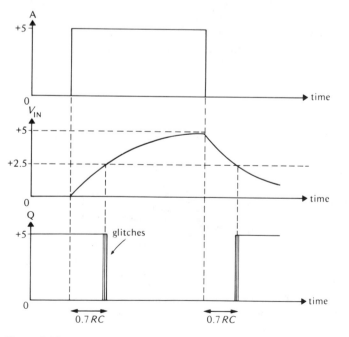

Figure 2.14

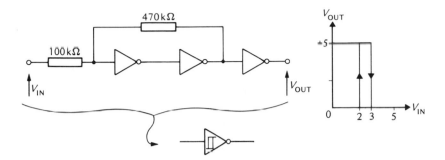

Figure 2.15 A Schmitt trigger NOT gate

into its new state. These glitches arise because V_{IN} (see figure 2.13) is a slowly changing signal. As it approaches the 1/0 threshold of $+2.5\,V$, any noise that it picks up will temporarily push it over the threshold. So as V_{IN} sweeps across $+2.5\,V$, Q will be wiggled up and down. You will find out in later chapters that there are many important logic systems which cannot tolerate glitches. They require clean, fast logic transitions if they are to function properly.

You can avoid dirty transitions by using Schmitt trigger NOT gates. Figure 2.15 shows how such a gate can be assembled from a trio of CMOS NOT gates. The two gates on the left, with their feedback loop, make the basic Schmitt trigger and the gate on the right inverts its output. If you look at the graphs of figure 2.16 you will see how the Schmitt trigger NOT gate irons out the glitches. Initially Q is logical 1, so the 1/0 threshold will be $+3\,V$. When A is raised to logical 1, V_{IN} rises slowly towards that threshold. As V_{IN} gets near to $+3\,V$, it is inevitable that sooner or later some noise will push it over the threshold. Q immediately dives down to logical 0, pulling the 1/0 threshold down to $+2\,V$ as it goes. So unless a noise spike is bigger than $1\,V$ it will be unable to pull V_{IN} below this new threshold, and Q will stay at logical 0 without any glitches.

The time delay system of figure 2.16 is an open loop one. That is, an external input (A) controls the signal fed out at Q; there is no way in which the output can control the input. If we introduce a feedback path, so that the output generates the input, something very interesting happens; the system oscillates.

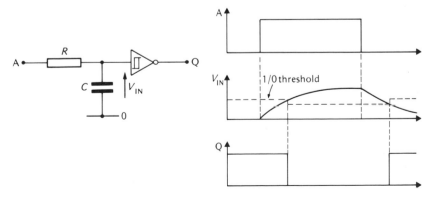

Figure 2.16 A better signal delay system

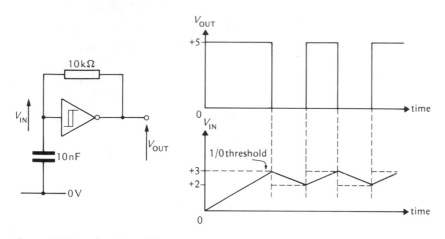

Figure 2.17 A relaxation oscillator

OSCILLATORS

Figure 2.17 shows the structure of a relaxation oscillator. At its heart is a Schmitt trigger NOT gate, with an *RC* network connecting the output to the input. This feedback makes the output continually change state at a rate fixed by the sizes of *R* and *C*.

In order to understand why the system behaves this way suppose that the power supply has just been connected. The capacitor will be completely discharged, so V_{IN} will start off at 0 V. Since the Schmitt trigger is an inverter, the initial value of V_{OUT} will be +5 V. There will be a voltage drop of 5 V across the 10 kΩ resistor, so current will flow through it. So V_{IN} will slowly rise as charge accumulates on the top plate of the capacitor. Eventually V_{IN} will reach the 1/0 threshold of +3 V (look at figure 2.16) and V_{OUT} will drop from +5 V to 0 V. Now the current will have to flow the other way through the 10 kΩ resistor, discharging the capacitor. V_{IN} will slowly fall towards the new 1/0 threshold of +2 V. Once it gets there, of course, V_{OUT} shoots back up to +5 V and the capacitor starts to charge up again. The system is permanently unstable and V_{OUT} regularly changes state.

How do we set about calculating how fast the system oscillates? Start off by assuming that V_{OUT} has just shot up to +5 V. Then V_{IN} will be +2 V. A glance at figure 2.18 should make it clear that we are dealing with a standard *RC* network. One end is held at +5 V, the other at 0 V, with an initial voltage of +2 V in the middle. Thinking about the resistor, $V_0 = 5 - 2 = 3$ V,

Figure 2.18   **Figure 2.19**

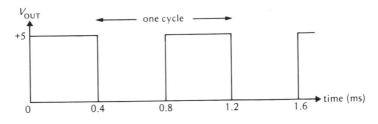

Figure 2.20

and the system will remain stable until $V = 5 - 3 = 2\,V$ (i.e. when $V_{IN} = +3\,V$). We can now use the usual formula to find the elapsed time;

$$t = RC \log_e (V_0/V)$$
$$= 10 \times 0.1 \times \log_e (3/2)$$

therefore $\quad t = 0.4\,ms$

So V_{OUT} stays at $+5\,V$ for 0.4 ms. We can do a similar calculation to find out how long V_{OUT} remains at 0 V. Figure 2.19 illustrates the initial state of the RC network; V_{IN} starts off at $+3\,V$ and slowly falls to $+2\,V$. The values of V_0 and V will be 3 V and 2 V, respectively. If these are substituted into the formula for t, we obtain $t = 0.4\,ms$ again.

So V_{OUT} spends as much time at $+5\,V$ as at 0 V, staying in each state for only 0.4 ms before changing. This sort of output signal is known as a **square wave**, despite the fact that it always looks more like a rectangle wave when you look at it with a **CRO** (**Cathode Ray Oscilloscope**). Figure 2.20 illustrates the shape of the waveform fed out by our system. Each cycle of the signal takes place in a time of 0.8 ms; this is called the **period** of the waveform. It is more usual to specify the **frequency** of the waveform i.e. the number of cycles that it goes through in one second. Frequency and period are linked by this formula;

$$f = 1/T$$

where f is the frequency of the waveform and
$\quad\quad T$ is the period of the waveform.

The usual units for f and T are hertz and seconds; in electronics it is more convenient to use kilohertz (kHz) and milliseconds (ms) instead. So the frequency of our oscillator is $1/0.8 = 1.25\,kHz$.

The 555 IC

The system which we have been considering is an example of a **relaxation oscillator**. You will meet another relaxation oscillator when you find out about operational amplifiers. Indeed, there are many ways in which the Schmitt trigger part of the circuit can be assembled. Unlike some of the other types of oscillator that you will meet in this chapter, the relaxation oscillator is a sure-start system. That is, it **has** to oscillate and will do so automatically when switched on. Furthermore, its frequency is utterly predictable, depending only on the values of three resistors and a capacitor; if the supply rails wobble up and down in voltage there is no effect on the frequency (see the problems) so it makes quite a good clock.

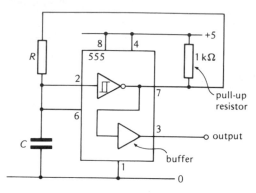

Figure 2.21 A 555 relaxation oscillator

A popular package for building a relaxation oscillator is the 555 IC. This eight pin package is widely used for generating general purpose square waves. It contains a Schmitt trigger NOT gate and a buffer, as shown in figure 2.21. The buffer just feeds out what is fed into it i.e. the logic state of pin 3 is always the same as the state of pin 7. The buffer can sink or source up to 200 mA with ease. The NOT gate, however, has an open collector output which is incapable of sourcing any current at all. The term open collector belongs to the realm of transistors, so we shall not discuss what it means until Chapter Five. Until then just remember that the NOT gate can sink current easily, but needs a 1 kΩ pull-up resistor to source current. The external RC network makes the system oscillate. Since the switching thresholds of the NOT gate are fixed at $\frac{2}{3}V_{CC}$ and $\frac{1}{3}V_{CC}$, the frequency of the output waveform is 1/(1.4RC). The frequency is independent of the supply voltage V_{CC}; the value of R has to be much larger than that of the pull-up resistor for the waveform to be a true square wave.

The 555 is versatile because its supply voltage, V_{CC}, can range from 4.5 V to 16 V. Also, it can be arranged to feed out a single pulse rather than a train of them. But it has some shortcomings, notably a tendency to inject glitches into the supply rails. If you want an alternative to the 555 for generating 5 V square waves, use one of the six Schmitt trigger NOT gates on a 7414 IC; you can only use a 1 kΩ resistor between input and output and the waveform will not be very square, but the system is easily assembled and trouble-free.

EXPERIMENT
Assembling a relaxation oscillator based on three of the
NOT gates in a 4069 IC

Start off by putting together the Schmitt trigger NOT gate shown in figure 2.22. Gates 1 and 2 are the trigger part of the system; gate 3 just inverts the output. The TTL NAND gate is the usual buffer needed to persuade a CMOS output to run an LED. When V_{OUT} is 0 V (i.e. logical 0) the LED will glow.

Use a 5 kΩ potentiometer to generate a variable voltage V_{IN} (look at figure 1.64 if in doubt). Measure, with a voltmeter, the switching thresholds of the system i.e. the values of V_{IN} which will make the LED change state.

Calculate what the thresholds ought to be, assuming that your CMOS gates have 1/0 thresholds of +2.5 V.

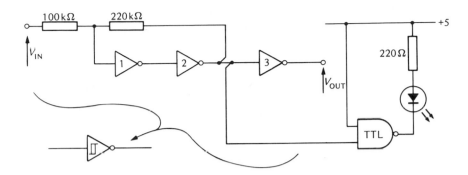

Figure 2.22 A Schmitt trigger NOT gate

Now use V_{OUT} to generate V_{IN} as shown in figure 2.23. You will have to remove the potentiometer of course, but leave the voltmeter so that it monitors V_{IN}. The capacitor will be an electrolytic one, so you will have to make sure that you get it round the right way.

The system ought to be oscillating with a period of about 10 s. Measure the period. Have a look at what V_{IN} is doing as a function of time, and make sure that you understand why it behaves as it does. Calculate the period of oscillation of the system.

Finally, replace the 1000 µF capacitor with a 100 nF one. The system will now oscillate with a frequency of about 1 kHz, so fast that the LED will appear to be continually lit. So you will have to use an oscilloscope (CRO) to look at the waveforms of V_{IN} and V_{OUT}. Trigger the CRO with V_{OUT}; use the CRO to look at the shape of the waveforms V_{IN} and V_{OUT}. With the help of the waveforms displayed on the screen, draw graphs of V_{OUT} and V_{IN} for a couple of cycles; put both curves on the same graph.

If you have time, find out how fast you can make the system oscillate. You can use any size capacitor, but the resistor must be at least 4.7 kΩ (remember that a CMOS output cannot source or sink more than about 1 mA). Measure the fastest frequency with the help of the CRO; it will be about 500 kHz.

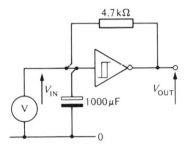

Figure 2.23 A Schmitt trigger NOT gate at the heart of an oscillator

PROBLEMS

1 Calculate the frequency of oscillation of the relaxation oscillator shown in figure 2.24. Assume that the gates are CMOS run off +5 V with 1/0 thresholds of +2.5 V.

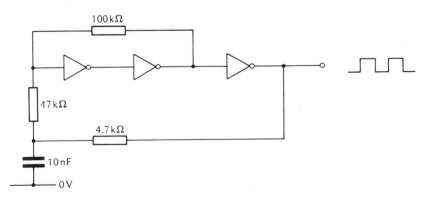

2 By calculating the frequency of oscillation of the system shown in figure 2.24 when it is run off a +15 V supply, show that its frequency is independent of the supply voltage. (The 1/0 thresholds of CMOS gates are halfway between the supply rails.)

3 Look at figure 2.25. A square wave is fed into an RC network. V_{IN} spends equal times at +5 V and 0 V, so the average value of V_{OUT} is +2.5 V. If the maximum value of V_{OUT} is $2.5 + X$ and its minimum value is $2.5 - X$, show that;

$$\frac{2.5 - X}{2.5 + X} = e^{-5/f}$$

where f is the frequency of V_{IN}. (The graph of V_{OUT} shown in figure 2.25 is for $f = 5$ kHz.)

Calculate the maximum and minimum values of V_{OUT} for frequencies of

a) 0.5 kHz

b) 5 kHz

c) 50 kHz.

Draw graphs of V_{OUT} as a function of time for all three frequencies.

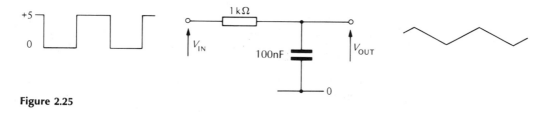

Figure 2.25

THE MONOSTABLE

A **monostable** is a device which feeds out a single pulse when it is triggered. If you look at figure 2.26, you will see that by briefly pulling a pin ($\overline{T}$) down to logical 0, the two outputs of the system change state for a while. A monostable has one stable state in which it will stay quite happily, and another state that it can only stay in for a fixed length of time. When the system has been triggered it enters its temporary state, stays there for a bit and then returns to its permanent state to await the next triggering. It should be fairly clear why an alternative name for the device is the **one-shot**!

Every time that $\overline{T}$ falls from logical 1 to logical 0, the monostable is triggered into pushing Q up to logical 1. Q will stay there for a fixed length

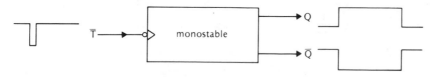

Figure 2.26 Properties of a monostable

of time, regardless of the state of $\overline{T}$ i.e. once you have triggered the system, it ignores its input. Then Q is lowered back to logical 0, and the system is ready to be triggered again. The triangle at the $\overline{T}$ input of the monostable tells you that it is triggered by a logic transition; the circle next to the triangle tells you that 1 to 0 transitions set the system into action. Using the standard jargon, the monostable of figure 2.26 could be described as a falling-edge triggered one-shot with complementary outputs.

SPIKE GENERATORS

There are many ways in which a monostable can be constructed. You are about to find out how they can be built out of logic gates; in Chapter Seven you will meet a transistor version of the monostable. Somewhere inside each of them is a timing element. This is a bit of circuitry which tells the monostable how long to keep Q at logical 1. The timing element is actually provided by the *RC* network shown in figure 2.27; it is sometimes called a **differentiator**. The graphs in figure 2.27 show how the output changes when the input is changed. As you can see, any 1 to 0 transitions (falling edges) generate negative spikes at the output. Similarly, a rising edge which is fed into the circuit generates a positive spike.

Why does the network behave in this way? Suppose that V_{IN} is $+5\,V$ and that it has not changed for some time. Then V_{OUT} will have been pulled down to $0\,V$ by the resistor. One plate of the capacitor will be at $+5\,V$, the other at $0\,V$. Now let V_{IN} suddenly fall from $+5\,V$ to $0\,V$ and stay there. There will be no time for any charge to flow on or off the capacitor plates, so the voltage difference between the plates must remain at $5\,V$ (remember that $C = \text{constant} = Q/V$). The left hand plate is pushed from $+5\,V$ to $0\,V$, so the right hand plate must be pushed from $0\,V$ to $-5\,V$ at the same instant. Thereafter, current will flow through the resistor as one end will be at a lower voltage than the other. That current will dump charge onto the capacitor, so that V_{OUT} slowly rises up to $0\,V$. It should be fairly obvious that $0.7RC$ ms after the change of V_{IN}, V_{OUT} will have crept back up to $-2.5\,V$ and that after $3RC$ ms, V_{OUT} will be almost back to $0\,V$.

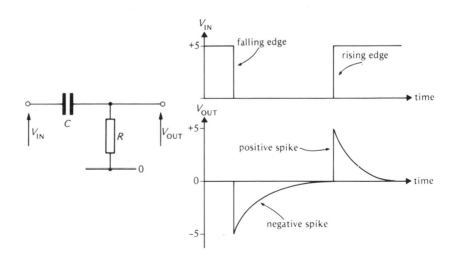

Figure 2.27 A spike generator

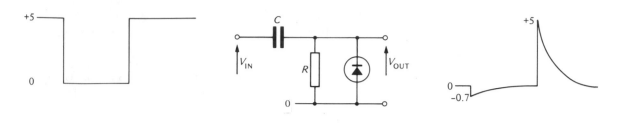

If V_{IN} is subsequently raised back up to $+5\,V$, V_{OUT} will also be pushed up to $+5\,V$ and will then decay back down to $0\,V$ with a half-life of $0.7RC\,ms$.

Figure 2.28 A clamp diode suppressing negative spikes

Clamping diodes

The version of the monostable which we are about to introduce involves the use of logic gates. As you know, these are designed to sample input signals whose voltage lies between the two supply rails. An acceptable input for a logic gate lies between $+5\,V$ and $0\,V$. The output of the differentiator shown in figure 2.27 clearly does not satisfy this criterion; the positive spikes are satisfactory, but the negative spikes are not. Indeed, the negative spikes would destroy the IC!

We can swallow up the negative spikes by adding a diode to the circuit. This is shown in figure 2.28. The diode prevents V_{OUT} from going below $-0.7\,V$, but leaves all of the positive spikes intact. This use of a **clamp diode** is widely used to protect the inputs of systems against signals which are potentially harmful. When V_{OUT} is above $0\,V$ the diode is reverse biased, so it behaves like an enormous resistor. But when V_{OUT} is pushed below $0\,V$, the diode becomes forward biased. Once V_{OUT} gets down to $-0.7\,V$ the diode behaves almost like a short circuit. So as far as rising edges are concerned, the circuits of figure 2.27 and 2.28 are identical; when V_{OUT} is pushed up to $+5\,V$, the capacitor has to discharge to ground via the resistor. But when a falling edge is fed into the system the capacitor will very rapidly charge up via the diode; V_{OUT} will only fall by $0.7\,V$ before it gets clamped there. The two graphs of figure 2.28 summarise the effect of the diode.

If you have a look at figure 1.58 you will see that both TTL and CMOS gates are equipped with clamp diodes at their inputs. So it is quite safe to feed the output of an unclamped spike generator into a CMOS NOT gate, as shown in figure 2.29. The diodes within the gate will prevent its input from rising above $+5.7\,V$ or falling below $-0.7\,V$. The spike generator is being driven by another CMOS NOT gate because the internal clamp diodes are quite fragile. They will burn out if more than $10\,mA$ flows through them. The output of a CMOS gate cannot sink or source much more than $1\,mA$, so this limits the current that can flow on or off the capacitor. When V_{OUT} falls below $-0.7\,V$, no more than $1\,mA$ will be drawn out of the clamp diode of gate 2.

PULSE GENERATORS

The circuit of figure 2.29 converts a falling edge into a pulse. It is not a true monostable, although it has most of the characteristics of one. If you look at the graphs of figure 2.29 you will see that $\bar{Q}$ is a logical 0 for $0.7RC\,ms$

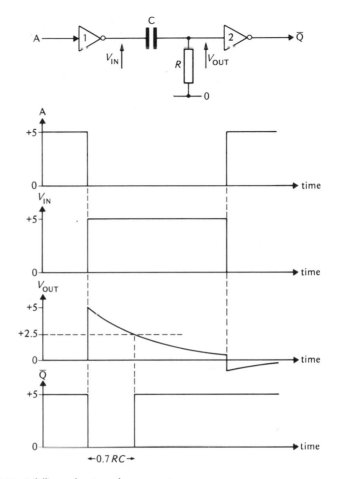

Figure 2.29 A falling edge to pulse converter

every time that a falling edge is fed in at A, but that it remains a steady logical 1 when rising edges are fed in. However, there are two major defects in the circuit.

Firstly, the output is noisy. More specifically, the rising edge of $\bar{Q}$ is likely to be covered with glitches as V_{OUT} falls slowly past the 1/0 threshold of $+2.5\,V$.

Secondly, the output pulse can be terminated by raising A back up to logical 1 before the $0.7RC$ ms are up. If V_{IN} drops by 5 V while the capacitor is discharging then V_{OUT} will also drop by 5 V crossing the 1/0 threshold on the way. So the system is sensitive to rising edges if they follow too hard on the heels of the triggering falling edges.

You could cure the first fault by using Schmitt trigger NOT gates. Both faults, however, can be eradicated by adding a feedback path.

True monostables

A true monostable is illustrated in figure 2.30. It looks very similar to the pulse generator of figure 2.29, except that the output $\bar{Q}$ is fed back to the input where a NAND gate mixes it with the triggering signal $\bar{T}$.

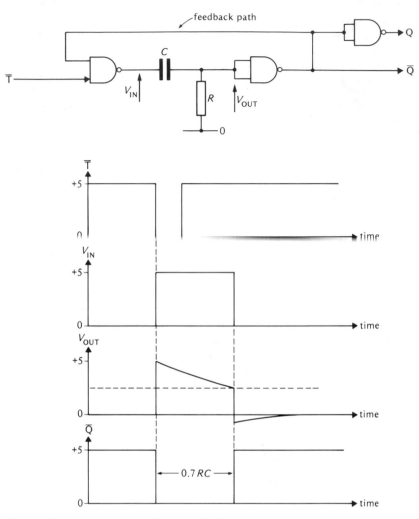

Figure 2.30 A monostable made from NAND gates

The circuit works like this. Starting off with $\overline{T}$ at logical 1, $\overline{Q}$ will be logical 1 as V_{OUT} is pulled down to 0 V by the resistor R. When a falling edge is fed into $\overline{T}$ it goes from logical 1 to logical 0, V_{IN} rises to $+5$ V and V_{OUT} is pulled up with it. So $\overline{Q}$ dives down to logical 0. Since a logical 0 at either input of a NAND gate will hold its output at logical 1, the feedback path ensures that V_{IN} will remain at $+5$ V until $\overline{Q}$ goes back up to logical 1. So $\overline{T}$ can be raised back up to logical 1 at any time without affecting the steady fall of V_{OUT} towards the 0 V rail. Once V_{OUT} gets down to $+2.5$ V, $\overline{Q}$ will rise to logical 1. If $\overline{T}$ is also logical 1, then this forces V_{IN} down from $+5$ V to 0 V as both inputs to the NAND gate are logical 1. So V_{OUT} is pushed down below the ground supply rail and clamped by the diode in the NOT gate.

Provided that the triggering pulse is shorter than the output pulse then the system works very well. Once it has been triggered it ignores the state of its input, and the output pulse is well defined and clean. You need to use Schmitt trigger NOT gates if you want a clean output pulse for long triggering pulses. You can purchase monostables in IC form (the 74123 contains two independent ones in a single package) which only require the addition of a resistor and a capacitor to set the pulse width, but our NAND gate version is cheaper and just as reliable.

EXPERIMENT
Exploring the properties of a monostable

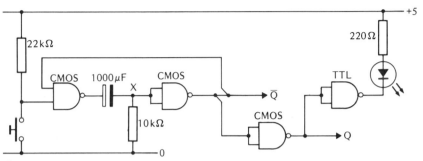

Figure 2.31

Assemble the system shown in figure 2.31. The push switch is used to trigger the system and a TTL NAND gate buffers the output so that its state can be monitored with an LED. Use a 4011 CMOS IC for the NAND gates of the monostable.

When the switch is pressed and released the LED should come on for about 7 s before going off again. Take a look at the voltage at X with a voltmeter. Check that releasing the switch at any time during the seven seconds does not affect the length of the output pulse.

Measure the length of the output pulse when the pull-down resistor has the values 10 kΩ, 22 kΩ, 47 kΩ and 100 kΩ. Do not have the voltmeter attached. Is the pulse width equal to 0.7RC every time?

EXPERIMENT
*Designing a couple of systems which consist of two
monostables connected together*

When you have designed each system, assemble it and see if it does what it is supposed to do. Then draw a circuit diagram of it, showing all of the component values. Finally, explain how the system works.

1) A delayed monostable. The system is triggered by briefly pressing a push switch. There is then a delay of about 30 seconds before an LED flashes on for about one second.

2) An oscillator. The system has no external trigger input, but has two outputs, both LEDs. The LEDs alternate on and off continuously, with one on while the other is off. Each LED is only on for one second at a time.

PROBLEMS

1 The circuit of figure 2.32 (overleaf) is a CMOS monostable which triggers on a rising edge. Draw graphs of the voltage at T, X, Y, Q and $\bar{Q}$ to show what happens when T is briefly raised to logical 1. Draw the graphs under each other in the fashion of figure 2.30. Explain, in detail, how the system works.

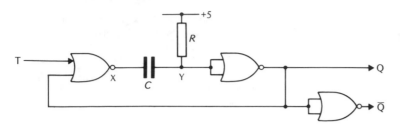

Figure 2.32 A rising edge triggered monostable

2 Figure 2.33 shows a TTL monostable. If a typical TTL logical 1 is +4 V and the 1/0 threshold of a TTL input is +1.4 V, show that the output pulse has a duration of C ms when C is measured in μF.

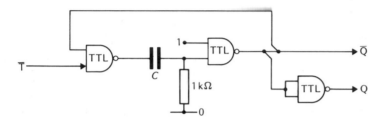

Figure 2.33 A TTL monostable

3 Figure 2.34 shows a spike generator which is being fed with a square wave. The average value of V_{OUT} will always be 0 V, but its shape will depend on the frequency of the square wave. Draw graphs of V_{OUT} as a function of time if the frequency of V_{IN} is

a) 0.1 kHz
b) 1.0 kHz
c) 10 kHz.

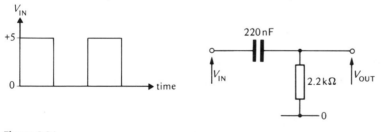

Figure 2.34

THE ASTABLE

An **astable system** is one which continually oscillates between two states. It cannot permanently settle into just one state. Such systems have outputs but no inputs. This is because the oscillation is caused by feedback from the output to the input via a network that introduces a time delay.

The relaxation oscillator is an example of an astable system. You are about to meet another example, the **astable multivibrator**. This is a classic circuit for generating square waves. Its frequency is fixed by a pair of resistors and a pair of capacitors, the rising and falling edges of the waveform are clean and fast, it has complementary outputs and can be started and stopped by a logic signal. Furthermore, with a little modification, it can be converted into a system whose frequency depends on the size of a voltage fed into it; this permits signals to be frequency modulated onto a carrier, as you will find out in Chapter Thirteen.

Two NOT gates

A version of the astable multivibrator assembled from a pair of NOT gates is shown in figure 2.35. The graphs show how the voltages at the various points in the system change with time as it oscillates. We have assumed that A_1 is initially at $+5V$ and that A_2 is at $0V$; we shall discuss this assumption later. As A_1 is above the CMOS 1/0 threshold, Q_1 is logical 0 and stays there for the next $0.7R_1C_1$ ms. When A_1 eventually dips below $+2.5V$, Q_1 rushes up to logical 1, pushing A_2 up to $+5V$ as it goes.

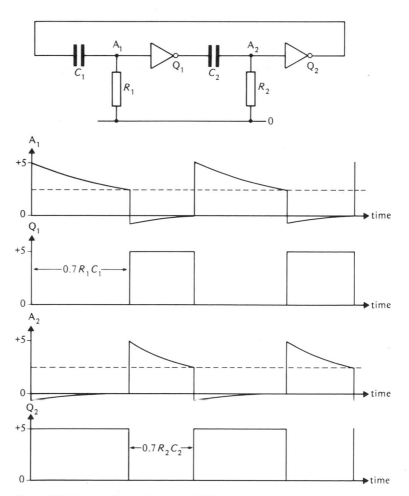

Figure 2.35 An astable made out of NOT gates

(Remember that if the voltage of one plate of a capacitor changes suddenly the other plate must change by the same amount.) Once A_2 is a logical 1, Q_2 is pushed down to logical 0, and as it goes down it firmly rams A_1 down below the ground supply rail. So $0.7R_1C_1$ ms after A_1 was $+5$ V there is a trigger action which moves Q_1 from 0 to 1 quickly and cleanly; at the same time, Q_2 is changed from 1 to 0.

It is now the turn of A_2 to slowly fall towards 0 V. It takes $0.7R_2C_2$ ms to get from $+5$ V to $+2.5$ V. The instant that it crosses the 1/0 threshold there is a chain of events around the feedback loop which ultimately rams A_2 down to -0.7 V. Q_2 goes from 0 to 1, pushing A_1 up to $+5$ V which, in turn, causes Q_1 to fall from logical 1 to logical 0. This last event forces A_2 below the ground supply rail.

We are now back where we started in our description of the action of the circuit. You can see that the system must continue to oscillate indefinitely, repeating the same cycle over and over again.

Two NOR gates

Once the two outputs of an astable multivibrator are in different states, the system will oscillate. Unfortunately, there is nothing in the circuit which will guarantee that this situation will arise when it is switched on. Should both Q_1 and Q_2 be at logical 1 at the same instant then the system will get stuck in that state! So an astable made out of a pair of NOT gates is not a sure-start system; it does not automatically oscillate when you switch it on.

If you use NOR gates as the inverters in the astable circuit, then you can arrange to have an **enable** input to the system. When that input is held low, the system oscillates. Take a look at figure 2.36. Since a NOR gate is equivalent to a NOT gate when one of its inputs is held at logical 0, the circuit is identical to the one of figure 2.35 when $\bar{E}$ is low. When $\bar{E}$ is high the system cannot oscillate; Q will be 0 and $\bar{Q}$ will be 1. However, if $\bar{E}$ starts at logical 1 and is suddenly brought down to logical 0 and left there, Q rises to logical 1 and the system starts to oscillate.

The graphs of figure 2.36 show how the states of Q and $\bar{Q}$ depend on the state of $\bar{E}$. When $\bar{E}$ is 1, Q is always 0. When $\bar{E}$ is 0, Q oscillates with a frequency of $1/(1.4RC)$ (assuming CMOS gates).

EXPERIMENT

Assembling and testing an astable made out of a couple of
TTL NOR gates

The circuit is shown in figure 2.37; assemble it. Make sure that you know which LED monitors which output.

The system should oscillate at 1 Hz when you press the switch. (Note that because $\bar{E}$ is a TTL input there is no need for a pull-up resistor.) Furthermore, Q should return to logical 0 as soon as the switch is released.

Replace the two capacitors with 1 µF ones. Trigger a CRO with the signal at Q. Look at the waveforms at Q, Y, $\bar{Q}$ and X as shown in figure 2.37. Use the oscilloscope traces to draw graphs of the voltages at these points as a function of time; arrange the graphs one under the other and ensure they are drawn to the same scale, similar to those in figure 2.35.

Use the CRO to measure the frequency of the signal at Q for the following values of capacitors; 10 µF, 1 µF, 100 nF, 10 nF and 1 nF. Use your results to confirm the rule of thumb:—

Frequency (kHz) = 1/Capacitor (µF)

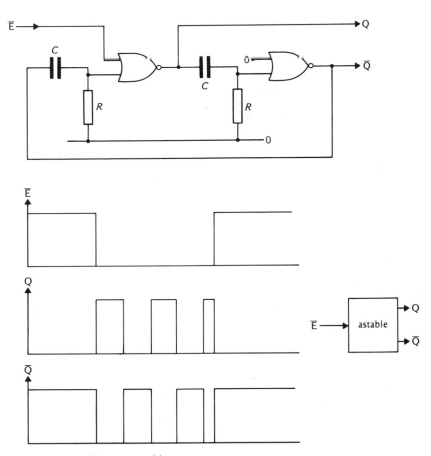

Figure 2.36 A NOR gate astable

Finally, see if you can work out how to make the astable self-starting. You will need to take out the push switch and replace it with a network consisting of a 1 kΩ resistor and a 1 μF capacitor. The network has to pull Ē up to logical 1 for a short while every time that the circuit is powered up. It then lets it down to 0 and leaves it there. Try out your design. When you have got it right, draw the circuit diagram of a self-starting astable and explain how the self-start works.

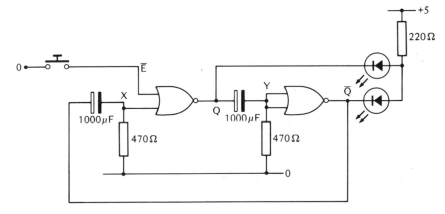

Figure 2.37

PROBLEMS

1 Draw a voltage–time graph for the output Q of the astable shown in figure 2.38. Assume ideal CMOS gates (i.e. that the 1/0 thresholds are +2.5 V) and that the system is oscillating. Make sure that you label the scales on both axes carefully.

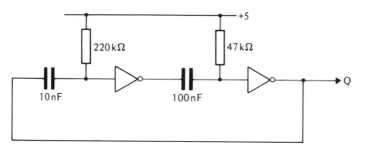

Figure 2.38

2 The two NAND gates in figure 2.39 are ideal CMOS ones. Draw voltage–time graphs, one under the other and to the same scale, of the waveforms at points A, B, C and D to show what happens when E is raised from 0 to 1.

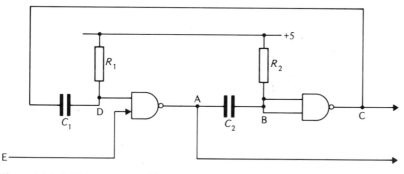

Figure 2.39 A NAND gate astable

DRIVING SPEAKERS

The bulk of this chapter has been concerned with explaining how capacitors can be used to create systems which oscillate. One of the major uses of such oscillators is to generate signals that can be heard. So this final section is going to deal with the problem of persuading a TTL logic gate (or any other system for that matter) to drive a loudspeaker.

POWER

Figure 2.40 shows an example of a system which drives a loudspeaker. When the switch is pressed the monostable output goes low for 10 s. This **enables** the oscillator, which therefore feeds a 1 kHz square wave into the final NOT gate. The output of that gate is pulled up to logical 1 by a loudspeaker; when it tries to oscillate a buzz comes out of the speaker.

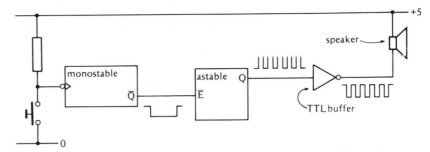

Figure 2.40 A logic system with a loudspeaker output

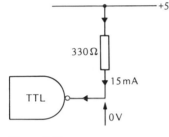

Figure 2.41

In order to get a lot of noise out of the speaker, the NOT gate has to feed a lot of energy into it. We shall start off by calculating how much energy a single TTL output can provide in a second. Then we shall compare this with the rate at which the loudspeaker in figure 2.40 is being fed energy.

Consider the system shown in figure 2.41. A TTL output is sinking current through a resistor. The resistor has 5 V across it, so a current of $5/0.33 = 15.2$ mA flows through it. This is just about the maximum current that a TTL output can manage to sink whilst remaining at 0 V; if you try to get it to swallow more current its output voltage will rise. Now, whenever current flows from one voltage to another in a circuit electrical energy is converted into another sort of energy. The rate at which that energy conversion takes place is called **power**. It can be calculated with this formula;

$$W = VI$$

where W is the power,
V is the voltage drop,
I is the current flowing.

The standard units for W, V and I are watts, volts and amps. For a lot of electronic applications it is very convenient to use milliwatts (mW), volts and milliamps instead. In the case of a resistor the converted energy takes the form of heat. For the resistor of figure 2.41, the power is $5 \times 15.2 = 76$ mW.

So a TTL output at logical 0 can pump energy into a resistor at a rate of 76 mW. If the output of the gate oscillates so that it only spends half of its time at logical 0 (i.e. it is feeding out a square wave), it will only put half that power into the resistor i.e. 38 mW. This represents the maximum power that a TTL output can push into a loudspeaker. Yet if the speaker is connected as shown in figure 2.40, only about 1 mW is pumped into it.

The low energy transfer to the speaker, when it is connected directly to a TTL output, is a consequence of its very low resistance. A speaker typically behaves like an 8 ohm resistor. When the output of the gate tries to go to logical 0 it has to stop when it gets to about 0.12 V below the +5 V supply rail. The output cannot sink much more than 15 mA, so the voltage drop across the speaker will be at most $15 \times 0.008 = 0.12$ V. The power of the speaker is given by the usual formula;

$$W = VI$$

therefore $W = 0.12 \times 15$

therefore $W = 1.8$ mW

Of course, as the output only spends half its time pulling current through the speaker, the rate at which energy is pumped into it will be 0.9 mW.

In practice you will get a buzz out of an 8 ohm speaker if it is directly connected between a TTL output and the +5 V rail, but it will not be as loud as it could be. In order to get 100% of the possible energy (i.e. 38 mW) we need to persuade the TTL gate that it is feeding its square wave into a 330 ohm load.

TRANSFORMERS

If you put the right sort of **transformer** between the speaker and the logic gate, you can get virtually all of the 38 mW fed out of the gate pumped into the speaker.

The symbol for a transformer is shown in figure 2.42. Its construction is very simple, consisting of a pair of coils of wire wound around a common slab of iron. The coils are insulated from each other so that there can be no direct current flow between them. The coils are given special names, the **primary winding** and the **secondary winding**. The transformer of figure 2.42 has a **turns ratio** of $n : 1$ i.e. it has n turns of wire in the primary winding for each turn of wire in the secondary winding.

The electrical behaviour of a transformer is as simple as its construction. A well designed transformer obeys three rules.

The first rule is that any waveform fed into the primary appears across the secondary without any change of shape. In practice there is an upper and lower frequency beyond which this ceases to be true, but it is certainly true for waveforms which can be heard by the human ear.

The second rule is that if the turns ratio is $n : 1$, then the voltage of the waveform put into the primary is n times larger than that which comes out of the secondary.

The third rule is that no energy conversion takes place within the transformer. So all of the electrical energy that is fed into the primary comes out of the secondary and is passed on to the next element in the circuit. Referring to the circuit in figure 2.42, this means that if the gate pumps 38 mW into the primary, the power transferred to the speaker is also 38 mW. In practice a small amount of energy is converted to heat within the transformer, but it should not be a significant amount.

Matching the load

We are now going to calculate the optimum value of the turns ratio which will allow 38 mW to get transferred to the speaker shown in figure 2.42. In order for this to happen, the gate will have to see the primary winding as behaving like a 330 ohm resistor. The transformer will take the 5 V square wave out of the gate, make it smaller and deliver it to the speaker. That

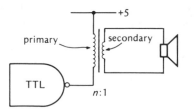

Figure 2.42 How to connect a speaker to a TTL output via a transformer

voltage reduction allows the speaker to draw more current from the secondary than the gate delivers into the primary.

Let us start off by considering the speaker. It will see a square wave of amplitude $5/n$ V, and since it has a resistance of $0.008\,k\Omega$, we can easily calculate how much current flows through it.

$$R = \frac{V}{I}$$

$$\text{therefore} \quad 0.008 = \frac{5/n}{I}$$

$$\text{therefore} \quad I = \frac{625}{n}\,mA$$

The square wave fed out by the secondary will only drive current through the speaker for half the time. The power developed by the speaker while current flows through it is given by;

$$W = VI$$

$$\text{therefore} \quad W = \frac{5}{n} \times \frac{625}{n}$$

$$\text{therefore} \quad W = \frac{3125}{n^2}\,mW$$

So the rate at which energy is pumped into the speaker when driven by a square wave will be half of this. We want this power to equal 38 mW, our previously calculated maximum power from an oscillating TTL output. So

$$38 = \frac{1}{2}\left(\frac{3125}{n^2}\right)$$

$$\text{therefore} \quad n = 6.4$$

Thus a transformer which has 6.4 times more turns on the primary winding than the secondary can make the TTL gate think that it is pushing energy into a 330 ohm load, whereas that energy is actually being delivered to an 8 ohm load. Transformers are widely used as intermediaries between electronic systems and loudspeakers. Electronics is usually good at driving small currents through large voltages, but loudspeakers (because of their low resistance) take high currents at low voltages. A transformer that is being fed with an oscillating waveform can be used to multiply the resistance of the load by a factor of n^2 i.e. the source of the waveform thinks that it is pumping energy into a resistance that is n^2 larger than the true load resistance.

When they are used this way transformers allow energy to be efficiently transferred from the electronics into the load. They do not amplify the signal in energy terms, although they can be used to change the voltage of the signal. We shall be returning to the problem of energy transfer between a circuit and a load in Chapter Four. It is an important facet of the subject since an electronic system is virtually useless unless it is capable of translating its output signals into some physical signal such as light, sound, heat or motion.

PROBLEMS

1 Figure 2.43 shows a transformer being used to mediate between the source of the square wave V_{IN} and the 100 ohm load. The turns ratio is 4 : 1 and the square wave has an amplitude of 12 V.

a) What is the amplitude of the square wave across the load?

b) How much current flows through the load?

c) What is the average power of the load?

d) How much current flows in the primary?

e) What is the apparent resistance of the primary winding as far as the source of the square wave is concerned?

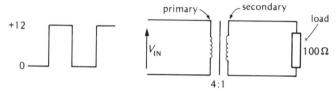

Figure 2.43

2 You are going to prove that a transformer multiplies the load resistance by a factor of n^2. For a system like the one shown in figure 2.43 with a turns ratio of $n : 1$, a load resistance R and an input square wave of amplitude V, obtain expressions for

a) the power developed in the load

b) the current flowing in the primary

c) the apparent resistance of the primary winding.

3 Calculate the turns ratio which would be needed for a CMOS output to drive an 8 ohm speaker with the maximum power. Assume that a CMOS gate can sink a maximum current of 1 mA from a 5 V supply.

EXPERIMENT

Here are a number of problems for you to solve. You will have to interconnect the systems which are shown in figure 2.44 to make circuits which generate interesting output displays.

Tackle each of the problems in turn. When you have designed a solution, assemble it and check that it works. Then draw its circuit diagram and explain how it works.

1) This one is easy, but it is a subsystem of the next two problems so it has to be got right first. The system has a switch and an LED; when the switch is pressed briefly the LED comes on for 30 seconds and then goes off again.

2) This system also has a press switch and an LED. When the switch is pressed briefly the LED flashes on and off at a rate of 1 Hz; after 15 seconds it goes off again and stops flashing.

3) Now adapt the system so that it has an audible output. Press the switch briefly and a 1 kHz buzz comes out of the speaker which is pulsed on and off at a rate of 1 Hz; after a minute the noise ceases.

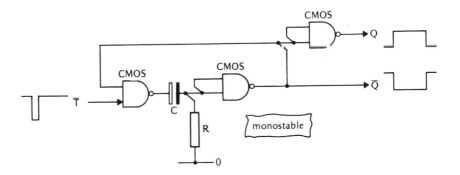

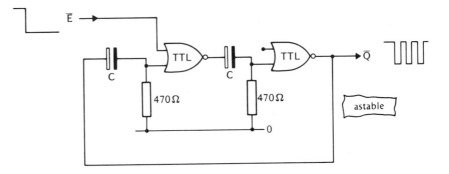

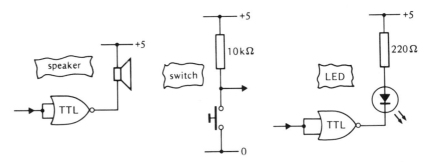

Figure 2.44

4) The final system has a continuous audible output. The sound alternates between 0.5 kHz and 1 kHz at a rate of 1 Hz. It should sound a bit like a two-tone police siren. You will have to combine two signals before feeding them into a speaker; don't just connect two TTL outputs to each other.

REVISION QUESTIONS

1 A 100 µF capacitor is connected to a 16 V supply. How much charge is stored on each plate?

2 The same capacitor is now disconnected from the supply. If its terminals are connected by a 220 kΩ resistor, how much time has to elapse before there is only 1 V across the resistor?

3 A 22 µF capacitor starts off completely discharged. The bottom plate is grounded and the top plate pulled up to a +5 V supply line by a 47 kΩ resistor. Sketch a graph to show how the voltage across the capacitor changes with time. Calculate

a) the time taken for the voltage across the capacitor to get to 4.5 V

b) the voltage across the capacitor one second after it has started to charge up.

4 Explain what the formula $V = V_0 e^{-(t/RC)}$ means.

5 Draw the circuit diagram of a relaxation oscillator. You may assume a Schmitt trigger at the heart of the system. Explain how the circuit works, including a couple of sketch graphs to show how voltages evolve with time.

6 A relaxation oscillator has a 10 nF capacitor being charged and discharged via a 4.7 kΩ resistor. The Schmitt trigger is based on CMOS gates run off a 5 V supply, and has switching thresholds of +1 V and +4 V. Work out the frequency of the square wave produced by the oscillator.

7 Show how a resistor, a capacitor and a diode can be connected to make a system which will convert the rising edge of a square wave into a brief pulse.

8 Draw the circuit diagram of a monostable based on a pair of TTL NAND gates. State what happens to the output when

a) a rising edge

b) a falling edge is fed into the input.

9 Sketch some graphs to show how the voltages within a monostable change with time when it is triggered.

10 Draw the circuit diagram of an astable based on a pair of TTL NOR gates. State what values of resistors can be used and roughly how the frequency depends on the size of the capacitors.

11 Explain what a clamp diode is and what it does.

12 Explain why a transformer needs to be placed between a loudspeaker and a logic gate. Show that a 6 : 1 turns ratio is suitable for interfacing an 8 ohm speaker to a TTL output.

3

Analogue systems

SIGNALS

Electronic systems can be divided into three broad categories. There are systems which generate **electrical signals**, systems which process them and systems which convert electrical signals into other sorts of signals. A signal, of course, is something that carries information from one place to another. An electrical signal uses voltages and currents to transfer information from one part of a circuit to another. Thus in the circuit of figure 3.1 information travels from left to right as an electrical signal in the two cables that connect the three systems. We shall now look at those three systems in some detail and discuss how information is passed from one to the other.

The switch/resistor network on the left of the figure is an example of an electrical signal generator, or **signal source**. It contains a **transducer** which converts a physical property into an electrical one; the switch "senses" the presence of a finger and reacts by being either a conductor or an insulator. When the switch is open there is a voltage difference of 5 V between the wires in the cable that carries the signal to the monostable in the centre of the diagram. When the switch is closed the electrical signal being fed into the cable changes i.e. the voltage difference between the two wires becomes 0 V. That signal change travels down the two wires close to the speed of light; the exact speed depends on the geometry of the cables and the nature of the dielectric that separates them. As a rule of thumb, signals travel down cables at a speed of about 30 cm per nanosecond. So if there is 1 m of cable between the switch/resistor network and the monostable, the latter will be triggered about 3 ns after the switch has been closed.

The signal (from now on the word "signal" refers to electrical signals) from the signal source is **processed** by the monostable. That is, the monostable does something to the signal which goes into it and feeds out a different signal. Thus when the input signal goes from logical 1 to logical 0, its output signal goes down to logical 0 for $0.7RC$ ms. In general, a processing system has a number of inputs and outputs; the signals pumped

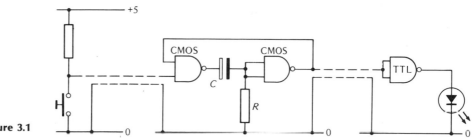

Figure 3.1

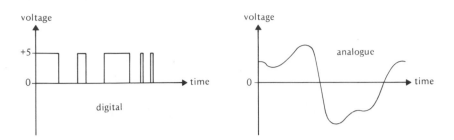

Figure 3.2 Typical analogue and digital signals

out of the outputs are a consequence of the signals going into the inputs. Most of the systems that you met in Chapter One are digital processing systems

The final block in the circuit of figure 3.1 contains the **output transducer**. A **power amplifier** takes the signal that arrives from the processing system and, with the help of an LED, converts it into a visual signal. A TTL NAND gate acts as the power amplifier, boosting the energy of the signal before it is fed into the transducer (the LED) where it is converted into light (or a lack of light). (Do not worry about the missing resistor. As the TTL output goes up to logical 1 it finds that it has to pump out its maximum current (a few mA) once the LED has become forward biased. If it tries to rise any higher than $+2\,V$ (the voltage drop across a lit LED) it cannot provide the extra current, and is therefore effectively clamped to that voltage.)

The signals that travel down the two cables of the circuit that we have been considering are **digital**. This means that only two types of signal go down the cable, known as logical 1 and logical 0. There is a second class of signal known as an **analogue signal**; whereas the wires of a cable carrying a digital signal can only have two voltage differences, those carrying an analogue signal can have any voltage difference. The graphs of figure 3.2 should make clear the distinction between the two classes of signal; note that changes of digital signal are very sudden (a few nanoseconds to change from one logic state to another) whereas analogue signals can vary more slowly with time.

This chapter is going to be concerned with the generation and processing of analogue signals, concentrating mainly on the use of a particular IC (the 741) to assemble different sorts of processing systems. The next chapter will deal with power amplifiers, the systems which take analogue signals and boost their energy so that they can drive output transducers.

SIGNAL SOURCES

In Chapter One you learnt how to use a switch to generate a digital signal. Since switches are either open or closed it is fairly obvious that one switch can only generate two sorts of signal. Switches are almost unique among input transducers, most of which have electrical properties that vary smoothly as their physical properties are changed. So most transducers generate analogue signals; you will be introduced to some in this section.

Phototransistors

The circuit of figure 3.3 generates an electrical signal in response to the amount of light that falls on the **phototransistor**. We shall leave a dis-

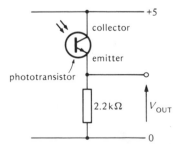

Figure 3.3 Using a phototransistor

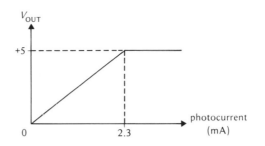

Figure 3.4 How the signal from the phototransistor source changes with illumination

cussion of transistor behaviour until Chapter Five, so there is no need to worry too much about the jargon that we are about to introduce. A phototransistor acts like a light controlled current source. The amount of current that flows from the **collector** to the **emitter** only depends on the amount of light shining on the transistor (which is a small slab of silicon housed in a clear piece of plastic). More precisely, the size of the **photocurrent** is proportional to the rate at which light energy is absorbed by the transistor base. The photocurrent flows through a resistor to generate a signal that can be sent down a cable to a processor.

So if the light intensity is such that the photocurrent is 0.5 mA, there will be a voltage drop of 1.1 V across the 2.2 kΩ resistor (use Ohm's Law). The remaining $5 - 1.1 = 2.9$ V is the voltage drop across the transistor. Provided that the collector is at a higher voltage than the emitter the photocurrent only depends on the amount of illumination. Once the voltage drop across the transistor is zero it **saturates** i.e. no further increase of photocurrent is possible. In our example, the photocurrent will saturate when it gets to 2.3 mA i.e. when V_{OUT} is $+5$ V. (Take a look at the graph of figure 3.4.)

Photodiodes

Although their interior is somewhat different, **photodiodes** behave just like phototransistors. Figure 3.5 shows a photodiode set up so that it behaves like a light controlled current sink i.e. the amount of current which flows from **cathode** to **anode** is proportional to the rate at which light energy is absorbed by the diode. The current for the photodiode is sourced by the resistor and the voltage drop across it is the signal. For example, if the photocurrent is 0.2 mA, the voltage drop across the 10 kΩ resistor will be $0.2 \times 10 = 2$ V. This leaves $5 - 2 = 3$ V across the photodiode, so the signal sent down the cable will be $+3$ V. The graph of figure 3.6 shows how the size of the signal depends on the amount of light; notice the saturation when the voltage across the diode becomes zero.

The diode in the circuit of figure 3.5 is reverse biased. In other words, if it was a normal silicon diode no current would flow through it. In practice there is always a small **leakage current** through a diode when it is reverse biased (about 25 nA for a small signal diode of the sort that you will be using in practical work). The size of the leakage current depends on the size of the diode and on its temperature (it increases very rapidly as the diode is heated up). When light energy is absorbed by a photodiode, the energy is used to break up the structure of the silicon, allowing it to leak more. The damage is only temporary; photodiodes can recover in a few nanoseconds.

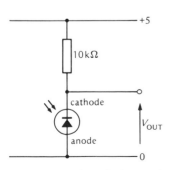

Figure 3.5 A photodiode signal source

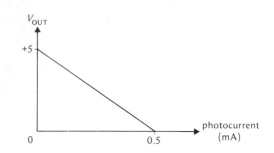

Figure 3.6 How the signal of figure 3.5 changes with illumination

Phototransistors use **transistor action** to amplify the leakage current through a photodiode that is built into them. This means that their photo-current is larger than that of the photodiode, but they take a few micro-seconds to recover. Otherwise, the two devices are identical in behaviour; indeed, they could replace each other in the circuits of figures 3.5 and 3.3.

Light dependent resistors

The two sources that we have looked at so far have generated **linear signals** i.e. the value of the signal (in volts) has been strictly proportional to the size of the physical property being sensed (look at the graphs of figures 3.4 and 3.6. Both are straight lines until saturation occurs). Figure 3.7 shows a non-linear light detector based on a **light dependent resistor** (usually called an **LDR**). The graph of figure 3.8 shows quite clearly that V_{OUT} is not proportional to the amount of light hitting the circuit. As we shall see below, this is not a consequence of any non-linearity of the LDR but arises from the fact that the circuit contains a pair of resistors connected between two supply rails.

As you might expect from its name, an LDR is a resistor (i.e. it obeys Ohm's Law) whose resistance depends on the amount of light hitting it. The resistance is very high in the dark (a few MΩ) and falls to a low value (about 0.1 kΩ) in strong sunlight. LDRs respond quite slowly to changes in illumination (about 100 ms) but are very rugged compared with photo-transistors or photodiodes; you can put much larger voltages across an LDR.

Voltage dividers

The circuit of figure 3.7 is an arrangement known as a **voltage divider**. There are five volts across a pair of resistors (one fixed, the other variable in

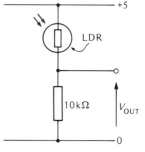

Figure 3.7 A signal source with an LDR

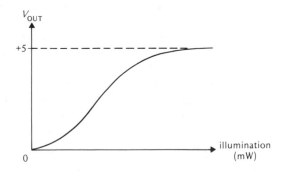

Figure 3.8 The non-linear response of a signal source with an LDR

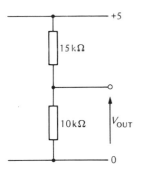

Figure 3.9

size) and the output signal is the voltage across just one of the resistors (the 10 kΩ fixed resistor). How do we work out the size of the output signal? Let us suppose that the LDR is illuminated such that its resistance is 15 kΩ. The situation is then that shown in figure 3.9; we want to calculate the value of V_{OUT}.

The total resistance between the supply rails is $15 + 10 = 25$ kΩ. (When resistors are connected in line the total resistance is the sum of the individual resistances in the line.) Since there is 5 V across the 25 kΩ, we can use Ohm's Law to calculate the current that flows through the resistors.

$$R = \frac{V}{I}$$

therefore $$25 = \frac{5}{I}$$

or $$I = \frac{5}{25} = 0.2$$

So 0.2 mA flows through both resistors from the 5 V rail down to the ground rail. The signal V_{OUT} is the voltage across just the 10 kΩ resistor; using Ohm's Law again:

$$R = \frac{V}{I}$$

therefore $$10 = \frac{V_{OUT}}{0.2}$$

or $$V_{OUT} = 10 \times 0.2 = 2$$

therefore $$V_{OUT} = +2\,V$$

There is another, more instructive, way of working out the value of V_{OUT}. We know that there is a drop of 5 V across 25 kΩ in the circuit. If we replaced the two resistors with twenty-five 1 kΩ resistors strung in line (in series) between the supply rails the total resistance would still be 25 kΩ. As the current through each of the twenty-five resistors has exactly the same size, each resistor will have the same voltage across it. In other words, 5 V is shared equally among twenty-five resistors, giving 5/25 V across each one. Since ten of the resistors are needed to make 10 kΩ, V_{OUT} will be 10 × 5/25 V = 2 V.

In fact, the value of V_{OUT} depends on only two things. These are the voltage difference between the two supply rails and the ratio of the two resistors connected between them. More precisely:

$$V_{OUT} = V_{IN} \times \left(\frac{R_2}{R_1 + R_2} \right)$$

where the symbols are defined in figure 3.10. You will be asked to verify this piece of algebra in one of the problems at the end of this section. The name of voltage divider for this simple network refers to the way in which the output voltage is a fixed fraction of the input voltage, the fraction being decided by the size of the two resistors. In fact voltage dividers are widely used to process analogue signals; the system shown in figure 3.11 (overleaf) feeds out a signal that has half the size of the signal that went in.

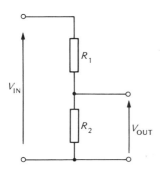

Figure 3.10 A voltage divider

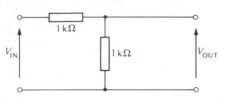

Figure 3.11 A divide-by-two network

Thermistors

There are quite a few devices whose resistance depends on a physical quantity; they can be used to generate signals if they are made one of the two resistors in a voltage divider. Just as an LDR has a resistance that depends upon its illumination a **thermistor** has a resistance that depends on its temperature. So a voltage divider which contains a thermistor can be used to generate a signal that is temperature dependent. Such a circuit is illustrated in figure 3.12. The resistance of a thermistor falls rapidly as it is heated (it is not a linear device at all) so V_{OUT} will rise as the temperature of the circuit is raised. Thermistors are widely used for monitoring temperatures because of their relatively large variation of resistance with temperature, typically about -4% per $°C$. They are available in a wide variety of sizes and average resistances at room temperature.

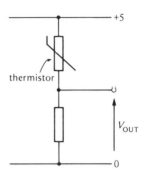

Figure 3.12 A signal source containing a thermistor

Potentiometers

You have already used one of these in the experiments to find the switching thresholds of Schmitt triggers. A **potentiometer** is a device with three terminals and a knob; figure 3.13 shows how it has to be connected to generate a variable voltage from a fixed voltage supply. As far as the supply rails are concerned, the potentiometer looks like a fixed resistor of value R. The position of the third terminal (the wiper) with respect to the other two depends on the angle of the potentiometer knob. The whole thing looks like a voltage divider where the value of R_2 is proportional to the angle of the knob, yet the value of $R_1 + R_2$ remains constant (R). So a potentiometer set up as in figure 3.13 produces an analogue signal proportional to the angle of rotation of the knob (provided that the potentiometer is a linear one). The relationship between V_{OUT} and the angle is shown in the graph of figure 3.14.

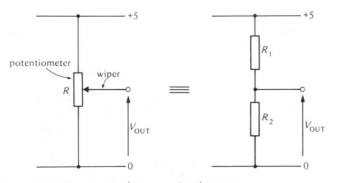

Figure 3.13 A potentiometer wired up as a signal source

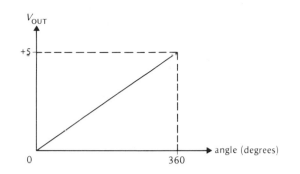

Figure 3.14

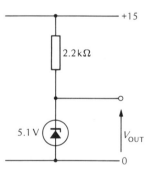

Figure 3.15 A zener diode

Zener diodes

All of the signal sources that we have looked at so far have converted a variable physical quantity into a signal which mirrors that variation. We have looked at systems which could be used to monitor light level, temperature and angle; variation in the size of the electrical signal coming out of the system would imply that the appropriate physical quantity had changed. Our last example of a signal source is a bit different; it is designed to feed out a constant signal that never changes! (It is in fact very useful, especially when designing power supplies—see Chapter Twelve.) The system is shown in figure 3.15; the diode with the extra line in its symbol is a **zener diode**, and it is the part of the circuit which forces the output to have a fixed voltage.

Zener diodes look just like ordinary silicon diodes from the outside and behave exactly like them in many respects. They have the usual 0.7 V drop across them when they are forward biased and current flows through them. When they are reverse biased they turn into insulators and no current flows through them, apart from the usual negligible leakage current. Leakage current is more or less independent of the size of the voltage across the diode when it is reverse biased, provided that the voltage is not too big. If the reverse bias voltage is gradually increased across a diode there comes a point where the leakage current suddenly increases dramatically. The diode stops looking like an insulator and becomes a short circuit. The change of resistance is so sudden that unless measures are taken to control the current that flows through the diode there is a strong possibility that it will be destroyed. The voltage at which the diode "**breaks down**" in this fashion is called the **zener voltage**. For most signal diodes it is quite high (for obvious reasons) but its value can be adjusted by altering the materials from which the diode is constructed. Zener diodes have zener voltages that are fixed in their manufacture and usually stamped on their sides.

The graph of figure 3.16 (overleaf) shows how the current through a 5.1 V zener diode varies with the voltage across it. The current increases rapidly when the diode breaks down. This clamps the value of V_{OUT} in figure 3.15 close to the zener voltage.

The 2.2 kΩ resistor limits the amount of current which flows through the diode to a safe level so that it doesn't overheat. Since the current which flows through the diode also flows through the resistor, we can use Ohm's Law to calculate its size. As the voltage across the diode is 5.1 V and the

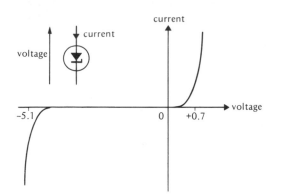

Figure 3.16 The current–voltage characteristics of a 5.1 V zener diode

supply rails are 15 V apart, the voltage across the resistor is $15 - 5.1 = 9.9$ V.

$$R = V/I$$

therefore $2.2 = 9.9/I$

or $I = 9.9/2.2 = 4.5$

So a current of 4.5 mA flows through the diode. The rate at which heat is generated in the diode can be calculated as follows:

$$W = VI$$

therefore $W = 5.1 \times 4.5 = 23$

A small diode can easily cope with 23 mW of heat; the zener diode in the circuit of figure 3.15 is in no danger self-destructing.

Now suppose that the supply voltage for the circuit falls from 15 V to 10 V. Assuming that V_{OUT} remains constant (we shall justify this later) at 5.1 V, the primary effect of this change is to reduce the voltage drop across the resistor and hence reduce the current flowing through it to about half of its previous value. Referring to the graph of figure 3.16, you will see that a factor of two change in the current has a very small effect on the voltage across the diode when it has "broken down". In fact, if the average current is big enough, the change in voltage becomes very small indeed. So a zener diode/resistor network generates a fixed signal whose value is very nearly independent of the supply voltage.

EXPERIMENT
Making a series of measurements on a zener diode and plotting a current–voltage graph

Start off by setting up the circuit shown in figure 3.17. The amount of current that flows through the diode (which is forward biased) is fixed by the size of the resistor R. The voltage drop across the diode is measured by the voltmeter; if its reading is V, Ohm's Law can be used to calculate the current that flows through the resistor.

$$R = \frac{15 - V}{I}$$

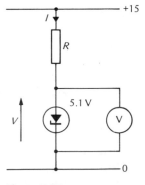

Figure 3.17

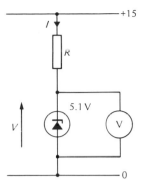

Figure 3.18

I, the current through the resistor, will equal the current which flows through the diode if a negligible amount is diverted into the voltmeter. Use values of *R* between 470 Ω and 47 kΩ, roughly doubling the value each time; record the value of *V* for each resistor and calculate the equivalent value of *I*.

Then turn the diode round so that it is reverse biased as shown in figure 3.18. Repeat the experiment with values of *R* between 220 Ω and 10 kΩ, so that you have a series of values for *V* and *I*.

Plot your results on a graph so that it looks like the graph of figure 3.16; the results of the first series of experiments give you data for the top right hand quadrant of the graph.

PROBLEMS

1 For the circuit shown in figure 3.19, calculate

    a) the current flowing through the two resistors

    b) the value of V_{OUT}.

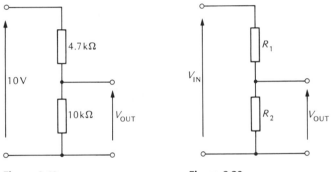

Figure 3.19   **Figure 3.20**

2 Repeat question **1** using algebraic symbols instead of numbers; figure 3.20 defines the symbols to be used. You should end up with an expression for V_{OUT} in terms of V_{IN}, R_1 and R_2.

3 Calculate the voltages at the points W, X, Y and Z in the circuits of figure 3.21. Remember that voltages are usually quoted with respect to 0 V i.e. ground.

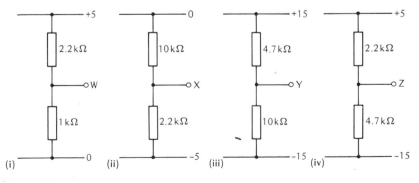

Figure 3.21

4 The diode of figure 3.22 has a zener voltage of 2.7 V and a maximum power rating of 500 mW.

a) What is the maximum current that can flow through the diode without damaging it?

b) What is the minimum value of R that can be used?

c) If R is 2.2 kΩ, at what rate is heat generated in the diode?

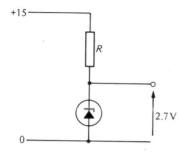

Figure 3.22

5 Figure 3.23 shows a waveform (a sine wave) that is fed into four different circuits i), ii), iii) and iv). Draw voltage–time graphs to show the shape and size of the signals fed out by each of the circuits.

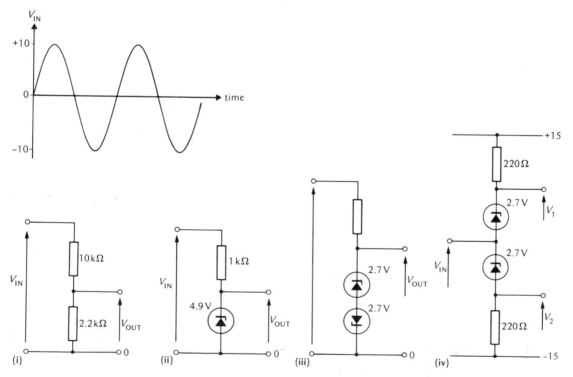

Figure 3.23

6 Each of the circuits shown in figure 3.24 contains a 10 kΩ potentiometer. What is the maximum and minimum value of the voltage of the wiper in each case? (Voltages with respect to 0 V.)

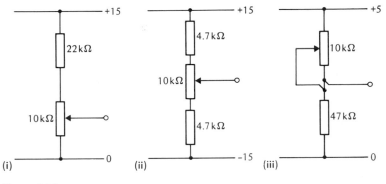

Figure 3.24

INPUT AND OUTPUT IMPEDANCES

There are many different ways in which an analogue signal source may be constructed, but they all have a common characteristic. The voltage of the signal that they send down a cable depends on the amount of current drawn from the signal source. More specifically, as more current is drawn out of the source the voltage of the signal gets smaller.

As an example, look at the circuit of figure 3.25. The voltage divider sends a signal down the cable to the voltmeter. If no current flows down the cable we can use the techniques of the previous section to work out the size of the signal.

$$V_{OUT} = \frac{V_{IN} \times R_2}{R_1 + R_2} = \frac{5 \times 47}{22 + 47} = 3.41$$

You will remember that the formula we have just used assumes that the same current flows through both resistors of the voltage divider. This will only be true in our example if the voltmeter behaves like an infinitely large resistor. In fact, a typical moving-coil voltmeter will look like a 100 kΩ resistor as far as the signal source is concerned. Using the language of electronics, we say that the meter has an **input impedance** of 100 kΩ; for

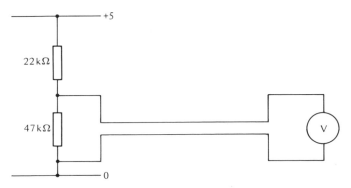

Figure 3.25 A voltage divider feeding a signal into a voltmeter

the moment you are quite safe if you think of resistance and impedance as alternative names for the same thing. So we can forget that the signal is being fed into a voltmeter and imagine that it is going into a 100 kΩ resistor instead. That resistor is the **load** into which the signal is fed.

We are now going to look in some detail at the way in which the load alters the signal generated by the voltage divider. The calculation will be a bit messy, but we shall use its results to check that some of the ideas we shall introduce later do work correctly. Look at figure 3.26. The current that flows through the 22 kΩ resistor splits into two when it gets to the junction (marked J). Some goes down the cable to the voltmeter and the rest goes through the 47 kΩ resistor. Since the amount of current coming out of the top supply rail must be equal to the current that flows into the bottom one, we can say that:

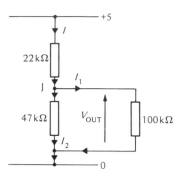

Figure 3.26 Another version of figure 3.25

$$I = I_1 + I_2 \tag{1}$$

By applying Ohm's Law to each of the three resistors in the circuit we can get three more equations:

$$22 = \frac{5 - V_{OUT}}{I} \quad \text{or} \quad I = \frac{5 - V_{OUT}}{22} \tag{2}$$

$$47 = \frac{V_{OUT}}{I_2} \quad \text{or} \quad I_2 = \frac{V_{OUT}}{47} \tag{3}$$

$$100 = \frac{V_{OUT}}{I_1} \quad \text{or} \quad I_1 = \frac{V_{OUT}}{100} \tag{4}$$

When we substitute (2), (3) and (4) into (1) we get an expression for V_{OUT}:

$$\frac{5 - V_{OUT}}{22} = \frac{V_{OUT}}{100} + \frac{V_{OUT}}{47}$$

therefore

$$\frac{5}{22} = \frac{V_{OUT}}{100} + \frac{V_{OUT}}{47} + \frac{V_{OUT}}{22}$$

therefore

$$\frac{5}{22} = V_{OUT}\left(\frac{1}{100} + \frac{1}{47} + \frac{1}{22}\right)$$

therefore

$$V_{OUT} = \frac{5}{22 \times (1/100 + 1/47 + 1/22)} = 2.96$$

So the output of the voltage divider falls from 3.41 V to 2.96 V when the voltmeter is connected to it. The current drawn out of the signal source (i.e. I_1) has pulled down the signal. In practice you rarely need to do a precise calculation of the effect of the current drawn from an analogue source; instead, you try to arrange things so that the current is small enough not to upset the signal source. In other words, you try to make the input impedance of the load sufficiently large that it draws a sufficiently small current out of the source. Obviously, the larger the input impedance is, the less it will **corrupt** the signal, but you need to know the minimum input impedance that a signal source will tolerate i.e. how big a load the source can feed its signal into without being corrupted.

Thévenin's theorem

For sources that are just assemblies of supply rails and resistors **Thévenin's theorem** comes to the rescue. This useful rule states that such a system can

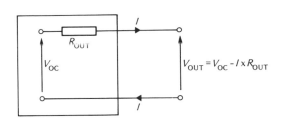

Figure 3.27 The equivalent circuit of a signal source

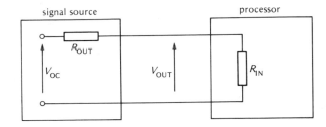

Figure 3.28 The equivalent circuit of figure 3.25

be replaced by a single pair of supply rails and a resistor. Any source, however complex, can be replaced (for the purposes of calculation) by its **equivalent circuit**, shown in figure 3.27. The equivalent circuit contains just two things; a pair of supply rails that are V_{OC} apart and a resistor R_{OUT}. V_{OC} is the **open-circuit voltage** of the source i.e. the signal that it feeds out when no current is being drawn from it. R_{OUT} is the **output impedance** of the source; any current that flows out of the source flows through this imaginary resistor, so that V_{OUT} must always be less than V_{OC}.

Figure 3.28 shows the equivalent circuits for an analogue signal source feeding a signal into a processor. The source has been replaced by its Thévenin equivalent of a pair of supply rails and an output resistor; the processor has similarly been replaced by an input resistor. The whole system looks like a voltage divider, with V_{OC} across R_{OUT} and R_{IN} in series. V_{OUT}, the signal carried by the wires connecting the source and the processor, can be calculated from the usual voltage divider formula:

$$V_{OUT} = \frac{V_{OC} \times R_{IN}}{R_{OUT} + R_{IN}}$$

Calculating output impedances

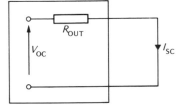

Figure 3.29

To illustrate the use of this idea we shall have another look at the circuit of figure 3.25. We need to calculate the **open-circuit voltage** and the **output impedance** of the voltage divider if we are to calculate the size of the signal being sent to the voltmeter. The value of V_{OC} is 3.41 V; we have already worked this out using the voltage divider formula. In order to get R_{OUT} we need to calculate I_{SC}, the **short-circuit current**. Figure 3.29 shows what happens when the source is short-circuited; the outputs of the source are connected by a wire, so that the signal fed out is zero. Since the voltage across R_{OUT} will be V_{OC} in this situation, Ohm's Law tells us that the short-circuit current will be given by:

$$R_{OUT} = \frac{V_{OC}}{I_{SC}}$$

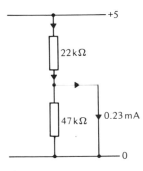

Figure 3.30

When the signal source of figure 3.25 is short-circuited a current of 0.23 mA flows through the short-circuit (look at figure 3.30). So its output impedance will be given by:

$$R_{OUT} = \frac{V_{OC}}{I_{SC}}$$

therefore $\quad R_{OUT} = \dfrac{3.41}{0.23} = 15\,k\Omega$

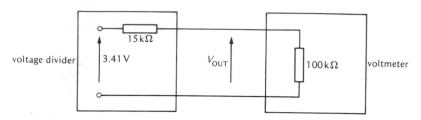

Figure 3.31 A more detailed equivalent circuit of figure 3.25

The equivalent circuit of figure 3.25 is shown in figure 3.31. A 3.41 V supply feeds current into a 15 kΩ and a 100 kΩ resistor in series. The signal being sent down the line is given by:

$$V_{OUT} = \frac{V_{OC} \times R_{IN}}{R_{OUT} + R_{IN}}$$

therefore $$V_{OUT} = \frac{3.41 \times 100}{15 + 100} = 2.96$$

Of course, we get the same answer as before! But the point of using the Thévenin equivalent circuit to do the calculation is that we can easily judge how high the input impedance of the load has to be if it is not to corrupt the signal from the source. By using the voltage divider formula on the circuit of figure 3.28, you can convince yourself that if R_{IN} is at least ten times larger than R_{OUT} then V_{OUT} will differ from V_{OC} by at most 10%. This sort of loss of signal is tolerable under most circumstances; if you want the loss to be less than 1% then you must arrange for R_{IN} to be a hundred times larger than R_{OUT}. If you are stuck with a certain value of R_{IN}, then you will have to amend the source so that its output impedance is sufficiently small. A **stiff source** is one whose output impedance is much smaller than the load into which it is feeding its signal. For example, the source of figure 3.32 is a stiff voltage divider as far as a voltmeter load is concerned. Its output impedance is 1.5 kΩ, much smaller than the 100 kΩ input impedance of the meter. The stiffness of the source is a consequence of the large current that flows through the two resistors of the divider network; the current drawn off into the voltmeter is negligible in comparison with it.

Thévenin's theorem does not apply to signal sources that contain non-resistive components such as transistors and diodes. Nevertheless, the idea of specifying an open-circuit signal and an output impedance for a source is very useful. In general the output impedance of a source will be a function of the size of the signal, but it tells you roughly how big you have to make the input impedance of the load for the signal to be more or less faithfully transmitted between the two. **The golden rule of the analogue world is that signals should be fed out of low impedances into high impedances.**

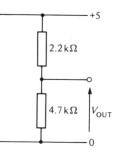

Figure 3.32 A stiff voltage divider

EXPERIMENT

You are going to find the input impedance of a voltmeter on its 5 V range. Then you will use the meter to measure the size of the signal being fed out by a number of different voltage dividers. By the time that you have finished you should appreciate that Thévenin's theorem works.

Set up the circuit of figure 3.33; it shows a voltmeter in series with a resistor R across a pair of supply rails at +15 V and 0 V. Starting off with

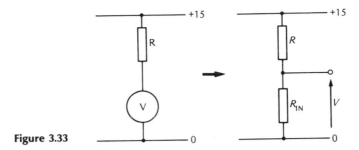

Figure 3.33

$R = 1\,\text{M}\Omega$, decrease it by factors of about two (i.e. try $470\,\text{k}\Omega$, $220\,\text{k}\Omega$ etc.) until the voltmeter reads a large value, but less than $5\,\text{V}$. The equivalent circuit is also shown in figure 3.33; use the values of V and R to find R_{IN}, the input impedance of the voltmeter.

Now calculate the values of the open-circuit voltage, short-circuit current and output impedance of the voltage divider shown in figure 3.34; the values of R_1 and R_2 are shown in the table. Since you know the value of the input impedance of the voltmeter you can also calculate the value of V_{OUT}, i.e. what the voltmeter ought to read in each case.

Finally, build each of the four voltage dividers in turn. Record the reading on the voltmeter and note how it agrees with your calculations.

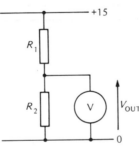

R_1	R_2	V_{OC}	I_{SC}	V_{OUT}
$1\,\text{M}\Omega$	$470\,\text{k}\Omega$			
$100\,\text{k}\Omega$	$47\,\text{k}\Omega$			
$10\,\text{k}\Omega$	$4.7\,\text{k}\Omega$			
$1\,\text{k}\Omega$	$470\,\Omega$			

Figure 3.34

EXPERIMENT

In this experiment you will explore the remarkable stiffness of a zener diode/resistor network

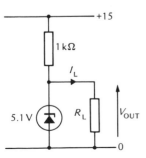

Figure 3.35

The circuit to be assembled is shown in figure 3.35. Use a voltmeter to measure the value of V_{OUT} for different values of the load R_{L}. Start off with $R_{\text{L}} = 47\,\text{k}\Omega$ and reduce it by factors of two until you get to $0.1\,\text{k}\Omega$.

Use your results to plot a graph of V_{OUT} as a function of I_{L}, the current drawn from the source by the load. Remember that $R_{\text{L}} = V_{\text{OUT}}/I_{\text{L}}$.

Try to explain why the source suddenly stops being stiff when more than $10\,\text{mA}$ is drawn from it.

OPERATIONAL AMPLIFIERS

The **operational amplifier** does for the analogue world what the NAND gate does for the digital world. It provides a cheap and easy-to-use basic component which can be used in a wide variety of ways to build systems that process analogue signals. The rest of this chapter will be devoted to showing you how operational amplifiers (or **op-amps**) can be made into a number of different devices with the addition of a few resistors, capacitors and diodes.

The circuit symbol of an op-amp is shown in figure 3.36. Beside it is shown the pin-out of a **741 IC**, the op-amp which has become the industry standard. The 741 is the Model T Ford of the op-amp world; it has been around for a long time, and was the first device on the market that was cheap, reliable, easy to use and versatile. Only recently has it had to contend with serious rivals (e.g. the 081) for the title of ' best all-rounder '. We shall be using the 741 as the basis of all our op-amp data in this book. Although op-amps differ in their capabilities and supply arrangements, their function and behaviour is the same; just as NAND gates can be built out of several technologies, op-amps can be made in many different ways.

Returning to figure 3.36, you will notice that op-amps run off **split supply rails**. That is, you need two supply rails with equal and opposite voltages. A 741 will tolerate supply rails that are 36 V apart but the usual arrangement is to have them at +15 V and −15 V. From now on, unless specifically stated, we shall assume that a 741 is run off such a 30 V split supply. This means that we will not bother to draw in the supply lines in a circuit diagram, just as we did with NAND gates. The 741 is quite flexible about supply arrangements; it will operate well with a pair of supply rails at +5 V and −5 V.

The **output** of an op-amp is controlled by a pair of **inputs** which have rather cumbersome, but standard names. The plus and minus signs beside each input tell you what they do. An op-amp takes the difference in voltage between its inputs and multiplies it by a large number before feeding it out. More specifically, referring to figure 3.37:

$$V_{OUT} = -100\,000 \times v$$

V_{OUT} is, as usual, the voltage of the output with respect to 0 V. The number 100 000 is the **DC gain** of a 741; v is the voltage difference of the two inputs. An op-amp is a **high gain differential amplifier.**

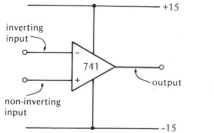

Figure 3.36 A 741 operational amplifier

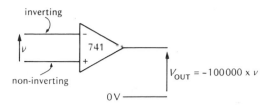

Figure 3.37

When it is run off the standard supply rails the inputs of a 741 can take any value between $+15\,V$ and $-15\,V$. The input impedance is high; each input draws a current of about 80 nA from a signal source. On the other hand the output of a 741 can only get to within $2\,V$ of the supply rails. So the maximum and minimum outputs are $+13\,V$ and $-13\,V$ with the usual supply rails. The output impedance is very low, about $75\,\Omega$, and a 741 can source or sink up to about 10 mA. The output is **internally protected**, so that the output voltage is automatically clamped if more than 13 mA is drawn from it.

In general there are three distinct ways in which an op-amp may be used. They may be used in **open-loop mode**, with **positive feedback** and with **negative feedback**. Negative feedback is the most flexible and widely used way of employing an op-amp, so we shall deal with the other two first and get them out of the way.

OPEN-LOOP MODE

When an op-amp is used as a processing unit on its own it acts as a **comparator**. It looks at the signals coming into its two inputs, decides which one has the higher voltage and then pushes its output as close as possible to one of the supply rails. The symbol against the input at the higher voltage fixes the sign of the output. So if the **non-inverting input** is higher than the **inverting** one the output is $+13\,V$ (look at figure 3.37). The output will always be $-13\,V$ when the inverting input is higher than the non-inverting one.

We shall look at two examples of op-amps acting as comparators. The first is illustrated in figure 3.38. On the left of the circuit are two voltage dividers, giving out signals X and Y. The size of X is fixed by the setting of the potentiometer; the size of Y is determined by the temperature of the thermistor. In the centre of the circuit is a 741 comparing the sizes of the signals from the two sources. On the right of the circuit are a pair of LEDs; the green one will only come on when V_{OUT} is above $+2\,V$, the red one only when V_{OUT} is below $-2\,V$.

The high input impedance of the 741 means that it does not interfere with the signals coming out of the two sources. Provided that the output impedance of a signal source is less than $1\,M\Omega$ you do not have to worry about the loading effects of a 741; if the output impedance is larger than $1\,M\Omega$ there are plenty of op-amps with input impedances greater than $1\,G\Omega$ (i.e. $10^9\,\Omega$)!

You can always assume that the output of a 741 in open-loop mode is **saturated** i.e. the output is trying to get as near as it can to one of the supply rails. This is just a consequence of the enormous DC gain of the

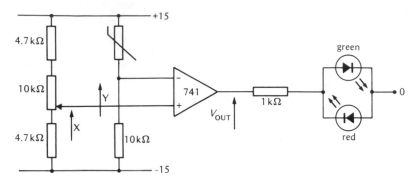

Figure 3.38 A hot/cold detector

op-amp; if the output is not to be saturated the two inputs must differ in voltage by less than about 130 µV. Two separate signal sources are unlikely to stay this close for very long.

Returning to the circuit of figure 3.38, we can now discuss what it does. Suppose that the thermistor is cold (i.e. Y is low) and that X has been set at some fixed value greater than Y. The input terminal with the plus sign will be higher than the other one so V_{OUT} will be +13 V. The green LED will be forward biased and a current of 11 mA will flow through it; the red LED is reverse biased so no current flows through it. If we now raise the temperature of the thermistor the value of Y will increase as the resistance of the thermistor goes down. The instant that Y becomes greater than X, V_{OUT} will go down to −13 V and the red LED will come on instead of the green one. The whole system acts as a hot/cold detector. The red LED indicates that the thermistor is hot, the green one that it is cold; the temperature at which the LEDs swap over is determined by the setting of the potentiometer.

Our second example is shown in figure 3.39. The LED only comes on if V_{IN} lies in the range +5 V to +10 V. When V_{IN} lies outside this range the outputs of the two op-amps have the same value i.e. +13 V if V_{IN} is below +5 V, and −13 V if V_{IN} is above +10 V. The LED can only come on when the outputs are different; the 2.2 kΩ resistor limits the current through it to about 10 mA.

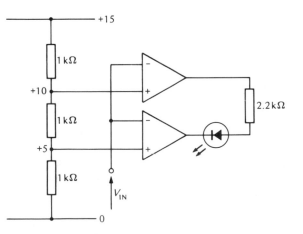

Figure 3.39 A +5 V to +10 V detector

PROBLEMS

1 For the circuit shown in figure 3.40:

a) Describe what happens to the LED as the LDR is exposed to light of gradually increasing intensity. Explain your answer.

b) What is the voltage at point G in the circuit?

c) What will the resistance of the LDR be when the LED is about to go off?

d) Most LEDs are very limited in their tolerance of reverse bias voltages. Usually five volts of reverse bias is enough to destroy most LEDs. Explain how the presence of the diode in the circuit protects the LED against destruction.

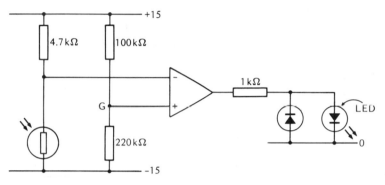

Figure 3.40

2 Devise a temperature sensing circuit built around a thermistor whose characteristics are shown in the graph of figure 3.41. The circuit must have three LEDs as its output. A green one shows that the temperature is less than 15 °C. A yellow one shows that the temperature is between 15 °C and 25 °C. A red one shows that the temperature is above 25 °C.

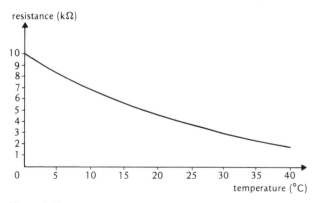

Figure 3.41

3 A simple digital voltmeter has a row of ten LEDs as its output. The number of LEDs that are lit tells you the size, in volts, of the signal that is being fed into the device. So if six LEDs are lit, the input is above +6 V and below +7 V. The whole thing is run off a 30 V split supply.

Draw a circuit diagram to indicate how such a device could be realised in practice.

POSITIVE FEEDBACK

When an op-amp is used in open-loop mode **information** only flows one way through the system. Signals arrive at the two inputs, are processed by the op-amp and a signal gets pushed out at the output. If you allow information to get from the output back to one or both of the inputs then the behaviour of the system is radically altered. The technique of allowing information to flow both ways through a system is called **feedback**. The simplest way in which this can be done with an op-amp is to link the output to one of the inputs with a wire as shown in figure 3.42. If the wire goes to the non-inverting input then **positive feedback** has been applied to the system. With this form of feedback the input signal is being compared with the output signal; the graph of figure 3.42 shows what effect this has on the response of the system.

To start with, suppose that the output of the 741 is saturated i.e. $V_{OUT} = +13V$. Then if V_{IN} has any voltage less than $+13V$ the system will be stable, with the non-inverting input at a higher voltage than the inverting one. But once V_{IN} goes above $+13V$, V_{OUT} will dive down to $-13V$. The positive feedback ensures that once V_{OUT} has begun to fall the initial small difference between the inputs is steadily increased so that it has to go all the way down to $-13V$ as quickly as it can (i.e. in about $50\,\mu s$ for a 741).

By now you will probably have realised that the system behaves like a **Schmitt trigger** with **thresholds** of $+13V$ and $-13V$. Furthermore, the system **inverts** the input signal; when V_{IN} is raised above the threshold, V_{OUT} dives down to $-13V$. By adding a pair of zener diodes we can cripple the output of the 741 and tailor the switching thresholds to any requirements. For instance, the circuit of figure 3.43 has switching thresholds of $+5V$ and $-5V$; the output of the 741 is clamped at these two voltages by the two 4.3 V zeners.

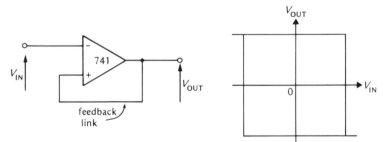

Figure 3.42 An op-amp with positive feedback

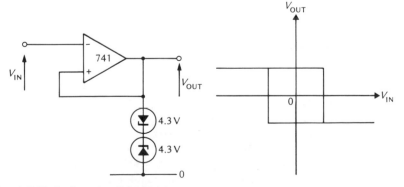

Figure 3.43 An inverting Schmitt trigger

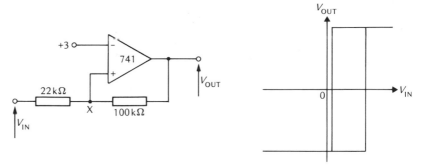

Figure 3.44 A non-inverting Schmitt trigger

Non-inverting Schmitt triggers

A somewhat more flexible approach to building a Schmitt trigger out of an op-amp is shown in figure 3.44. Perhaps the best way of understanding how it works is to jump in at the deep end and calculate its switching thresholds.

Since the feedback is positive we can safely say the output of the op-amp will be saturated; let us assume that V_{OUT} is initially -13 V. In order for the system to be stable, point X in the circuit must be at less than $+3$ V so that the inverting input is the one at the higher voltage. Once X goes above $+3$ V, V_{OUT} will start to rise and positive feedback will firmly push it up to $+13$ V. All we need to do is calculate what value of V_{IN} will get X up to $+3$ V. No current goes in or out of the inputs of a 741 (none worth worrying about) so we can forget that the op-amp is there and concentrate on the voltage divider shown in figure 3.45. Using the voltage divider formula:

$$V_{OUT} = \frac{V_{IN} \times R_2}{R_1 + R_2}$$

therefore $\quad 3 + 13 = \dfrac{(V_{IN} + 13) \times 100}{100 + 22}$

therefore $\quad V_{IN} = \dfrac{16 \times 122}{100} - 13 = +6.5$

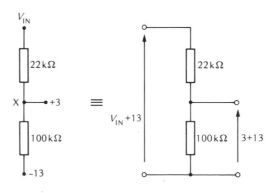

Figure 3.45

So V_{IN} has to get up to $+6.5\,V$ to trigger V_{OUT} into switching from $-13\,V$ to $+13\,V$. The $26\,V$ increase in the value of V_{OUT} is fed back to produce a rise in the voltage at X. In order to get X back below $+3\,V$ (i.e. to push V_{OUT} back down to $-13\,V$) V_{IN} will have to be lowered. Figure 3.46 shows the voltage divider part of the circuit when V_{IN} is just small enough to trigger it.

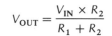

Figure 3.46

$$V_{OUT} = \frac{V_{IN} \times R_2}{R_1 + R_2}$$

therefore

$$3 - V_{IN} = \frac{(13 - V_{IN}) \times 22}{100 + 22}$$

therefore

$$V_{IN}(122 - 22) = 3 \times 122 - 13 \times 22$$

therefore

$$V_{IN} = \frac{366 - 286}{100} = +0.8$$

The behaviour of the whole system is summarised in the graph of figure 3.44; note that it is a non-inverting Schmitt trigger.

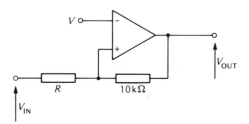

Figure 3.47 A general purpose Schmitt trigger

Asymmetrical Schmitt triggers

It should be clear that this sort of trigger circuit can be tailored to have any pair of switching thresholds. The general circuit is shown in figure 3.47. You choose the values of the voltage V and the resistor R to give you the thresholds that you want. As a rule of thumb, the value of V sits about halfway between the thresholds and the value of R fixes how far apart the thresholds are. If the thresholds are fairly close to $0\,V$, then they are approximately given by $V \pm 13 \times (R/10)$. In practice you use the rule of thumb to get the values of V and R approximately correct then generate them with potentiometers as shown in figure 3.48. You then adjust the circuit until it has exactly the thresholds that you want.

A simple way of arranging the circuit so that one of the switching thresholds is at $0\,V$ is shown in figure 3.49. The addition of a diode into the feedback path introduces an asymmetry; the thresholds will be ground and some positive voltage.

Suppose, for example, that we want $+2\,V$ and $0\,V$ as the thresholds. When V_{OUT} is $+13\,V$ the diode is reverse biased so there is no way that the output of the 741 can feed current (i.e. information) back towards the non-inverting input. It is effectively operating open-loop. So once V_{IN} gets below $0\,V$, V_{OUT} shoots down to $-13\,V$, the diode turns into a short-circuit and positive feedback can take place. In order to get V_{OUT} back up to $+13\,V$, V_{IN} will have to be raised. Look at figure 3.50, which shows the

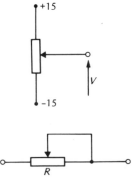

Figure 3.48 The adjustable components for a Schmitt trigger

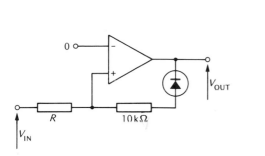

Figure 3.49 A Schmitt trigger with one threshold at 0 V

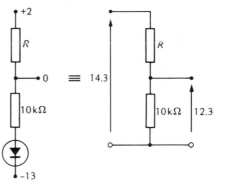

Figure 3.50

voltage divider part of the circuit when it is about to be triggered. Using the voltage divider formula:

$$12.3 = \frac{14.3 \times 10}{10 + R}$$

therefore $\quad R = \frac{14.3 \times 10}{12.3} - 10$

therefore $\quad R = 1.6\,\text{k}\Omega$

(Note how we have taken the voltage drop across the diode into account by assuming that the 741 output sits 0.7 V higher than it actually does.)

The relaxation oscillator

There is more than one way of making an op-amp into a Schmitt trigger. Look at the circuit of figure 3.51. It is similar to the one shown in figure 3.42, with the input being compared with a fraction of the output signal. The feedback is positive so V_{OUT} will be saturated. The voltage divider ensures that X sits at a fixed fraction of the value of V_{OUT} i.e. $+6.5\,\text{V}$ or $-6.5\,\text{V}$ for our example. A little thought should convince you that these two voltages are the switching thresholds of the whole system i.e. that any

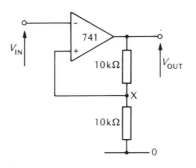

Figure 3.51 Another inverting Schmitt trigger

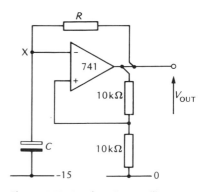

Figure 3.52 A relaxation oscillator

value of V_{IN} above $+6.5\,V$ will push V_{OUT} down to $-13\,V$, and that V_{IN} will have to subsequently go below $-6.5\,V$ to get V_{OUT} back up to $+13\,V$.

You will recall from the last chapter that the addition of a resistor and a capacitor to an inverting Schmitt trigger will make it generate square waves. Such an oscillator is shown in figure 3.52. The square wave that is fed out has a frequency of about $1/(2RC)\,kHz$ if R and C are in the usual units of $k\Omega$ and μF (see the problems at the end of this section).

A 741 will happily feed its output signal into an input impedance of $1\,k\Omega$. If you want to feed the output of a 741 based relaxation oscillator into a loudspeaker you will have to use a transformer to make the 8 ohms of the speaker look like $1\,k\Omega$. A transformer that has a turns ratio of 12:1 and is wired up as shown in figure 3.53 will make the 741 think that it is pushing its signal into $(12)^2 \times 8\,ohms$ i.e. $1.1\,k\Omega$. A 741 that is going flat out can push $13\,mA$ out at $13\,V$, so the speaker receives energy at a rate of about $13 \times 13 = 169\,mW$. This makes quite a respectable noise! Chapter Four will show you how the power fed out of a 741 can be boosted to much higher levels with the help of a pair of bipolar transistors.

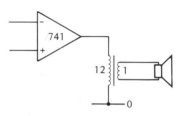

Figure 3.53 Feeding a 741 output into a speaker via a transformer

EXPERIMENT

Assembling and testing a resistance-to-frequency converter based on a 741 relaxation oscillator

Start off by assembling the Schmitt trigger of figure 3.54. The potentiometer generates a variable input signal and the two LEDs display the state of the output. Using a voltmeter to measure the value of V_{IN}, measure the switching thresholds of the circuit. Compare the measured values with your calculated ones.

Then remove the potentiometer and add a resistor and a capacitor as shown in figure 3.55. Measure the period of the square wave; calculate what it ought to be.

Now speed up the oscillator by replacing the $100\,\mu F$ with $100\,nF$ and the $47\,k\Omega$ with $22\,k\Omega$. Trigger a CRO on V_{OUT} and use it to look at the waveforms of V_{OUT} and V_{IN}. Draw graphs of V_{OUT} and V_{IN} as functions of time, putting both on the same graph.

Finally, replace the $22\,k\Omega$ with an LDR and listen to the square wave output with the help of a loudspeaker. The frequency of the sound coming

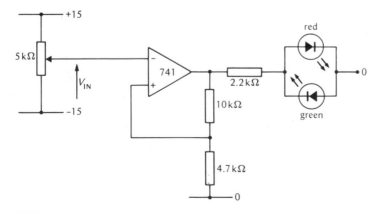

Figure 3.54

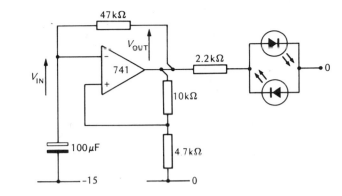

Figure 3.55

from the speaker will depend on the amount of light falling on the circuit. You might even care to replace the LDR with yourself as the variable resistor in the circuit!

PROBLEMS

1 Calculate the switching thresholds of the Schmitt triggers shown in figure 3.56. Draw a graph for each to show how its output is related to its input. Assume a 30 V split supply.

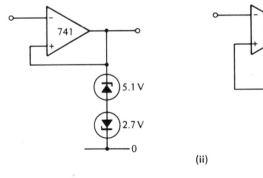

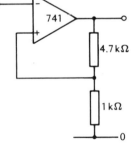

(i) (ii)

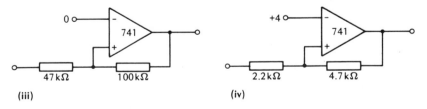

(iii) (iv)

Figure 3.56

2 By calculating how long it takes for point X in figure 3.52 to charge up from one switching threshold to the other, show that the system of figure 3.52 has a frequency of about $1/(2RC)$.

3 Draw graphs for the system shown in figure 3.57, to the same scale and one under the other, to show how the voltages at points X, Y and Z vary with time.

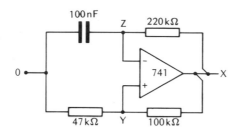

Figure 3.57

4 Design a non-inverting Schmitt trigger that has switching thresholds of roughly $+5\,V$ and $0\,V$. Do not use zener diodes, but feel free to use trial and error!

5 Figure 3.58 shows a 20 V peak-to-peak sine wave being fed into a Schmitt trigger. The thresholds of the trigger circuit depend on the value of V. Draw the waveforms of V_{IN} and V_{OUT} on the same graph for $V = -5\,V$, $0\,V$ and $+5\,V$.

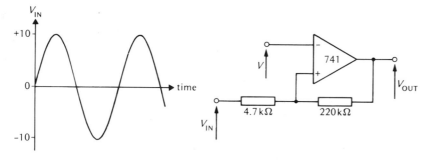

Figure 3.58

NEGATIVE FEEDBACK

Figure 3.59 shows the simplest way in which **negative feedback** can be applied to an op-amp. You can tell that feedback has been applied because the output is able to transfer information back to the inverting input via the wire that connects them. Why the feedback is called negative will become clear after we have discussed the behaviour of the system.

Look at the graph of figure 3.60; it shows the two conditions that the circuit of figure 3.59 has to obey. The curve labelled "open-loop" shows how V_{OUT} varies with F for a fixed value of V_{IN} when no feedback is applied. The other curve marked "closed-loop" shows the other condition $(V_{OUT} = F)$ that is imposed by the connection between input and output. There is only one point on the graph where both conditions can be satisfied at the same time i.e. when $V_{OUT} = V_{IN}$. So the voltage fed into the system is equal to the voltage fed out by it; it is a **follower**, as the output signal follows the input signal.

But is the output of the follower stable? Suppose that V_{OUT} is greater than V_{IN}. Then the inverting terminal will be at a higher voltage than the

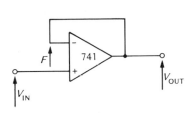

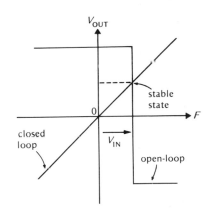

Figure 3.59 An op-amp with negative feedback

Figure 3.60 A graph to work out the stable state of an op-amp with negative feedback

non-inverting one and consequently the op-amp will start to push its output down towards $-13\,V$. This change in output will bring V_{OUT} back towards V_{IN}. Once V_{OUT} gets smaller than V_{IN} the situation changes and the 741 tries to push its output up to $+13\,V$; again, the negative feedback pushes V_{OUT} closer to the value of V_{IN}. So the output of the op-amp continually strives to make V_{OUT} the same as V_{IN}. This is typical of the behaviour of an electronic circuit when negative feedback is applied to it. The output is forced towards a certain value and is pushed back to that value whenever it strays from it. **Negative feedback stabilises the output of a circuit**.

There is one golden rule that will enable you to understand any op-amp circuit where negative feedback has been applied. **The output sits at exactly the right voltage for the two inputs of the op-amp to have the same voltage as each other**.

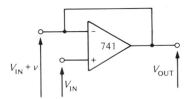

Figure 3.61

For the reasons that lie behind this invaluable rule refer to figure 3.61. It shows an op-amp with a small voltage difference v between its input terminals when negative feedback has been applied. The output will be stable and equal to the value of V_{IN}; V_{OUT} must lie between $+13\,V$ and $-13\,V$. Remembering that an op-amp amplifies v, $V_{OUT} = 100\,000 \times v$. If V_{OUT} is not to saturate then there is a maximum size that v can have:

$$v = \frac{V_{OUT}}{100\,000} < \frac{13}{100\,000} = 130\,\mu V$$

So we would expect that v will always be between $+130\,\mu V$ and $-130\,\mu V$. In other words, as the non-inverting terminal is varied over a range of $26\,V$ the other terminal follows it to within $130\,\mu V$ (i.e. to within 0.001%). This difference is so small that it might as well be called no difference at all for a lot of applications. Since an op-amp with a DC gain of infinity would always have both inputs at exactly the same value when negative feedback was applied, the golden rule is sometimes called the **infinite gain approximation**.

We shall spend the rest of the chapter looking at some of the ways in which negative feedback can be used to create useful systems from op-amps. We shall start off with three **linear amplifiers**, then look at some **non-linear circuits** that involve diodes and switches, and end up with a glance at some of the applications of **ramp generators**.

The non-inverting amplifier

Negative feedback can be used to transform an op-amp into a device which multiplies the signal fed into it by a fixed number. The circuit of figure 3.62, for example, has an output that is $+3.1$ times the size of its input. The behaviour of the circuit is easily understood with the help of the golden rule. V_{OUT} will always have the right value to make F the same as V_{IN}. The two resistors form a voltage divider whose input is V_{OUT} and whose output is F. So $F = 0.32 \times V_{OUT} = V_{IN}$. It is customary to quote a value for the **voltage gain** of this sort of system:

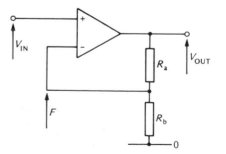

Figure 3.62 A non-inverting amplifier

$$G = \frac{V_{OUT}}{V_{IN}} = \frac{V_{OUT}}{0.32 \times V_{OUT}} = +3.1$$

The graph of figure 3.63 illustrates the behaviour of the system. Provided that it is not saturated, V_{OUT} is 3.1 times larger than V_{IN}; the circuit behaves like an amplifier whose gain is decided by the relative sizes of the resistors used in the feedback loop. The circuit is so useful that it is worth working out a formula for its gain; referring to figure 3.64 for the definitions of the symbols:

$$V_{IN} = F = \frac{V_{OUT} \times R_b}{R_a + R_b}$$

therefore
$$G = \frac{V_{OUT}}{V_{IN}} = \frac{R_a + R_b}{R_b}$$

The usefulness of the amplifier arise from three of its properties.

Firstly, it is **linear** i.e. the output is an exact copy of the input, only a different size.

Secondly, the gain can be easily and reliably calculated; you only need to know the sizes of the two resistors. (Wait until you meet transistor amplifiers in Chapter Five!)

Thirdly, the circuit has a very large input impedance. A 741 has an input impedance of $2\,M\Omega$ between the input terminals, but if they are always at

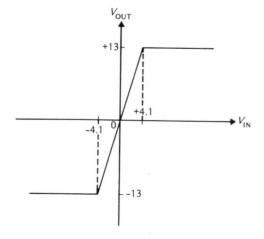

Figure 3.63 The input-output characteristics for figure 3.62

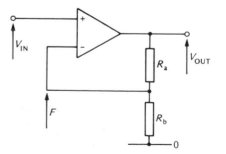

Figure 3.64

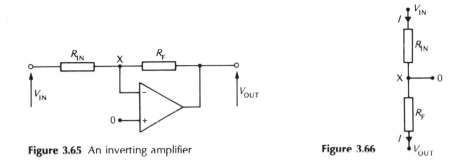

Figure 3.65 An inverting amplifier   **Figure 3.66**

the same voltage no current can flow through that equivalent resistor. If no current flows into the non-inverting terminal, the whole circuit must appear to have an infinite input impedance.

The inverting amplifier

Figure 3.65 shows the second way in which negative feedback can be used to convert an op-amp into an amplifier. Since the feedback is negative we can use the golden rule and state that the point marked X in the circuit will always sit at 0 V (provided that V_{OUT} is not saturated of course). For this reason X is sometimes referred to as a **virtual earth**; it behaves as though it is connected to 0 V. Furthermore, since no current will flow into the inverting input of the op-amp, we need only consider the voltage divider formed by the two resistors to calculate the gain of the system.

Consider the input resistor R_{IN} (look at figure 3.66). Applying Ohm's Law:

$$R = \frac{V}{I}$$

therefore $R_{IN} = \frac{V_{IN} - 0}{I}$

therefore $I = \frac{V_{IN}}{R_{IN}}$

All of that current must flow through the feedback resistor to the output of the op-amp. Remembering that one end is held at 0 V by the virtual earth, we can calculate the voltage of the other end of the resistor:

$$R = \frac{V}{I}$$

therefore $R_F = \frac{0 - V_{OUT}}{I}$

therefore $V_{OUT} = -IR_F$

Combining our two equations we get:

$$V_{OUT} = \frac{-V_{IN}}{R_{IN}} R_F = \frac{-R_F}{R_{IN}} V_{IN}$$

So the voltage gain of the system is given by:

$$G = \frac{V_{OUT}}{V_{IN}} = -\frac{R_F}{R_{IN}}$$

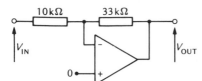

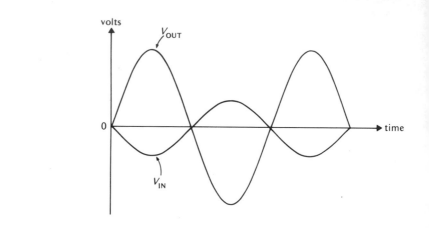

Figure 3.67 An inverting amplifier in action

Figure 3.67 shows an **inverting amplifier** in action. The gain is -3.3, so the output waveform is 3.3 times larger than the input and has the opposite sign. The input impedance of the amplifier is considerably lower than that of its non-inverting cousin (R_{IN} is 10 kΩ in this case), so you have to be fairly careful about the output impedance of the source of the signal that is being amplified.

The summing amplifier

The last of our trio of linear amplifiers is a useful building block that can be used to add two analogue signals together; you will meet its alter ego, the **difference amplifier**, in the problems at the end of this section.

As you can see from figure 3.68, the circuit has a virtual earth at X. This allows us to use Ohm's Law to calculate the currents I_1 and I_2 (0.2 mA and 0.3 mA, respectively). Then the current that flows through the feedback resistor must be equal to $I_1 + I_2 = 0.5$ mA. So the voltage drop across that resistor must be 5 V; since one end is held at 0 V (it is a virtual earth), the other end must sit at -5 V. The output is equal to the sum of the two inputs multiplied by -1.

There is no reason why the resistors in a **summing amplifier** have to be the same as each other. If they are different the two input signals are amplified before they are added to each other. For example, the circuit of figure 3.69 has an output that is roughly $-(2X + 5Y)$.

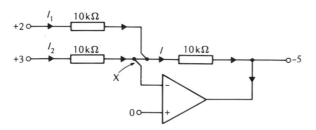

Figure 3.68 A summing amplifier

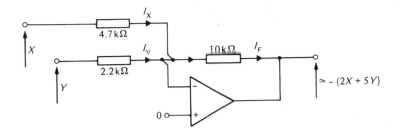

Figure 3.69

To work out what it does you apply Ohm's Law to every resistor in turn:

$$4.7 = \frac{X - 0}{I_X} \qquad \text{therefore} \quad I_X = \frac{X}{4.7}$$

$$2.2 = \frac{Y - 0}{I_Y} \qquad \text{therefore} \quad I_Y = \frac{Y}{2.2}$$

$$10 = \frac{0 - V_{OUT}}{I_F} \qquad \text{therefore} \quad I_F = \frac{-V_{OUT}}{10}$$

Knowing that $I_F = I_X + I_Y$, we can combine these three expressions to get:

$$\frac{-V_{OUT}}{10} = \frac{X}{4.7} + \frac{Y}{2.2}$$

$$\text{therefore} \qquad V_{OUT} = -(2.1X + 4.5Y)$$

EXPERIMENT

Exploring the characteristics of the inverting amplifier
in figure 3.70

The potentiometer allows you to generate values of V_{IN} that can range from $+15\,V$ to $-15\,V$. Choosing values of V_{IN} that cover the whole of this range, measure the voltages at X and the output of the 741. Draw graphs of V_{OUT} and V_X as functions of V_{IN}; explain their shapes.

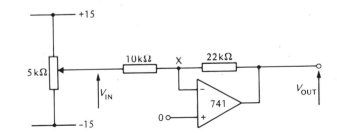

Figure 3.70 A circuit for investigating an inverting amplifier

EXPERIMENT

The system shown in figure 3.71 generates three analogue
signals, called R, S and T. You are going to design and
assemble systems that will mix the three signals to give each
of the waveforms shown in figure 3.72.

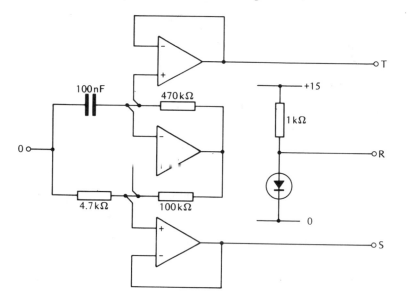

Figure 3.71

Start off by assembling the circuitry of figure 3.71, remembering to leave
plenty of room for the other bits that you will add later. It consists of a
relaxation oscillator and a forward biased diode, with a couple of followers
to buffer the time-varying signals. Trigger a CRO on the signal at S and look
at R, S and T in turn; draw graphs of each waveform, to the same scale
and one under the other.

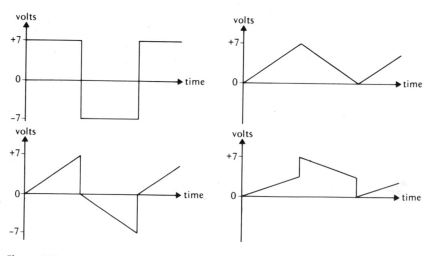

Figure 3.72

Each of the waveforms shown in figure 3.72 can be generated from those at R, S and T with appropriate combinations of inverting, non-inverting and summing amplifiers. Try to generate each of the waveforms in turn. When you have successfully built and tested it, draw the circuit diagram of the mixer and indicate the component values used.

PROBLEMS

1 What is the voltage of the output terminal of the op-amp in each of the circuits of figure 3.73?

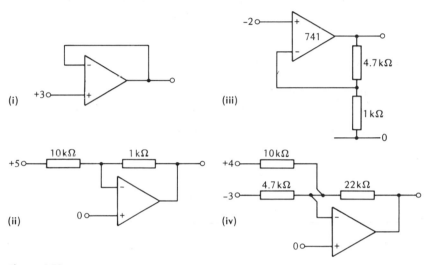

Figure 3.73

2 Draw circuit diagrams of amplifiers, based on op-amps, that have gains of
  a) −1
  b) +23
  c) −47
  d) +3

3 Show that the amplifier in figure 3.74 has an output voltage of −(A/8 + B/4 + C/2 + D); A, B, C and D are the analogue signals fed into it.

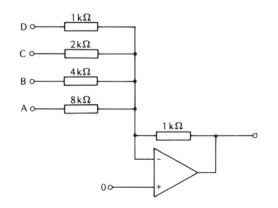

Figure 3.74

4 A digital-to-analogue converter has four inputs A, B, C and D. Each input is to be held at 0 V or +5 V by the digital system feeding signals into the converter; the output of the converter must equal half of the decimal equivalent of the binary word DCBA. So if DCBA = 0000_2, then the output = 0 V; if DCBA = $1000_2 = 8_{10}$, the output = +4 V.

Draw a circuit diagram of a system that could do this, based on the system of figure 3.74.

5 An analogue averager has four inputs (W, X, Y and Z) and an output which is equal to the average value of the four input signals. Design a suitable circuit (it can be done with just two op-amps).

6 For the circuits of figure 3.75, calculate

a) the voltage of the non-inverting terminal,

b) the voltage of the output terminal of the op-amp.

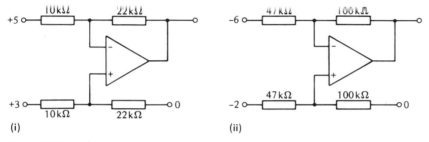

(i)   (ii)

Figure 3.75

7 Show that the circuit of figure 3.76 obeys the equation $V_{OUT} = B - A$ i.e. it subtracts analogue signals from each other.

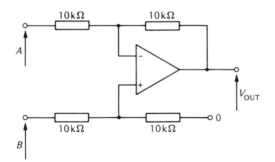

Figure 3.76 A difference amplifier

NON-LINEAR AMPLIFIERS

This section will introduce you to five applications of negative feedback to op-amps. The circuits are non-linear in the sense that you cannot simply write down a formula for their outputs as a function of their inputs; the non-linearity arises from the addition of switches and diodes to the feed-back path.

The first system is shown in figure 3.77. It always has a gain of 1, but the sign of the gain depends on whether the switch is open or closed. When the switch is closed, the non-inverting input is connected to 0 V and the circuit becomes an inverting amplifier. If the switch is open, both input terminals float up to V_{IN} (negative feedback) so no current can flow through

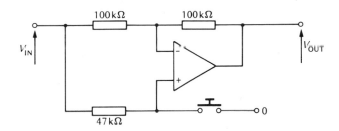

Figure 3.77 A plus-minus follower

the feedback resistor and V_{OUT} has to equal V_{IN}. This **plus-minus follower** will appear in one of the circuits to be discussed a little later on.

The second system uses a diode in the feedback path, so that it becomes a perfect **half-wave rectifier**. Look at figure 3.78. On the left is a sine wave that is fed into the circuit and on the right is the waveform that emerges from it; the circuit clamps the output so that it cannot go below ground, but otherwise is a faithful copy of the input signal. The negative feedback will always try and arrange for the current flow through the 4.7 kΩ to be enough for V_{OUT} to be equal to V_{IN}; the op-amp can only do this if the diode is forward biased i.e. V_{IN} is above ground. Once V_{IN} goes below 0 V,

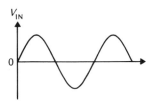

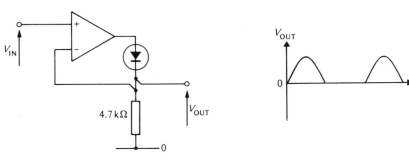

Figure 3.78 A perfect half-wave rectifier

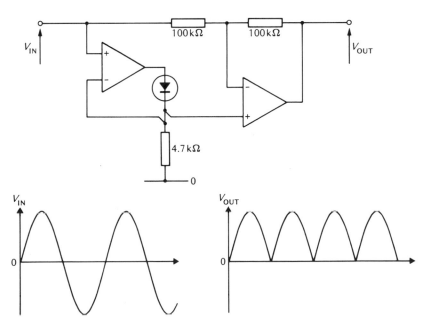

Figure 3.79 A perfect full-wave rectifier

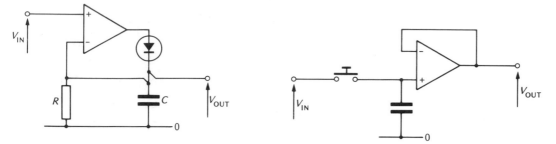

Figure 3.80 A peak-value detector   **Figure 3.81** A sample-and-hold circuit

the op-amp output dives down to $-13\,$V in an attempt to pull current through the (now reverse biased) diode so that V_{OUT} can follow V_{IN}. The diode defeats the op-amp and V_{OUT} is left at $0\,$V, with no current flowing through the resistor.

The third system is a perfect **full-wave rectifier** and combines the two systems that have already been introduced; it is shown, together with typical input and output waveforms, in figure 3.79. The op-amp on the right is a plus-minus follower, the one on the left is a half-wave rectifier. When V_{IN} is positive the whole system is a follower, with a gain of $+1$. Negative values of V_{IN} leave the output of the half-wave rectifier at ground, so that the plus-minus follower becomes an inverter.

The next circuit is shown in figure 3.80; it is a **peak-value detector**. Suppose that V_{OUT} is initially $0\,$V and that V_{IN} rises to $+3\,$V; the negative feedback will push the output of the op-amp up so that current flows onto the capacitor via the diode and V_{OUT} rises up to $+3\,$V. If V_{IN} now falls to $+2\,$V, the op-amp output dives down to $-13\,$V and tries to sink current from the capacitor and lower V_{OUT} from $+3\,$V to $+2\,$V. Since the diode is reverse biased the charge on the capacitor can only leak off via the resistor; V_{OUT} is stuck at $+3\,$V for a time short compared with RC.

The final circuit is an example of a **sample-and-hold system**. Look at figure 3.81. When the switch is closed, V_{OUT} follows V_{IN}, but when the switch is open V_{OUT} stays frozen at the value it had just before the switch was opened. In practice, V_{OUT} will steadily drift away from its initial value because of the small current that has to flow into the non-inverting terminal of the op-amp, but if you choose a low current op-amp and a large capacitor you can minimise the problem.

Ramp generators

Our final example of negative feedback with an op-amp shows what happens when you put a capacitor in the feedback path. Instead of forcing the output of the op-amp to have a certain value (as is usual with negative feedback), the presence of the capacitor makes the output change at a certain rate. The output **ramps** up or down at a rate fixed by the input signal and the components in the feedback network.

Figure 3.82 illustrates such a **ramp generator**. Since the feedback is negative, X is a virtual earth and we can easily get an expression for the current flowing through the resistor:

$$R = \frac{V_{IN} - 0}{I}$$

$$\text{therefore} \quad I = \frac{V_{IN}}{R} \quad\quad\quad\quad (1)$$

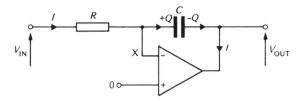

Figure 3.82 A ramp generator

All of that current must flow onto the capacitor, increasing the charge on its plates. If we consider a time interval t, then the amount of charge dumped on the capacitor in that time is given by:

$$Q = I \times t \qquad (2)$$

Suppose that the capacitor starts off with no charge. Then as the charge on the capacitor increases, so does the voltage across it; the virtual earth at X means that this voltage change appears at V_{OUT}. Using the equation linking the charge on a capacitor to the voltage across it, we can say that:

$$C = \frac{Q}{0 - V_{OUT}} = \frac{-Q}{V_{OUT}} \qquad (3)$$

Combining equations (1), (2) and (3) we obtain:

$$C = -\frac{Q}{V_{OUT}} = -\frac{I \times t}{V_{OUT}} = -\frac{V_{IN} \times t}{R \times V_{OUT}}$$

So if V_{OUT} starts off at 0 when $t = 0$, then at time t its value will be given by:

$$V_{OUT} = -V_{IN} \times \left(\frac{t}{RC}\right)$$

The output of the op-amp changes at a constant rate of V_{IN}/RC volts per millisecond. V_{IN}, R and C have the usual units.

For example, look at the circuit of figure 3.83. When the switch is closed, the capacitor is short-circuited and the op-amp acts like a follower i.e. $V_{OUT} = 0\,V$. When the switch is opened the output steadily increases at 7.0 volts per second until it saturates. The output stays at $+13\,V$ until the switch is briefly pressed; the capacitor discharges through the switch, so that V_{OUT} returns to $0\,V$. Notice that a negative input signal makes the output ramp upwards.

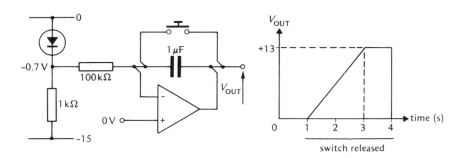

Figure 3.83 A ramp generator in action

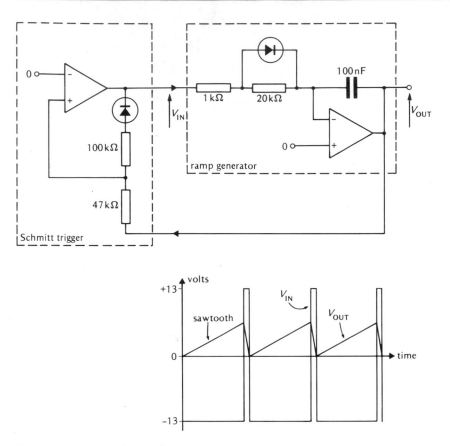

Figure 3.84 A sawtooth waveform generator

Generating sawtooth waveforms

Ramp generators are very useful building blocks. One of their applications is the generation of **sawtooth waveforms** for **CRO timebases**. An example of a sawtooth waveform is shown in figure 3.84, together with a circuit that could be used to generate it. The signal ramps up from 0 V to +6 V once every millisecond, diving down quickly from +6 V to 0 V when it has finished ramping up. On the right of the circuit is the ramp generator and on the left is the Schmitt trigger that senses the output waveform and makes it ramp up or down accordingly. The Schmitt trigger has thresholds of 0 V and +6 V. When V_{OUT} gets up to +6 V, V_{IN} will fly up to +13 V and a current of about 12 mA will flow onto the capacitor (the diode "short-circuits" the 20 kΩ resistor so that we can forget that it is there). As 12 μC of charge are dumped on the 0.1 μF capacitor in each millisecond, V_{OUT} will ramp down at 12/0.1 = 120 volts per millisecond. After 0.05 ms V_{OUT} will get down to 0 V, the Schmitt trigger will quickly change state and V_{IN} will become −13 V. The diode in parallel with the 20 kΩ resistor is now reverse biased so we can treat it as an infinite resistor. The current flowing onto the capacitor will now be 0.61 mA, so V_{OUT} rises at 0.61/0.1 = 6.1 V/ms. After about a millisecond V_{OUT} gets up to +6 V, the Schmitt trigger changes the state of V_{IN}, etc.

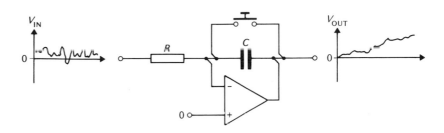

Figure 3.85 Integrating a signal

Integrators

Ramp generators are sometimes called **integrators**. This is because they can be used to **integrate** an analogue signal i.e. find out its average value. Suppose that you need to measure the size of a small analogue signal, and that the signal has a lot of noise on it; this happens frequently when analogue signals are sent down long, noisy cables. The signal will wobble up and down about its average value in a random fashion, as shown in figure 3.85. If the signal is fed into a ramp generator, for a fixed length of time, then the output is proportional to the average value of the signal.

To understand why this happens, suppose that the signal has the value V_i during the *i*th millisecond. During that millisecond the output will change by $-V_i/RC$. In each millisecond the value of V_{IN} will be different, so the output changes by a different amount each time. However, if we start off with $V_{OUT} = 0\,V$ (i.e. switch closed), and let it ramp up and down freely for 10 000 ms (i.e. 10 s) then it will end up with a value given by:

$$V_{OUT} = \sum_i - \frac{V_i}{RC}$$

Now, the time average of V_{IN} is

$$\sum_i V_i/10\,000 = \langle V_{IN} \rangle$$

So after ten seconds integrating the input signal, V_{OUT} is given by;

$$V_{OUT} = -\frac{10\,000}{RC} \times \langle V_{IN} \rangle$$

If you choose RC to be equal to 10 000, then V_{OUT} will have the same size as $\langle V_{IN} \rangle$ after ten seconds of integration.

Time measurement

Our final example of an application of a ramp generator is shown in figure 3.86 (overleaf). Because the value of the output of a ramp generator changes steadily with time, it can be used to measure the duration of single pulses. The capacitor is initially shorted by pressing the switch, so that V_{OUT}

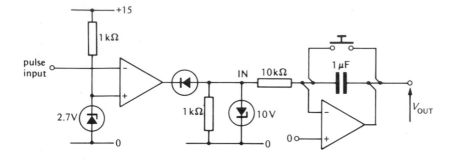

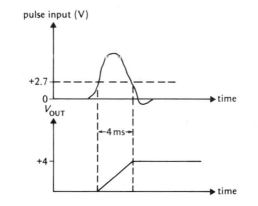

Figure 3.86 A pulse width to voltage converter

is 0 V. When a short pulse is subsequently fed into the system, V_{IN} goes to -10 V for the duration of the pulse before returning to 0 V. So V_{OUT} rises to a new value; if the pulse was above $+2.7$ V for t ms, V_{OUT} will be $+10 \times t/10 \times 1 = +t$. So the value of V_{OUT}, in volts, is equal to the duration of the pulse in milliseconds.

EXPERIMENT
Assembling a ramp generator; after looking at its behaviour you will combine it with a Schmitt trigger to make a triangle waveform generator

The ramp generator is shown in figure 3.87; the potentiometer allows you to set the value of V_{IN} anywhere between about $+1.5$ V and -1.5 V. Set V_{IN} to several values in this range, set V_{OUT} to zero with the switch and measure the time taken for V_{OUT} to ramp up to $+10$ V (or down to -10 V). Check that the time taken agrees with the theoretical value.

Now connect the ramp generator to a Schmitt trigger as shown in figure 3.88; the output of the ramp generator should ramp up and down continually to generate a triangle waveform. Draw graphs of V_{OUT} and V_{IN} as functions of time, one under the other and to the same scale.

Finally, amend the system so that it produces the waveforms shown in figure 3.89; you will need to use a CRO to monitor the output. Draw the circuit diagrams of the two systems, complete with component values.

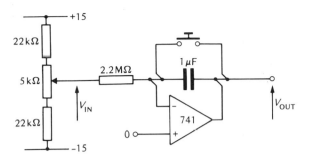

Figure 3.87 A circuit for investigating a ramp generator

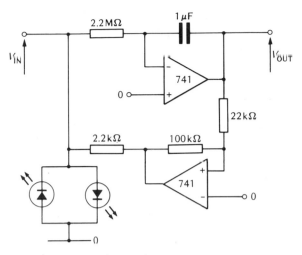

Figure 3.88 A triangle waveform generator

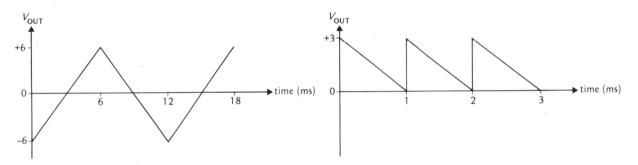

Figure 3.89

PROBLEMS

1 At what rate will the outputs of the circuits shown in figure 3.90 change?

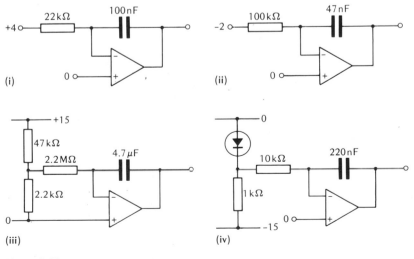

Figure 3.90

2 The waveform shown in figure 3.91 is fed into the circuit shown there. If the output is initially at 0 V, draw a graph to show how it changes with time.

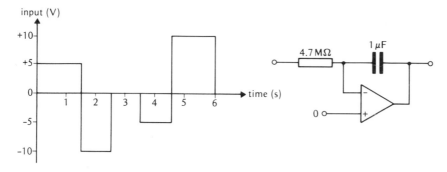

input (V)

Figure 3.91

3 The system of figure 3.92 is designed to give an analogue signal that is proportional to the number of logic pulses which enter the input. Describe how it works and state how to convert the value of the output voltage into the number of input pulses.

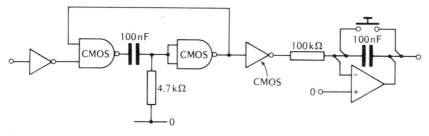

Figure 3.92

REVISION QUESTIONS

1 Explain the difference between analogue and digital signals.

2 Describe the electrical properties of the following devices; phototransistor, photodiode, LDR, thermistor, potentiometer and zener diode.

3 A zener diode is rated at 4.3 V, 500 mW. If it is connected to a 15 V supply with a resistor in series what is

a) the maximum permissible current through the diode,

b) the smallest permissible value of resistor?

4 Explain what is meant by the terms input impedance and output impedance.

5 A system feeds out a 9 V signal when no current is drawn from it. The signal drops to 1.5 V when a 2.2 kΩ resistor is placed across the system's output terminals. What is the output impedance of the system?

6 A signal source has an output impedance of 10 kΩ. By how much is the signal attenuated if it is fed into a system which has an input impedance of

a) $1\,k\Omega$,

b) $10\,k\Omega$,

c) $100\,k\Omega$?

7 Describe the electrical properties of a 741 op-amp which is run off $+15\,V$ and $-15\,V$ supply rails.

8 A circuit has an LED as its output. A signal is fed into it. If the signal lies between $+0.7\,V$ and $-0.7\,V$ the LED is lit, otherwise the LED is off. Draw a suitable circuit for the system using op-amps, resistors and diodes.

9 Draw the circuit diagram of an op-amp set up as

a) a non-inverting Schmitt trigger,

b) an inverting amplifier,

c) an inverting Schmitt trigger,

d) a non-inverting amplifier.

10 Explain what is meant by the term "virtual earth". Illustrate your answer by referring to an op-amp arranged as an inverting amplifier.

11 Show how an op-amp can be made into a system with a gain of

a) -22

b) $+48$

c) -0.1

d) $+1$.

12 Design a non-inverting Schmitt trigger which has switching thresholds of $+5\,V$ and $-5\,V$.

13 Draw a relaxation oscillator based on an op-amp. Sketch graphs to show how the voltages at various places in the system change with time. Use them to explain why the system oscillates. If the oscillator contains three $22\,k\Omega$ resistors and a $100\,nF$ capacitor, calculate the frequency of the square wave it generates.

14 Draw the circuit diagram of a ramp generator based on a 741 op-amp. State the formula for the rate at which the output of the system ramps up.

15 Draw the circuit diagram of a system which produces a sawtooth waveform with the following properties. It ramps slowly from $-6.5\,V$ to $+6.5\,V$ before rapidly returning to $-6.5\,V$. The waveform has a frequency of $500\,Hz$. Show all of the component values, with reasons. Explain how the circuit works.

16 Draw the circuit diagram of a digital to analogue converter based on a summing amplifier. The output voltage of the system should equal the decimal value of the three bit binary word fed into it. So if 001 is fed in, the output is $+1\,V$, and if 100 is fed in the output is $+4\,V$. Show all component values, assuming that logical 1 is $+5\,V$.

4

Power amplifiers

You cannot tell what a circuit is doing by simply looking at it or touching it; a wire at $+5$ V looks the same as one at 0 V. If there are high voltages around in the circuit you might get a shock when you touched it, and you might burn yourself on a hot component, but as a rule human beings cannot sense electronic signals directly. They have to rely on **output transducers** instead, devices which convert electrical signals into things that human beings can sense. So the output of a circuit has to be fed into an energy converter such as a voltmeter, a loudspeaker or an LED if you want to find out what it is doing.

In this chapter we shall be concerned with some of the techniques that are employed to enable electronic signals to successfully drive output transducers. With a few exceptions, most transducers pull a lot of current out of the signal source that is driving them, i.e. they need a lot of electrical energy. You will remember from the last chapter that whenever you draw current from a source you corrupt its signal; indeed, if you pull enough current out of it the signal virtually disappears!

For example, look at the circuit of figure 4.1. The potentiometer feeds out a signal that is converted into light by the resistor/LED network; the brightness of the LED indicates the size of the voltage of the wiper. Suppose that the wiper is placed so that the signal fed out by the potentiometer in the absence of the LED is $+10$ V. A little calculation should convince you that the output impedance of the source under these conditions is close to 1 kΩ. Now, if we ignore the voltage drop across the LED, the input impedance of the transducer network will also be 1 kΩ. So the signal from the potentiometer is halved when the transducer is connected to it; too much current is pulled out of the source by the transducer.

What we need is a device that will allow the source to control the transducer without having its signal corrupted in the process. Such a device is called a **power amplifier**, or **power amp**. A power amp will obviously have to present a high input impedance to the signal source and yet will have to have a low output impedance so that it can drive the transducer.

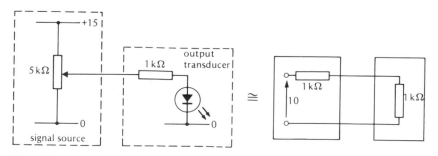

Figure 4.1

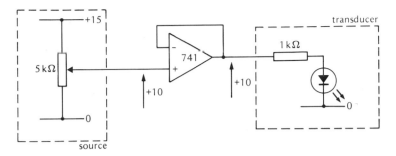

Figure 4.2 An op-amp follower in action

OP-AMP FOLLOWERS

Figure 4.2 shows how an op-amp follower can be used as a power amp. The op-amp has a very large input impedance (its inputs draw very little current) so it hardly affects the signal fed out by the potentiometer. The signal goes straight through the op-amp into the transducer; the op-amp has a very low output impedance $(0.08\,k\Omega)$ so it has no trouble putting most of the signal across the resistor/LED network. In effect, the op-amp makes the source think that it is feeding its signal into a very high resistance. At the same time the transducer thinks that it is being fed a signal out of a very low resistance. It may appear that the op-amp is performing the same trick as a transformer when it mediates between a loudspeaker and an oscillator i.e. it changes the apparent input impedance of the load. However, a more detailed analysis shows that the op-amp does something that a transformer never can do, namely, pump out more energy into the load than it receives from the signal source.

The currents flowing in the op-amp are shown in figure 4.3. A small current of about 80 nA will flow into the input; this is so small compared with the other currents which flow in the circuit that we shall ignore it. So virtually no energy enters the op-amp from the signal source (it is actually $10 \times 80 = 800\,nW$ or 0.0008 mW). On the other hand, the output of the op-amp sources 8 mA which falls through 10 V in the load, so the power fed out by the power amp is $8 \times 10 = 80\,mW$. This energy is not generated inside the op-amp itself, but comes from the supply rails that run the 741. It is instructive to work out how much of the energy that is taken from those rails by the op-amp is passed onto the load.

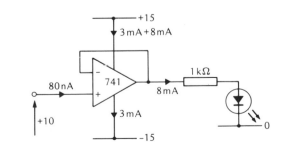

Figure 4.3

To start with, we need to consider the **quiescent current** of the op-amp. This is the current that it draws from the supply when it is not pushing current into a load; the quiescent (or **standing**) current keeps the 741 working. As you can see from figure 4.3, 3 mA flows from the +15 V rail, through the op-amp and into the −15 V rail; since it falls through 30 V, the quiescent current generates heat at a rate of 30 × 3 = 90 mW inside the IC itself. This amount of heat is generated even when no energy is being pushed into the load.

The other source of waste heat in the IC is the current that gets pushed into the load. Since 8 mA comes out of the op-amp output at +10 V and leaves the supply rail at +15 V, it must generate heat at a rate of (15 − 10) × 8 = 40 mW inside the op-amp.

So in our particular example 90 + 40 = 130 mW of heat is generated in the 741 when it dumps 80 mW into the load. Far more heat is wasted in the power amp than is transmitted to the load. Since a 741 can cope with up to 600 mW of waste heat, this inefficiency is no great problem. Indeed, since a 741 has internal current-limiting (i.e. it refuses to source or sink more than about 13 mA) you can never overheat one by trying to push too much current into its load. (This is not to say that a 741 is indestructible; it will evaporate spectacularly if you reverse the voltage of its supply rails!) This immunity against too much heat is certainly not the case with the transistor power amps that we shall be dealing with later on in this chapter.

DIGITAL POWER AMPS

A digital signal has only two states, so a power amp for such signals need only have two output states. One of those states can cause current to flow through the load, the other state can stop that current flowing. A power amp with this behaviour acts as an electrically controlled switch. We shall look at two devices that are used as digital power amps; they permit ordinary logic signals to control the current flowing through large loads such as lamps, motors and heaters.

RELAYS

Figure 4.4 shows how a relay can be used to amplify the power of a logic signal. A logical 0 fed out by the gate will sink current from the relay coil so that its contacts close and current can flow through the load. The relay contacts will be open when the gate output is at logical 1.

There are several things worth noting about the circuit of figure 4.4. Firstly, since there is no direct connection between the relay coil and its contacts, the two supply rails for the load can have any voltage. So a logic signal going between 0 and +5 V could control the current between rails that were a few hundred volts apart. Secondly, the relay is a very efficient power amp. Because there is no voltage drop across the contacts when they are closed, there will be no heat generated in the relay by the current that flows through the load. Finally, it is very slow; the fastest relays have response times of about 1 ms, but they slow up dramatically if they are designed to control large currents through the load.

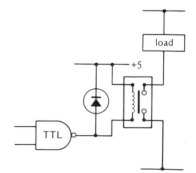

Figure 4.4 Connecting a relay to a TTL gate

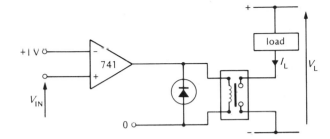

Figure 4.5 Connecting a relay to a 741

You may be wondering why the diode has been included in the circuit of figure 4.4. After all, since a TTL output does not go above $+5\,V$, the diode is always going to be reverse biased and act like an open circuit. In fact, the diode is there to clamp the output so that it cannot rise far above the $+5\,V$ supply rail when it goes from 0 to 1. Whenever there is a rapid drop of current flowing through a coil, a voltage develops across the ends of the coil; this **back emf** is particularly large if the current changes very quickly. A TTL output can go from logical 0 to logical 1 in a few nanoseconds, so there is likely to be quite a large voltage spike across the relay coil when this happens. Since the spike will always try to raise the TTL output above the $+5\,V$ supply rail, there is a good chance that the gate will destroy itself when it tries to turn the relay off. If a diode is wired in parallel with the coil this cannot happen as the back emf is clamped at about $+5.7\,V$. It is for this reason that the diode is called a **safety diode**; it should always be included whenever an IC is used to control a relay.

Figure 4.5 shows how a 741 comparator can be used to control a relay. A little thought should convince you that the current will only flow through the load when V_{IN} is greater than $+1\,V$; the safety diode performs the dual function of protecting the 741 output from the relay coil's back emf, and preventing the output from going below $-0.7\,V$.

Relays have been around for a long time and their use in controlling very large loads is well established. They are very rugged and can put up with very large voltages and currents if necessary. They do have disadvantages which are not shared by their solid state equivalents. They are mournfully slow, heavy, bulky, expensive to make and contain moving parts that must eventually wear out. **Solid state devices** (i.e. digital power amps that are manufactured like ICs) on the other hand can respond very quickly, be light and compact, and contain no moving parts that can fail or wear out. We shall defer discussion of FET and transistor power amps until later on (they can double up as analogue power amps as well as digital ones), and concentrate for the moment on solid state power amps that are designed to control alternating currents.

TRIACS

A **triac** is a solid state AC switch. That is, it is an integrated circuit version of a relay that can only effectively control alternating currents. (An alternating current (AC) flows in a load when it is connected across the live and neutral rails of the mains supply.) Triacs are very useful as they allow logic signals to control large loads without many of the disadvantages of relays.

Figure 4.6 shows a triac controlling the current which flows through a load. Initially, the triac will behave like an open-circuit i.e. no current will flow through the load. It will remain in this state if its **gate** is kept within about a volt of the lower supply rail. The triac can be triggered into becoming an effective short-circuit by raising V_G above a certain threshold (usually between 1 and 2 V) when the triac changes state very quickly, so that the current is very rapidly switched on. That current will continue to flow through the load, regardless of the subsequent value of V_G. Once you have triggered the triac into conduction with a signal fed into the gate you lose all control of it. The only way that you can switch the triac off again is to stop the current flowing through it. Once the current flow is zero the triac rapidly goes back to its non-conducting state.

Triacs are very flexible because they can be triggered by positive or negative signals fed into the gate; furthermore, they do not mind which way the current flows through them, so they can be used for controlling AC. As you can see from figure 4.6, a triac is not much use for controlling DC; once you have switched the triac on you cannot switch it off again without switching off the DC supply! This loss of control over a triggered triac is no problem if the current is AC because such currents regularly switch themselves off.

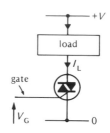

Figure 4.6 A triac

Switching AC

Look at the circuit of figure 4.7; it shows a triac controlling the current through a load that has been connected across the **live** (L) and **neutral** (N) rails of the **mains supply**. The top graph shows how V_{LN}, the voltage difference between the supply rails varies with time; it goes through zero every

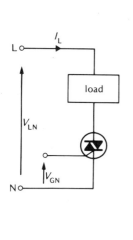

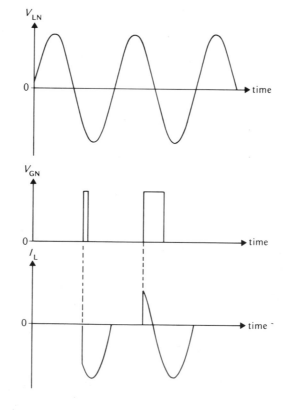

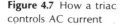

Figure 4.7 How a triac controls AC current

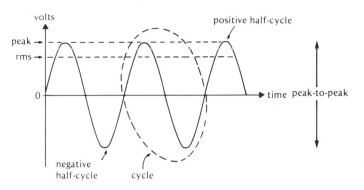

Figure 4.8 Various terms used in connection with AC signals

10 ms (assuming a 50 Hz supply). If the triac is triggered into conduction by a pulse fed into its gate, current will flow through the load until the next **zero-crossing point** of the supply; as soon as the current through the triac is zero it reverts back to being an open-circuit. If you want to keep the triac on continually you will either have to feed it pulses regularly or keep its gate a constant couple of volts above or below the neutral rail.

When using a triac to control AC currents it is good practice always to switch the triac on at the zero-crossing points of the mains supply. This is because a triac changes its state very rapidly, so the current through the load can increase very quickly when the triac is triggered. The sudden appearance of a large current causes a fair amount of **radio interference**; indeed, a carelessly used triac can wreak havoc with sensitive circuitry located nearby! If you ensure that the triac is only fed gate pulses when V_{LN} is small this problem can be avoided. We shall now look at how such "zero-crossing" pulses may be generated and administered to the gate of the triac, but we shall need to discuss some of the finer points of the mains power supply first.

The mains supply

Look at the graph of figure 4.8. It shows how the voltage difference between the live and the neutral rails of the mains supply varies with time. The neutral line stays at a fixed voltage, somewhere near to local ground voltage; the live line is the one whose voltage is continually changing. As you can see, the graph consists of repeated **cycles**, and each cycle contains a **positive half-cycle** followed by a **negative half-cycle**. In Britain the mains supply has 50 cycles in each second (i.e. it has a frequency of 50 Hz), so each half-cycle lasts for 10 ms. The size of the mains supply is usually quoted as **240 V rms**, where rms is short for "**root-mean square**". The rms voltage of an AC signal is an attempt to specify the average size of the signal during a half-cycle; if the signal is a sine wave (i.e. like the one shown in figure 4.8) the peak value of the voltage is $\sqrt{2}$ times its rms value. So the peak value of the mains voltage is 1.414 × 240 = 340 V.

Generating gate pulses

A circuit that will only allow a triac to switch on at the instants when the mains voltage is zero is shown in figure 4.9. On the right-hand side of the circuit is the triac and the load that it controls (e.g. a light bulb or a heater). On the left is the circuitry which generates the firing pulses and feeds them into the gate of the triac. Note that the two supply rails shown in the

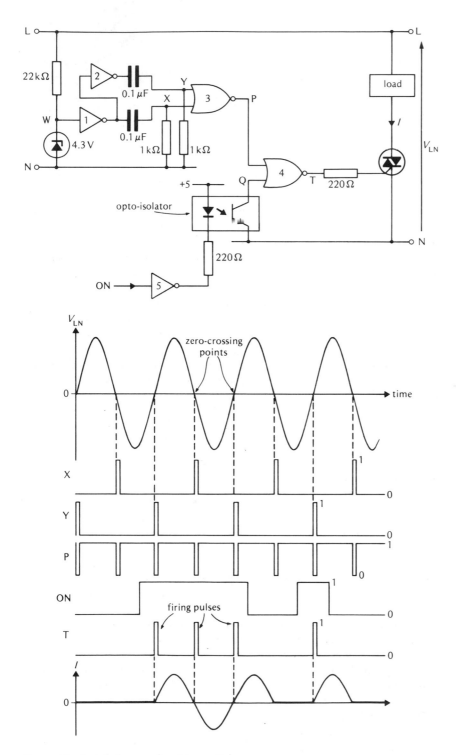

Figure 4.9 An interference-free triac switch

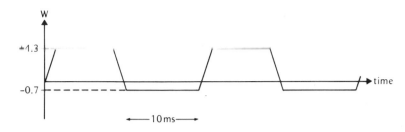

Figure 4.10

diagram are the live (L) and neutral (N) rails of the mains supply; the neutral line acts as a reference logical 0 for the gates marked 1, 2, 3 and 4, whereas gate 5 uses ground as its reference logical 0. Gates 1 to 4 inclusive are therefore run off a supply rail (not drawn) that is kept five volts above the neutral one; we shall discuss how this can be done after we have described the operation of the circuit.

Consider the point marked W in the circuit. During most of a positive half-cycle of the mains supply, W will be clamped 4.3 V above the neutral rail by the zener diode. Similarly, W will be 0.7 V below the neutral rail for most of the negative half-cycle. So the waveform at W will be a square wave that alternates between logical 1 and logical 0. Its rising and falling edges will coincide with the zero-crossing points of the mains supply. Figure 4.10 shows how the waveform at W varies with time.

The square wave at W goes through a pair of NOT gates to produce a pair of square waves that are opposites of each other i.e. when one rises, the other falls. Points X and Y spend most of the time being held at logical 0 by the 1 kΩ resistors, but they are briefly pushed up to logical 1 by the rising edges that come out of the NOT gates. Since a NOR gate is only at a logical 1 when both of its inputs are at logical 0, those rising edges will cause P to briefly dip down to logical 0 (for about 0.1 ms each time). As you can see from the graphs of figure 4.9, P goes to logical 0 every time the mains supply goes through zero.

Now consider gate 4, the one that controls the triac. A logical 1 fed out by that gate should manage to trigger the triac into conduction (big triacs may need an additional transitor to feed enough current into the gate). T is only going to be logical 1 when both P and Q are logical 0. If Q stays at 0 all the time, then the pulses fed in by P will cause T to briefly go up to logical 1 at each zero-crossing point of the mains supply. On the other hand, if Q is at logical 1, T remains fixed at logical 0 and the triac blocks off the current through the load. So the state of Q at the start of a half-cycle decides whether or not the triac is going to conduct during that half-cycle.

If gate 4 is a TTL NOR gate then a simple push switch could be used to short-circuit Q to the neutral rail (gate 4's logical 0) every time you wanted current to flow through the load. However, you usually want to feed Q with a signal from an external processor (e.g. a computer). Unfortunately, the logic gates of that processor (gate 5 in our example) are run off supply rails that are 0 V and +5 V with respect to the ground rail; gate 5 and gate 4 recognise different things as logical 0. Gate 5 thinks that the ground rail is logical 0, gate 4 thinks that the neutral rail is logical 0, so you cannot let one feed a signal directly into the other. The unsafe answer is to connect the ground and neutral rails together so that they are forced to be at the same voltage. A somewhat better solution is to use an **opto-isolator**, as shown in figure 4.9.

As you can see from its circuit symbol, this device consists of an LED and a phototransistor placed beside each other in the same package; from the outside it looks very similar to a 741. When current flows through the LED it emits infra-red light which crosses to the phototransistor where it makes current leak from the collector to the emitter. So when gate 5 goes to logical 0, the phototransistor sinks current from Q and pulls it down to the neutral rail. Although digital signals can pass from one side of the opto-isolator to the other, there is no direct electrical path. This means that one side can **float** up or down in voltage (easily up to 1 kV) without any effect on the other side. So if the triac unit fails, there is no chance that high voltages will be fed back to the processor that is driving it.

The operation of the whole triac system is illustrated with the graphs of figure 4.9. Note that the triac will only let current flow through the load in whole half-cycles when ON is a logical 1.

Floating supply rails

Although we shall discuss the business of generating power supply voltages in Chapter Thirteen, we shall spend a little time illustrating how a supply rail for the four "floating" logic gates in the triac controller could be provided; it will emphasise the need for the opto-isolator. Since all four gates can be provided by a single 7402 IC only a few mA will be drawn from the supply rail. So a simple power supply will suffice.

Figure 4.11 shows a circuit that generates a 5 V supply with respect to the neutral rail. S and N are the two supply rails for the 7402 IC so V_{SN} needs to be kept close to +5 V. L is the live rail of the mains supply; the

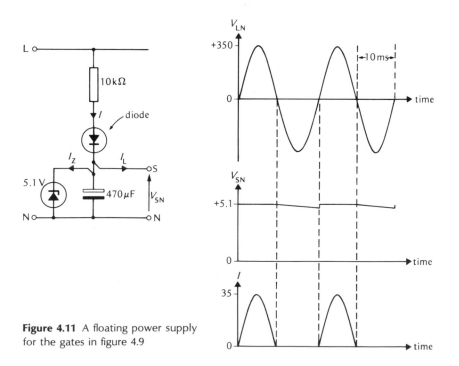

Figure 4.11 A floating power supply for the gates in figure 4.9

whole circuit is designed to convert the 240 V AC of this rail into a 5 V DC signal.

The diode rectifies the mains supply by cutting off the current I during negative half-cycles. During that time the capacitor acts as a battery and keeps V_{SN} at $+5$ V. During the positive half-cycles the diode is forward biased so a current flows towards the capacitor. The zener diode prevents the voltage across the capacitor rising above 5.1 V, so V_{SN} is clamped at $+5.1$ V for most of the positive half-cycle. I_L is the current drawn from S by the IC connected to it; during positive half-cycles this comes from L via the diode, otherwise it flows off the capacitor. If the capacitor is large enough, the voltage across it will not drop very much during the negative half-cycles; in any case, it will be rapidly charged back up to $+5.1$ V at the start of each positive half-cycle.

Note that the 10 kΩ resistor will have to be a high power type. When calculating the rate at which heat is generated in a component carrying AC you have to use the rms values of the current and the voltage. So during each positive half-cycle, the resistor dissipates heat at a rate of $240 \times 24 = 5760$ mW. As no heat is generated in the negative half-cycle, the average dissipation will be $\frac{1}{2} \times 5.76 = 2.88$ W. To be on the safe side and avoid a very hot component, the resistor needs to be able to cope with a least 4 W of heat!

EXPERIMENT
Assembling a simple triac control system that feeds pulses into a triac gate at the start of every negative half-cycle

Assemble the circuit shown in figure 4.12. The 12 V AC supply must have one of its supply rails grounded so that the logic gates can be run off the normal $+5$ V and 0 V supply rails. The bulb should light when Q is held at logical 0, but not when it is left to float up to logical 1.

Trigger a CRO on the 12 V supply. Look at the waveforms at the points V, W, X, Y and Z in the circuit when Q is held at logical 0. Draw the waveforms one underneath the other and to the same time scale.

Now modify the system so that the bulb is only lit for the positive half-cycles of the AC supply.

Finally, assemble a system that will let current flow in the bulb for both positive and negative half-cycles of the AC supply when Q is logical 0.

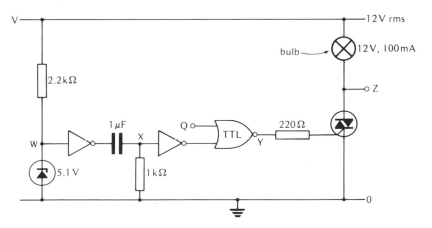

Figure 4.12

PROBLEMS

1 Figure 4.13 shows a triac set up as **over-voltage protector** for an electronic system. The fuse is supposed to protect the five volt power supply if the system develops a fault. For example, the two supply rails might get shorted to each other. Then a very large current would be drawn from the supply, enough to damage it. The fuse is designed to melt quickly when a lot of current flows through it, so that the supply is rapidly disconnected from the system when something goes wrong.

   The fuse will *not* protect the system if the power supply develops a fault. A faulty power supply may well raise the supply rail above 5 V and cause extensive damage to the system before the fuse has time to melt. The fuse will only melt when the system is sufficiently damaged so that it looks like a short-circuit. Indeed, the power supply only has to go above about 7.5 V for digital ICs of the TTL family to be irreparably damaged.

   The triac shown in figure 4.13 is designed to protect the system against power supply failure. Describe how the triac gives that protection; it will conduct if its gate is raised above +1 V.

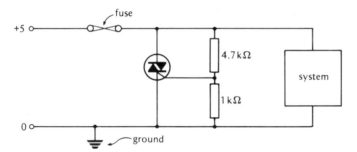

Figure 4.13 An over-voltage protection system

2 This problem concerns the use of a triac to control a DC motor. A DC motor is shown in figure 4.14. When current flows from L to R it drives an engine forwards. The engine can be driven backwards by sending current the other way through the motor, from R to L. The processor shown in the diagram has to send pulses into the gate of the triac, such that the engine goes forwards when F is pressed and backwards when B is pressed.

   Draw a suitable circuit diagram for the processor. Explain how the processor works.

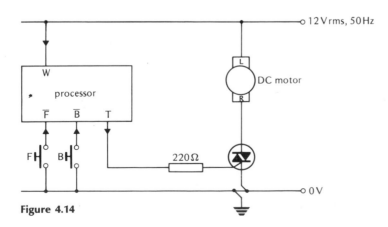

Figure 4.14

TRANSISTORS

If you want to assemble a power amp that is capable of pushing analogue signals into a big load then you will have to use **transistors**. The word "transistor" will be used in a rather specialised sense in this chapter. We shall only discuss **bipolar transistors** which are made out of silicon, the sort that are widely used for power amp work. The other sort of transistor (the **field effect transistor**, or FET) will have a chapter to itself later on. Furthermore, we shall only be dealing with the business of using transistors in power amps, and leave all of the other applications of transistors to Chapter Five.

Figure 4.15 shows the circuit symbol for an npn transistor. Its name hints at the physics behind its operation, as do the names of its three terminals. Since we are not concerned with understanding what goes on inside a transistor we will not be discussing what it is that the collector collects or why an emitter emits. Instead, we shall introduce a **model** of transistor behaviour. This model is a set of rules that tell you how the currents and voltages of the three terminals are related to each other. That model will enable you to use transistors effectively in a circuit without needing to know how a transistor works inside! This approach is similar to the one that we adopted towards the 741. By ignoring the contents of the eight-legged package and concentrating on its electrical behaviour, you were able to use op-amps as the basic components of a number of complex systems.

There is another similarity between our approach to transistors and the way in which we treated op-amps. The model that you met of the 741 used the infinite-gain approximation. It assumed an ideal sort of op-amp, i.e. an op-amp whose behaviour real op-amps aspire to, but do not quite achieve in practice. This did not really matter for the sort of system that you built in Chapter Three; you will meet some applications of op-amps where the model has to be elaborated on in Chapter Twelve. The infinite-gain approximation does not describe the complete behaviour of an op-amp, but it is very good for a lot of applications. The model that we are about to introduce for the npn transistor is likewise not the whole story, but is good enough for all of the applications that you will meet in this book. The model is an attempt to make a device which has complex behaviour easy to use i.e. it is a model of an ideal transistor. Real transistors fall far short of the ideal in many respects!

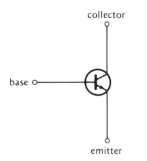

Figure 4.15 An npn transistor

THE MODEL OF AN NPN TRANSISTOR

The symbols that are used in the model are defined in figure 4.16, overleaf. Bear in mind that the rules which follow describe an ideal npn transistor; we shall discuss how real transistor behaviour deviates from the ideal after the model has been explained.

1) V_{CE} **is greater, or equal to 0 V**
The **collector** has got to be higher than the **emitter**. This means that current will always flow into the collector and flow out of the emitter when the transistor is used correctly. When V_{CE} is equal to 0 V, the transistor is said to be **saturated**.

2) $I_C = h_{FE} I_B$

The collector current of the transistor is a fixed multiple of its base current. This is only the case when the transistor is not saturated. The fixed multiple is called the **current gain** of the transistor. Its symbol is h_{FE}; it is one of a number of **h-parameters** that can be used to describe a transistor more fully than our model does. The current gain of a transistor is usually very big, at least 100 for the small ones you will be meeting in this book.

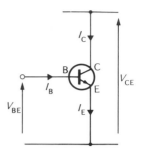

Figure 4.16

3) $I_E = I_C$

This rule is a fudge, but quite a realistic fudge. Since current cannot be lost or gained in an electronic component we can say that $I_E = I_C + I_B$ if the currents flow as shown in figure 4.16. However, as the collector current will normally be at least a hundred times larger than the base current (rule 2), we can ignore the latter's contribution to the emitter current.

4) $V_{BE} = 0.7 \text{ V}$ when the transistor is on

When the transistor is switched on, current flows into its base. The base will always be 0.7 V higher than the emitter when this happens. You can think of the transistor as having a diode between the base and the emitter; the arrow in the emitter leg of the symbol is a useful reminder of this.

5) **The transistor is switched off when V_{BE} is less than 0.7 V**

No collector current flows into the transistor when it is switched off, so there is no base current either.

Real transistors

Before we use the model to show how transistors can be used to make power amps we shall say a few words about the behaviour of real transistors. At this stage you are probably still unsure about the behaviour of ideal transistors, so we will simply state the ways in which transistors are non-ideal so that we can refer back to them later.

The current gain of a transistor is a highly variable quantity. That is, if you select transistors at random from a particular batch, their current gains may differ by up to a factor of four. So a BC107 (the type that you will be using extensively in practical work) might have a value for h_{FE} that lies between 100 and 400! Furthermore, the h_{FE} of a particular transistor ceases to have a constant value when it gets near to saturation. Finally, it is also a function of the temperature of the transistor.

Leakage currents abound in a transistor. At room temperature they are very small, but they grow by leaps and bounds as the transistor warms up. So you cannot expect a hot transistor to have very ideal behaviour; indeed, it will have a tendency to self-destruct.

Rule 4 is only approximate. The value of V_{BE} depends on the size of I_B, being proportional to the logarithm of the base current. This means that it rises slowly as the base current is increased; in practice the value of V_{BE} could lie between 0.5 and 0.9 volts.

Finally, you have to bear in mind that transistors are inherently fragile; they are easily destroyed. This fragility is a consequence of their versatility; if they were built so that all of the inputs were internally protected (as 741 and 7400 ones are) they could not be used in so many systems. When you put a transistor in a circuit you must ensure that you do not stress it too

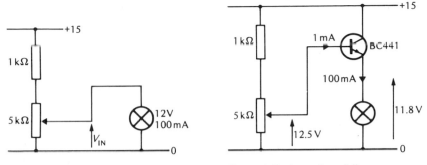

Figure 4.17

Figure 4.18 An emitter-follower

much. For example, a BC107 must not have a collector current of more than 100 mA, nor must its collector lie more than 45 V above its emitter. The base must not go more than 50 V below collector and the power dissipated as heat must always be less than 360 mW. If you exceed these limits you will destroy the transistor.

THE EMITTER-FOLLOWER

Figure 4.17 illustrates a situation which requires a power amp that can handle analogue signals. The signal source is a potentiometer whose open-circuit voltage can range from 0 V to +12.5 V. The signal is to be fed into a light bulb, so that its brightness indicates the size of the signal. It should be clear that the signal source and the transducer are incompatible. The bulb will take 100 mA from the source when it is fully lit, yet the current flowing down the voltage divider is only 2.5 mA. The value of V_{IN} will be severely reduced when the bulb is connected to the wiper of the potentiometer.

The simplest type of power amp that can be used to mediate between source and transducer is shown in figure 4.18. The transistor is being used as an **emitter-follower**. Remembering rule 4, the emitter of the BC441 will always be 0.7 V below its base. Since the signal source controls the voltage of the base, the voltage across the bulb will always be $V_{IN} - 0.7$. The voltage of the emitter "follows" the voltage of the base, hence the name "emitter-follower" for this sort of arrangement.

The current that flows through the bulb is provided by the emitter. A typical value for the h_{FE} of a BC441 is 100; the emitter current is about 100 times greater than the base current (rule 2). So the maximum current that the transistor will draw from the potentiometer is 1 mA, as the bulb draws 100 mA when it is fully lit. So the transistor makes the source think that the bulb has a resistance that is 100 times greater than its true value.

The current drawn from the source is too large for comfort because it is comparable to the current that flows inside the source itself (i.e. 2.5 mA). V_{IN} is likely to be pulled down a bit by the base current, but the potentiometer will be able to control the brightness of the bulb from fully off to fully on. Before we explain how the power amp can be improved by adding a second transistor, we shall run through a calculation of the power gain of an emitter-follower.

Power gain

The details of the system are shown in figure 4.19. The voltage of the base is +7 V, and the load has a resistance of 100 ohms. Given that the current gain of the BC441 is 100, we want to calculate the power gain of the follower.

We start off by considering the load. The voltage across it will be $7 - 0.7 = 6.3\,\text{V}$. Using Ohm's Law we can now calculate the emitter current:

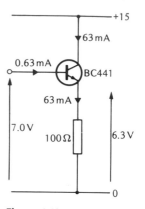

Figure 4.19

$$R = \frac{V}{I}$$

$$\text{therefore} \quad 0.1 = \frac{6.3}{I_\text{E}}$$

$$\text{therefore} \quad I_\text{E} = \frac{6.3}{0.1} = 63$$

So the emitter current is 63 mA. Rule 2 now allows us to calculate the base current:

$$I_\text{C} = h_\text{FE} I_\text{B}$$

$$\text{therefore} \quad 63 = 100 \times I_\text{B}$$

$$\text{therefore} \quad I_\text{B} = 0.63$$

The base current is therefore 0.63 mA.

(Note that we have used rule 3 and assumed that the emitter and collector currents are equal.)

The power drawn from the signal source is therefore $7 \times 0.63 = 4.4\,\text{mW}$. The emitter pushes power into the load; that power is $6.3 \times 63 = 370\,\text{mW}$. So the emitter-follower provides a power gain of $370/4.4 = 90$. In general, you can say that the power gain of an emitter-follower is more or less the same as its current gain.

One important quantity that needs to be calculated for an emitter-follower is the amount of heat generated within the transistor. Most of that heat is due to the collector current, as it is the only current that drops through an appreciable voltage. In the example of figure 4.19, the collector current of 63 mA falls from $+15\,\text{V}$ to $+6.3\,\text{V}$ as it goes through the BC441. So heat is generated at the rate of $(15 - 6.3) \times 63 = 548\,\text{mW}$. This has to be compared with the maximum power that the transistor can tolerate (2 000 mW for a BC441). If the transistor gets too hot it will be destroyed!

Darlington pairs

The power gain of an emitter-follower can be increased by using two transistors connected as a Darlington pair. This arrangement is shown in figure 4.20. The heavy-duty BC441 provides the current which flows through the load, and the smaller BC107 provides the base current for the power transistor. The base of the signal transistor is controlled by the signal source. A little thought should convince you that the emitter of the power transistor will always be $0.7 + 0.7 = 1.4\,\text{V}$ below the base of signal transistor. So the voltage across the load is 1.4 V less than the signal fed out by the signal source. Furthermore, the current drawn out of the source will be 10 000 times smaller than the current which flows through the load (assuming a current gain of 100 for both transistors). For example, suppose that the emitter current of the power transistor is 100 mA. Then the emitter current of the signal transistor will have to be 1 mA as it is the base current of the

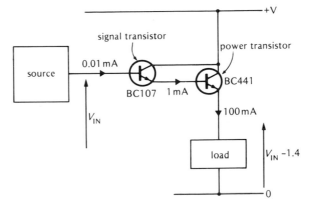

Figure 4.20 A Darlington pair

power transistor. The base current of the signal transistor will be 100 times smaller than its collector current, a meagre 0.01 mA.

A Darlington pair gives a power gain that is equal to the product of the current gains of the two transistors. Since most of the current flows through the power transistor, the signal transistor does not have to cope with much waste heat. Small transistors usually have much higher current gains than large ones, so there is a definite advantage to only using power transistors where you have to.

Compensating for V_{BE}

The circuit of figure 4.20 shows how a Darlington pair can act as a power amp that will draw a negligible current from the signal source. So the value of V_{IN} is unaffected by the presence of the power amp. Unfortunately, we have had to pay a price for this; the value of V_{OUT} is always 1.4 V below the value of V_{IN}. The ideal power amp would have $V_{OUT} = V_{IN}$ as well as presenting a very large input impedance to the signal source.

Figure 4.21 shows one way in which the voltage drop can be compensated for. The two diodes and the resistor placed before the Darlington pair

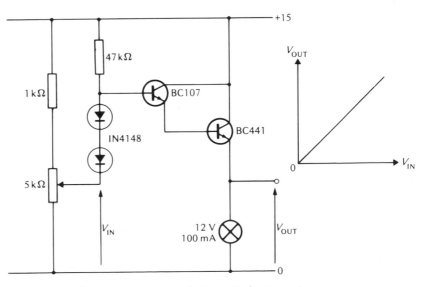

Figure 4.21 A linear power amp employing a Darlington pair

push the signal out of the source up by $0.7 + 0.7 = 1.4\,\text{V}$. When the signal goes through the transistors it drops by $1.4\,\text{V}$, so V_{OUT} is equal to V_{IN}. The compensation is not perfect as the voltage drop across a diode depends on its temperature as well as the current that flows through it. V_{OUT} is not going to differ from V_{IN} by more than $0.2\,\text{V}$, so the system is a better power amp than it was before. When we look at audio systems in Chapter Eight we shall show how the techniques of negative feedback can be used to eliminate **all** discrepancies between the values of V_{OUT} and V_{IN}.

PROBLEMS

1 Calculate the following quantities for the circuits shown in figure 4.22:

   a) the emitter voltage,

   b) the emitter current,

   c) the base current,

   d) the rate at which heat is generated in the transistor.

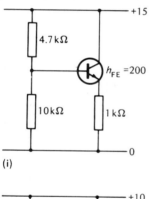

(i)

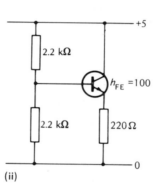

(ii)

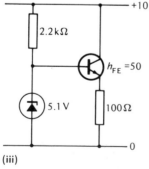

(iii)

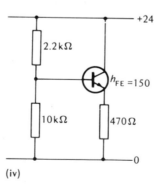

(iv)

Figure 4.22

2 For the system shown in figure 4.23, calculate:

   a) the current flowing through the load,

   b) the current drawn from the source,

   c) the rate at which heat is generated in each transistor.

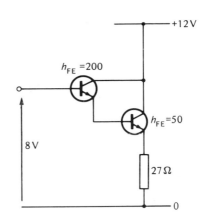

Figure 4.23

THE PNP TRANSISTOR

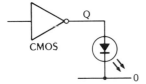

Figure 4.24

Figure 4.24 shows how you would hook an LED to a CMOS gate in order to get its output state displayed. When Q tries to be logical 1 it pumps a couple of mA into the LED so that it lights up feebly. You do not need a resistor to limit the current through the LED; the gate has internal current-limiting.

An improved system with the same function is shown in figure 4.25. The LED has been replaced with a light bulb, so that the display will be much brighter. The emitter-follower has been added to boost the current that the CMOS gate can source into the bulb. The transistor has a current gain of 100. So the gate only has to source 0.6 mA into the base for the emitter to source 60 mA into the bulb.

Suppose that we wanted the bulb to light up when Q was a logical 0? If we were using an LED as the output transducer, the answer would be simple. You would hold one end of the LED at +5 V and let the gate sink current from it when Q goes to logical 0 (figure 4.26). You can have a similar arrangement with a light bulb, but you must arrange for the current sinking capability of the gate to be increased. An npn emitter-follower will not do because its emitter can only source current. You need a **pnp emitter-follower**, as shown in figure 4.27.

pnp transistors are mirror images of npn transistors. They obey exactly the same rules except that all of the voltages have the opposite sign and

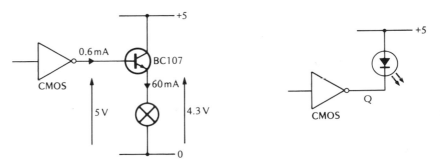

Figure 4.25   **Figure 4.26**

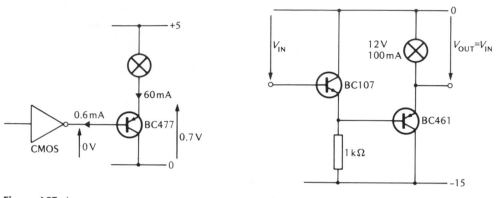

Figure 4.27 A pnp
emitter-follower in action

Figure 4.28

the currents flow the other way. Thus in figure 4.27 the emitter is 0.7 V higher than the base and the emitter is held at a higher voltage than the collector. The gate sinks 0.6 mA from the base, causing a current of 60 mA to flow into the emitter of the transistor.

You can use pnp emitter-followers whenever you need to sink current from a load. For example, the system shown in figure 4.28 shows a pnp emitter-follower sinking current from a light bulb; note the use of the npn signal transistor to compensate for the base-emitter voltage rise of the pnp power transistor.

PROBLEMS

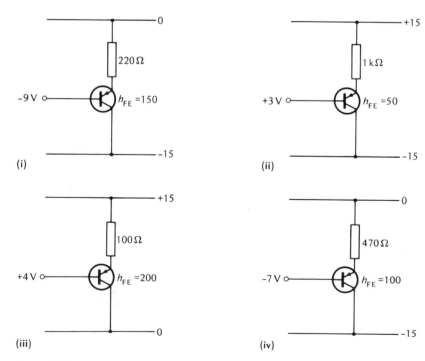

Figure 4.29

1 For the circuits shown in figure 4.29, calculate each of the following:

a) the emitter voltage,

b) the emitter current,

c) the base current,

d) the rate at which heat is being generated in the transistor.

2 Draw the circuit diagram of a power-amp that is similar to the one shown in figure 4.28 but has an npn power transistor. It has to be able to source current into a load.

THE COMPLEMENTARY FOLLOWER

How do we set about boosting the current output of a 741? If we use an npn transistor as an emitter-follower then we can only source current into a load, as shown in figure 4.30. Once V_{IN} goes below $+0.7$ V the transistor switches off so that no current can flow through it. If no current can be pulled out of the load by the emitter then V_{OUT} cannot drop below 0 V. Of course, when V_{IN} is above $+0.7$ V, V_{OUT} will be equal to $V_{IN} - 0.7$ and the system acts as a reasonable power amp.

A pnp emitter-follower has exactly the opposite defect. It will only work as a power amp if V_{IN} is smaller than -0.7 V. Once V_{IN} goes higher than this, the transistor switches off and its emitter is held at 0 V by the load. As you can see from figure 4.31, a pnp emitter-follower works for the values of V_{IN} that an npn emitter-follower cannot handle; the two devices **complement** one another.

A power amp that is suitable for boosting op-amp signals has to be capable of both sourcing and sinking current. The complementary follower,

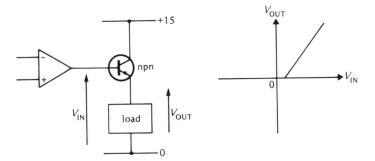

Figure 4.30 Boosting the current sourced by an op-amp

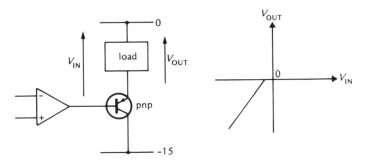

Figure 4.31 Boosting the current sunk by an op-amp

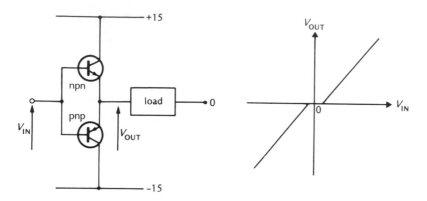

Figure 4.32 A complementary follower

shown in figure 4.32, is the simplest type of transistor power amp that will work effectively. The npn transistor does any sourcing of current that is required, and the pnp transistor does the sinking. In order to see how it works, we shall consider three different values of V_{IN} and work out the corresponding values of V_{OUT}. For the purposes of these examples we shall assume that the load has a resistance of $10\,\Omega$ and that both transistors have current gains of 100.

Figure 4.33 shows what happens when V_{IN} is $+8\,V$. The load tries to pull the transistor emitters down to $0\,V$ whereas their bases are held at $+8\,V$. So the npn transistor switches on, current is sourced into the load and V_{OUT} is clamped at $+7.3\,V$. Because its emitter is not $0.7\,V$ higher than its base the pnp transistor is switched off. The npn transistor sources $730\,mA$ into the load but the 741 only has to source $7.3\,mA$ into the npn base.

Our second example assumes that V_{IN} is $-5\,V$; this is illustrated in figure 4.34. The op-amp is trying to sink current from the load, so it pulls the bases of the transistors below their emitters. The pnp transistor switches on and its emitter sinks $430\,mA$ from the load; the op-amp sinks $4.3\,mA$ from its base. The npn transistor is switched off as its base is being held below its emitter. No current can therefore flow from the $+15\,V$ supply rail into the pnp transistor.

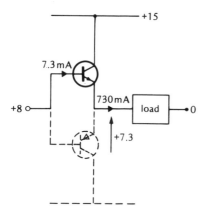

Figure 4.33 The npn transistor pushes

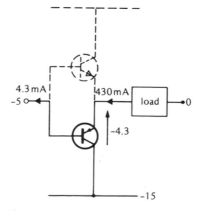

Figure 4.34 The pnp transistor pulls

Finally, we shall consider what happens when V_{IN} is $+0.5\,V$. The pnp transistor will be switched off because its base is being held above its emitter (remember that the load will pull the emitter towards $0\,V$). Similarly, the npn transistor will be switched off because its base is not the requisite $0.7\,V$ above the emitter. If both transistors are off there is no way that current can flow through the load, so V_{OUT} will be $0\,V$.

The complementary follower is sometimes referred to as a **push-pull output stage**. This name reflects the action of the two transistors; one pushes current into the load, the other pulls current out of it. The push-pull arrangement is quite efficient as far as waste heat is concerned. The quiescent current is zero when no signal is being pumped into the load. Furthermore, when an AC signal is fed into the load each of the transistors is only on for half of the time, so that they share the waste heat evenly.

Compensated push-pull followers

The complementary follower suffers from the defect that it is non-linear. This is because its output is zero when the input lies within the range $+0.7\,V$ to $-0.7\,V$. There are several ways in which the response of the system can be improved and we shall look at some of them now. The best method, however, will be left until Chapter Eight; there we shall show how the techniques of negative feedback can convert the worst power amp into a near-perfect one!

Look at the circuit of figure 4.35. It uses a diode/resistor network to generate two signals that are $0.7\,V$ above and below V_{IN}. These two signals control the bases of the npn and pnp transistors respectively. It follows that because of this V_{OUT} will always have the same value as V_{IN}. Both of the transistors will be on all of the time, although they do not have to carry the same current. When V_{OUT} is positive the npn transistor will source most of its current into the load, letting only a little flow through the pnp transistor to the $-15\,V$ supply rail.

So we appear to have designed a linear complementary follower, with $V_{OUT} = V_{IN}$ over the whole range of values of V_{IN}. Unfortunately, the circuit is not **thermally stable**; if you construct it, it is likely to heat itself

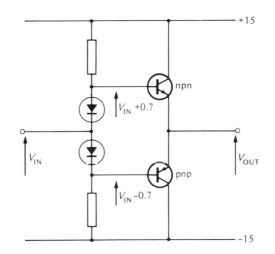

Figure 4.35 A bad push-pull output stage

into oblivion. In a moment we shall explain why the system will get uncontrollably hot, but we need to say a few words about the general problem of waste heat in transistor power amps.

Thermal runaway

When a transistor heats up, its properties change. To start with, all of the leakage currents increase. Then the collector current will rise, provided that the value of V_{BE} remains constant. Both of these effects mean that a hot transistor is going to generate more heat than a cold one, most of the heat being generated by the collector current. Thus if a transistor warms up a bit, more heat will appear within it, so it warms up a bit more, even more heat is generated etc. Once the temperature is so high that the leakage currents are comparable to the normal currents in the transistor you will have lost control of it. In a short while it will heat itself to destruction.

This process is called **thermal runaway**; it is less of a problem with silicon transistors than with the older germanium types, but it still has to be avoided. You can keep a transistor cool by fastening it to large bits of metal to conduct the heat away. These are called **heat sinks**; they are usually painted black and are finned to help convection currents carry away the heat.

Returning to the circuit of figure 4.35, we shall now explain why it is thermally unstable. The quiescent current of the system is fixed by the properties of the transistors themselves. Suppose that V_{IN} is 0 V. Then V_{OUT} will also be 0 V. Both transistors are switched on, so there will be a quiescent current flowing through both of them. Our model of transistor behaviour does not allow us to calculate the size of that current. In theory, we would expect the quiescent current to be about 100 times the current flowing through the diodes, but that does assume ideal transistors and diodes at the same temperature. In practice, the heat developed in the transistors will start them on the road to thermal runaway. There is a high probability that the power amp will self-destruct a few seconds after its supply rails are connected.

Thermally stable followers

Figure 4.36 shows a push-pull follower that has built in immunity to thermal runaway. The system is linear i.e. $V_{OUT} = V_{IN}$, because the diode/resistor network ensures that both transistors are on all of the time. The diodes keep the transistor bases 2.8 V apart, regardless of the value of V_{IN}. So the emitter of the npn transistor will be at $V_{IN} + 0.7$, that of the pnp transistor at $V_{IN} - 0.7$. Each of the 47 Ω resistors will therefore have 0.7 V across it when V_{IN} is 0 V. The quiescent current is fixed by the size of the resistor and the voltage across it; in our case it will be $0.7/0.047 = 15$ mA.

That current will, of course, generate heat in both of the transistors. In fact each transistor will gain heat at a rate of $15 \times 15 = 225$ mW when no current flows into the load. Although this represents a waste of electrical energy, it does not start the transistors on the road to thermal runaway. This is because the quiescent current is fixed by the size of the 47 Ω resistors, not by the properties of the transistors. Suppose that the current increases as a consequence of the rise in temperature of the transistors. The bases of the transistors are kept a fixed 2.8 V apart, so an increase of collector current through the resistors will tend to decrease the values of V_{BE} and hence start to turn the transistors off. Every time that the current starts to

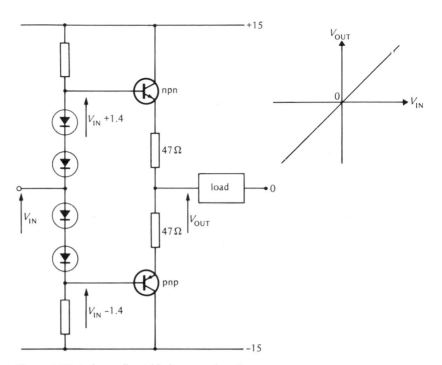

Figure 4.36 A thermally stable linear push-pull output stage

run away the transistors get shut off a bit to compensate. The transistors can run hot (up to 200 °C) without your losing control of them.

Short-circuit protection

One of the disadvantages of the simple push-pull follower is that it can be destroyed by connecting its output straight to ground. When you do this, the emitter is having to push or pull current through a load that has zero resistance. Of course, the emitter has to stay within 0.7 V of the base, so it will end up trying to maintain a finite voltage drop across a zero resistance. Ohm's Law predicts that this requires an infinite current to flow through the resistance, so the transistor will exceed its maximum collector current. The presence of a fuse between the emitters and the load is unlikely to save the transistors from their fate; they will be dead by the time that the fuse has melted.

On the other hand, the circuit of figure 4.36 has short-circuit protection. Let us suppose that V_{IN} is +6 V and that the output is shorted to ground; the situation is illustrated in figure 4.37. As V_{IN} is positive, we can forget about the pnp transistor as it will be switched off; its base will be at $6 - 1.4 = +4.6$ V and its emitter at 0 V. The npn transistor will hold its emitter at 6.7 V by sourcing 143 mA into the 47 Ω resistor. Provided that both the transistor and the resistor can cope with the heat generated (about 1 W each) the system will come to no harm.

The price that has to be paid for this protection is an increase in the output impedance of the follower. Since all of the current that flows through the load has to flow through one of the 47 Ω resistors, it follows that the output impedance of the system is also 47 Ω. You can make this smaller, but at the expense of an increase in quiescent current. As we shall show in Chapter Eight, the techniques of negative feedback can be used to

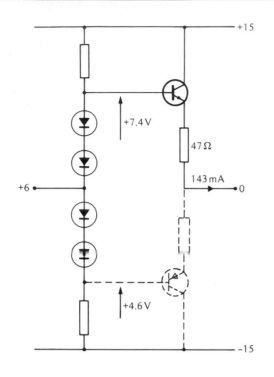

Figure 4.37

wipe out the output impedance of the follower, whilst maintaining its short-circuit protection.

EXPERIMENT

Assembling and evaluating two complementary followers

Assemble the system shown in figure 4.38. The value of V_{IN} can be varied from about $+12\,V$ to $-12\,V$ by rotating the potentiometer knob. Measure the values of V_{OUT} for values of V_{IN} that cover the whole of this range. Draw a graph of V_{OUT} as a function of V_{IN} for the system.

Then assemble the compensated follower shown in figure 4.39. As before, measure values of V_{OUT} for values of V_{IN} that span from $+12\,V$ to $-12\,V$ and use them to plot a graph of V_{OUT} as a function of V_{IN}.

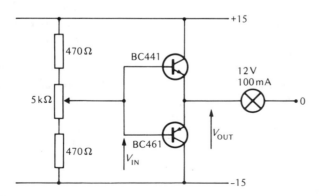

Figure 4.38

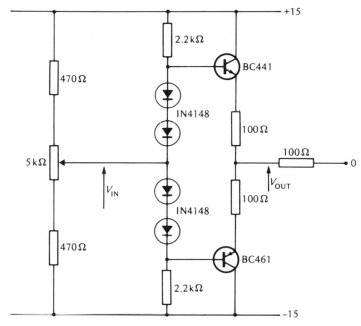

Figure 4.39

Comment on the shapes of the graphs, and try to explain their shapes as fully as you can.

PROBLEMS

This problem concerns the circuits shown in figure 4.40. A 3 V peak value sine wave is fed into each one; in each case draw a graph to show how V_{OUT} varies with time.

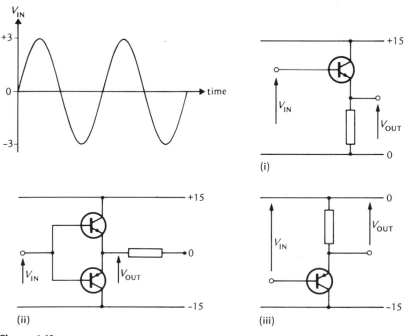

Figure 4.40

REVISION QUESTIONS

1 Explain what a power amplifier is supposed to do.

2 Show how a relay may be connected up to the output of a TTL logic gate. What is the function of the diode in your circuit?

3 Describe the electrical properties of a triac.

4 Why should a triac only be fired at the instants when the mains voltage is zero?

5 Describe the electrical properties of an npn transistor.

6 Draw the circuit diagram of an npn emitter follower. Describe how the output voltage depends on the input voltage.

7 An npn emitter follower is run off supply rails of 0 V and +15 V. If a signal source holds the base at +9 V, and the emitter is connected to the 0 V rail by a 220 ohm resistor what is

a) the emitter voltage,

b) the emitter current,

c) the base current,

d) the power drawn from the signal source,

e) the power fed into the resistor

f) the rate at which heat is generated within the transistor.

(Assume that $h_{FE} = 200$.)

8 What is a Darlington pair? In what way is it useful?

9 Draw the circuit diagram of a complementary follower. Explain why it is better than a single transistor emitter follower.

10 Draw the circuit diagram of a thermally stable compensated push-pull follower. Explain why it is thermally stable.

11 Explain what is meant by the terms "current gain" and "saturation" when applied to a bipolar transistor.

12 A 12 V DC motor is run off a 12 V rms AC supply via a triac. A single control pin makes the motor run backwards or forwards. When that pin is left to float up to logical 1 the motor goes forwards, and when the pin is held at logical 0 the motor goes backwards. Draw a circuit diagram of a suitable system for controlling the triac. Explain how it works; some sketch graphs showing how voltages change with time within your circuit might be useful.

5

Bipolar transistors

In every chapter of this book you have been introduced to at least one new component. So far you have met switches, resistors, relays, diodes, logic gates, capacitors, transformers, op-amps, triacs and transistors. You have learnt how to synthesise large and complex systems from these basic components. Each new component has opened up a new area of electronics for you. For example, once you had met the op-amp you were launched into the field of analogue signal processors such as amplifiers, integrators, rectifiers etc., all of which can be assembled from op-amps. Furthermore, each chapter has had more or less the same format. You have been introduced to a new basic component, and then shown how it can be used to build systems with a variety of different functions.

However, this chapter is going to be different. You are not going to learn about any new basic components as you go through it. This is because you have already become familiar with bipolar transistors from your study of power amps in Chapter Four. The chapter will show you how transistors can be put together to make various complex systems, but few of those systems will be directly useful or allow you to design devices that you could not design before.

The different flavour of this chapter is a consequence of the unique position that the transistor has in the scheme of electronics. It is the simplest component that can be used to amplify a signal. Before the invention of the transistor, the thermionic valve was the basic electronic component; virtually every electronic system was based on valves. Modern electronics relies on transistors. They were the first amplifying components to be made by integrated circuit technology. They could be built into the structure of a thin wafer of crystalline silicon. This made them cheap and plentiful, so that systems built out of large numbers of transistors became an economic possibility. Then, several transistors could be etched onto a single wafer of silicon, so that integrated circuits could be manufactured. In fact, it is the cheap and reliable IC that has made modern electronics possible.

Transistors are the fundamental building blocks inside all of the ICs that you have met in this book. ICs all consist of a group of transistors connected together in a single package. What makes the 741 behave differently from the 7400 is the arrangement of the transistors within the IC. This chapter will attempt to show you how transistors can be put together to make systems with the characteristics of op-amps, logic gates and AC amplifiers. That is, we shall try to show you how the ICs you use could be put together from more basic components. This ought to give you a better insight into the reasons behind the input and output idiosyncracies of the ICs. Furthermore, this appreciation of how the transistors are connected within the IC will be useful when you have to connect different types of IC together.

Very few of the circuits that you will meet in this chapter will be able to do things that you cannot already do with the ICs that you are familiar with. Indeed, the IC version of a transistor circuit is usually far superior in many respects. ICs are more reliable, cheaper, smaller, easier to use, and are generally superior in electronic behaviour. There are a few areas where discrete transistors reign supreme; i.e. where a circuit built out of separate transistors functions better than its IC equivalent. You have already met one of these areas, namely power amplification. Another area is that of high frequency amplifiers, but that sort of transistor circuit design is very complex; the IC versions are getting better rapidly, and you would be advised to use a pre-assembled circuit rather than design your own.

LOGIC GATES

Before the advent of integrated circuits, logic systems were assembled from gates that were built out of discrete transistors. NOT, NOR and NAND gates were put together from resistors, diodes and transistors in a variety of ways; we will be showing you one of those ways in this section.

Although the systems that we shall be discussing mimic the behaviour of the TTL logic gates that you are familiar with, they do not represent the actual arrangement of transistors in a TTL IC.

AN NPN NOT GATE

A simple NOT gate based on an npn transistor is shown in figure 5.1. We shall discuss its behaviour in detail, as this type of transistor connection (called the **common emitter** connection) is used for many other transistor applications.

First of all, we will explain how the value of V_{OUT} depends on the value of V_{IN}. The relationship between V_{OUT} and V_{IN} is summarised in the graph of figure 5.2; note that there are three distinct regions in the graph.

The first region has V_{OUT} at a constant $+5\,V$. For this to happen V_{IN} has to be less than $+0.7\,V$ so that the transistor is switched off. Its collector current will therefore be zero. As no current flows through the collector resistor there cannot be a voltage drop across it, and V_{OUT} is pulled up to the top supply rail.

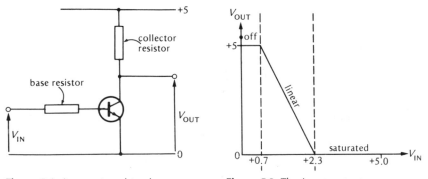

Figure 5.1 An npn transistor in common emitter mode

Figure 5.2 The input-output characteristics for figure 5.1

When V_{IN} goes above $+0.7\,V$ the transistor switches on. Current flows into its base, causing current to flow into the collector. That current has to flow through the collector resistor, pushing V_{OUT} down towards the bottom supply rail. This is the second region of the graph, the **linear region**, where V_{OUT} goes down as V_{IN} goes up.

As V_{IN} is raised the current flowing into the base increases. (As the emitter is held at $0\,V$, the base cannot rise above $+0.7\,V$ so that any increase in V_{IN} means an increase in the voltage across the base resistor.) If the base current rises there is a corresponding rise in the collector current. A larger collector current means a larger voltage drop across the collector resistor, pushing V_{OUT} further away from the top supply rail. So as V_{IN} goes up, V_{OUT} goes down.

The third region of the graph has V_{OUT} at a constant $0\,V$. The transistor has saturated, as its collector has got down to the same voltage as its emitter. When that happens, the collector current cannot be increased by an increase in base current. So as V_{IN} rises, it makes more current flow into the base, but the collector current does not change.

It should be clear that the circuit has the characteristic behaviour of a NOT gate. When V_{IN} is $0\,V$, V_{OUT} is $+5\,V$; when V_{IN} is $+5\,V$, V_{OUT} is $0\,V$. If a digital signal is fed into the circuit the inverse is fed out of it.

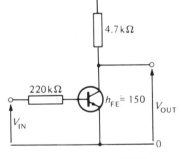

Figure 5.3 A transistor NOT gate

The saturation threshold

Before we proceed to show how more complex gates can be made out of npn transistors, we shall run through a detailed calculation of the saturation threshold of the NOT gate of figure 5.3. Not only will we need the result later on, but it will reinforce your understanding of how the NOT gate works.

We want to calculate the size of V_{IN} which will just get V_{OUT} down to $0\,V$. That is, we are going to calculate the value of V_{IN} which marks the transition between the linear and saturated regions of the graph of figure 5.2.

Consider the collector resistor in figure 5.3. Since V_{OUT} is $0\,V$, there must be $5\,V$ across the resistor. The current which flows through it will be $5/4.7 = 1.1\,mA$; that current flows into the collector.

Assuming that the current gain is 150 (typical for a BC107), the base current must equal $1.1/150 = 0.007\,mA$. That current has to flow through the $220\,k\Omega$ base resistor, causing a voltage drop of $0.007 \times 220\,V$. Since V_{IN} equals the sum of the voltage drop across the base resistor and the base-emitter voltage drop of the transistor it must be equal to $1.6 + 0.7 = 2.3\,V$.

So V_{IN} has to be at least $+2.3\,V$ for V_{OUT} to be $0\,V$.

NOR GATES

A transistor NOR gate is shown in figure 5.4. It is basically a pair of NOT gates that share a common collector resistor.

Start off by having both A and B at logical 0. Both of the transistors will be firmly switched off. No current will be able to flow into either of the collectors, so no current will flow through the collector resistor. Q will be pulled up to the top supply rail so that it sits at logical 1.

Now suppose that one of the inputs is raised above $+2.3\,V$. As we have already calculated, this feeds enough current into the base of the transistor for it to saturate. The other transistor is still firmly switched off, so we can

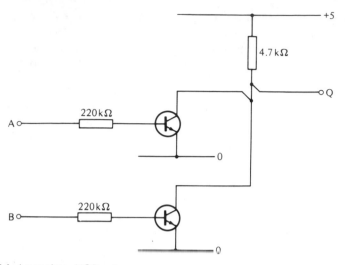

Figure 5.4 A transistor NOR gate

ignore any current flow into its collector. So 1.1 mA flows through the collector resistor into one of the collectors and Q is pushed down to logical 0.

Finally, if both inputs are raised above +2.3 V both of the transistors will be switched on. As each transistor is capable of sinking 1.1 mA from the collector resistor on its own, they will have no trouble in pulling Q down to logical 0 between them.

So if one or both of the inputs are at logical 1, the output is at logical 0. The circuit of figure 5.4 is therefore a NOR gate, the basic logic gate that can be used to make any other logic system. For example, figure 5.5 illustrates how npn NOT and NOR gates can be used to make a NAND gate.

Figure 5.5 An npn transistor NAND gate

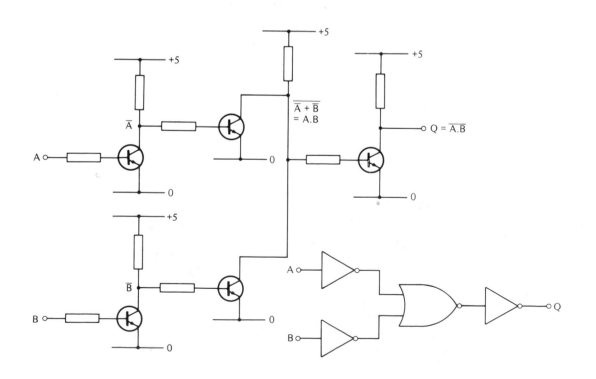
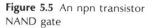

GATE CHARACTERISTICS

When you use a set of identical logic gates to make a logic system you need to bear three things in mind. Firstly, the range of signals that will be read in as logical 0. Secondly, the range of signals that will be read in as logical 1. Thirdly, the fanout of each gate i.e. the number of inputs that a single output can drive.

We are going to calculate each of these quantities for the NOT gate of figure 5.3; because the NOR gate of figure 5.4 is so similar, the characteristics will apply to it as well.

If V_{IN} is to be recognised as a logical 0 by the gate it must make V_{OUT} rise to logical 1 i.e. $+5$ V. V_{OUT} is only at $+5$ V when the transistor is switched off. This requires V_{IN} to be less than 0.7 V, so any signal that is less than $+0.7$ V will be read in as logical 0.

We have already calculated the value of V_{IN} that is just large enough to saturate the transistor; it is $+2.3$ V. When the transistor is saturated $V_{OUT} = 0$ V i.e. is a logical 0. So any signal which is above $+2.3$ V will be read in as logical 1.

In order to find the fanout of the gate we must assume that its output is connected to a number of inputs, as shown in figure 5.6. We need to calculate how many inputs are needed to stop the output working properly. As it happens, when the output is at logical 0 you can hook any number of inputs up to it. This is because transistors T_1 to T_n inclusive will be off so that they draw no current from T_0. But when the output is at logical 1 it has to source current into the inputs; if too much current is pulled out of the output it will fall below the logical 1 threshold. Suppose that n is just large enough to pull the output down to $+2.3$ V. Each input will sink 0.007 mA (the current through 220 kΩ when the voltage drop across it is $2.3 - 0.7$ V), so the total current pulled out of the output is $n \times 0.007$ mA. Bearing in mind that T_0 is switched off, the current through the 4.7 kΩ resistor will also be $n \times 0.007$ mA. It must produce a voltage drop of

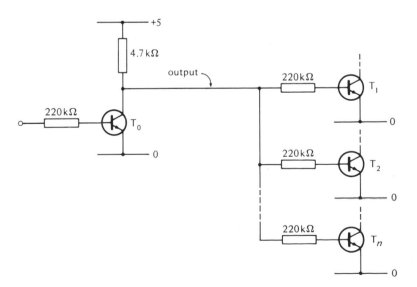

Figure 5.6 A NOT gate feeds a number of other NOT gates

$5 - 2.3 = 2.7\,\text{V}$ across the resistor. By applying Ohm's Law to that resistor we can calculate n:

$$R = \frac{V}{I}$$

$$\text{therefore} \quad 4.7 = \frac{2.7}{n \times 0.007}$$

$$\text{therefore} \quad n = \frac{2.7}{4.7 \times 0.007}$$

$$\text{therefore} \quad n = 82$$

So one output is capable of pulling up to 82 inputs up to logical 1. In practice, you would have to take account of the large variation in the value of h_{FE} for a particular type of transistor. Our calculated value of fanout has assumed a current gain of 150; to be on the safe side we would have to calculate the fanout for transistors which have the smallest value of h_{FE} that we would expect to find.

PROBLEMS

1 Calculate the minimum value of V_{IN} that will make the transistors in figure 5.7 saturate. Then draw a graph to show how V_{OUT} depends on V_{IN} for each circuit.

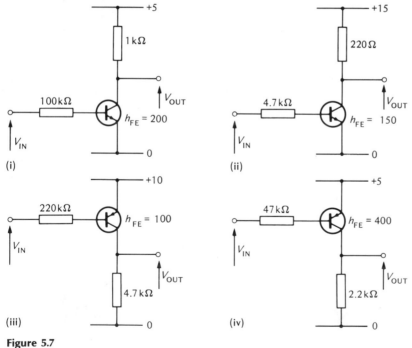

Figure 5.7

2 For each of the circuits in figure 5.8 calculate

a) the base current,

b) the collector current,

c) the current gain.

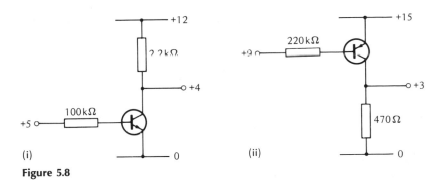

Figure 5.8

3 A system is to be assembled from logic gates of the type shown in figure 5.9. The transistors can have current gains in the range 50 to 200. Calculate the range of signals that will be read in as logical 0 and logical 1 for any of the gates. What is the value of the fanout?

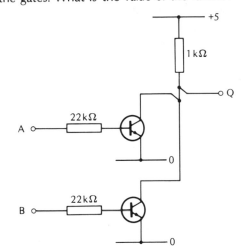

Figure 5.9

4 This question concerns the two circuits shown in figure 5.10.

a) State and explain what type of logic gate each circuit behaves like.

b) Calculate the range of signals that will be read in as logical 1 and logical 0 for this type of gate. Assume that the transistors have current gains of 100.

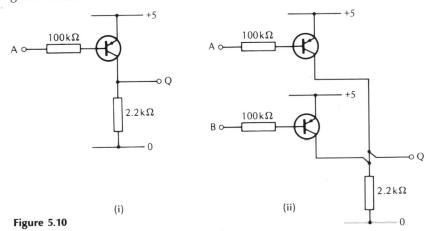

Figure 5.10

5 Draw the circuit diagram of an EOR gate based on transistors.

EXPERIMENT

Measuring the characteristics of a transistor NOT gate.
This will allow you to calculate its fanout, which you will
subsequently measure directly.

Assemble the NOT gate and signal source shown in figure 5.11. Use the potentiometer to generate values of V_{IN} between 0 and $+5\,V$; measure the values of V_{OUT} for values of V_{IN} that cover this range.

Use your results to plot a graph of V_{OUT} as a function of V_{IN}. It will look similar to the one shown in figure 5.12. Fit the best trio of straight lines through your points, and use them to read off the logical 1 threshold. Use this value to calculate the current gain of your transistor.

Now calculate the fanout of the gate. You can test the result of your calculation by hooking a dummy load onto the gate output when its input is held at logical 0. This is shown in figure 5.13. The $10\,k\Omega$ resistor and diode behave like ten inputs in parallel. By trial and error, find the value of dummy load resistor that will pull the output below logical 1 and hence work out the actual value of the fanout.

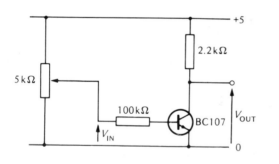

Figure 5.11

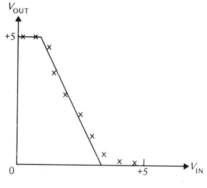

Figure 5.12

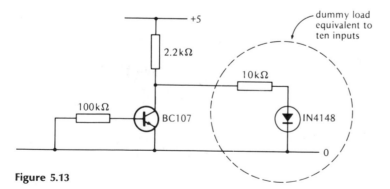

Figure 5.13

AC AMPLIFIERS

An amplifier is a device which feeds out a magnified copy of the signal which is fed into it. Figure 5.14 shows an example of a DC amplifier based on an op-amp. The voltage of the output is always twice as large as that of the input. So if we present a DC signal of $+2\,V$ at the input the output feeds out a DC signal of $+4\,V$ (a DC signal is one that does not change with time). Furthermore, if the input is wobbled up and down the output will wobble up and down twice as far. So any AC signals fed into the device are also amplified. In fact, the name of DC amplifier for the system is misleading as it will amplify both AC and DC signals.

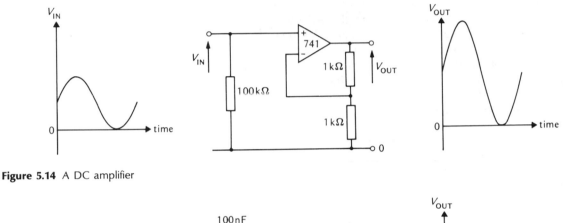

Figure 5.14 A DC amplifier

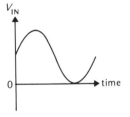

Figure 5.15 An AC amplifier

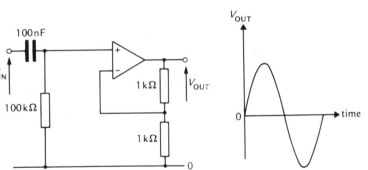

By adding a 100 nF capacitor the system can be made into an AC amplifier (figure 5.15). Then, only the AC components of a signal will be amplified. The DC component will be completely ignored so that the output will be a pure AC signal, with an average value of $0\,V$. We shall be explaining how the **coupling capacitor** cuts out the DC in the next section but first we need to discuss why AC amplifiers are so important in electronics.

AC amplifiers are very widely used in systems which process analogue signals. There are many transducers which generate small AC signals (microphones, aerials and gramophone pickups for example) which need amplifying. Consider the system of figure 5.16. The signal from the microphone will be small, a few millivolts at most. This must be raised to the level of at least a few volts to make a reasonable noise when fed into the speaker. So the amplifier needs to have an AC gain of about 10^4 i.e. make the AC components of the input signal 10 000 times larger.

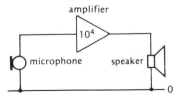

Figure 5.16

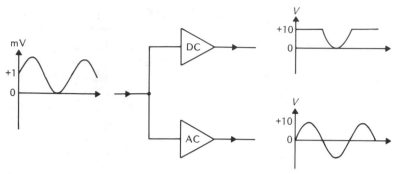

Figure 5.17

Whenever different metals are in contact, and whenever different parts of a circuit are at different temperatures, small DC signals are generated. In a number of cases, they will be larger than the AC signal that is to be amplified. There is a danger that these DC signals will saturate the amplifier so that it cannot respond properly to the AC signals. To understand how this can happen, look at figure 5.17. Both amplifiers have a gain of 10^4 and their outputs can go from $+10\,V$ to $-10\,V$. The AC amplifier is able to amplify faithfully the AC component of the signal. The DC amplifier has its output pushed too close to its top limit for it to be able to amplify the AC signal properly.

So when you are only interested in amplifying AC signals it is a good idea to use an AC amplifier. We shall now show how the addition of a coupling capacitor can convert a DC amplifier into an AC one.

Coupling capacitors

When a signal is fed through a capacitor only the AC components will get through. Look at figure 5.18. There is no way that a DC current can flow through the capacitor, so the DC voltage across the resistor must be zero. The average value of V_{OUT} will be $0\,V$. On the other hand if V_{IN} is wobbled up and down, V_{OUT} will wobble up and down with it. The AC signal at V_{OUT} is the same as the AC signal at V_{IN}.

AC signals will only pass through the network unaffected if they change on a timescale that is fast compared with $0.7RC$. If a cycle of the signal takes longer than this, the voltage difference between the plates will be able to change by an appreciable amount. If one cycle takes much less than $0.7RC$ then there is no time for the capacitor to change the charge on its plates.

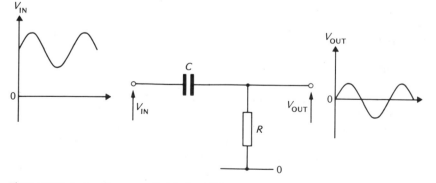

Figure 5.18 A coupling capacitor losing DC

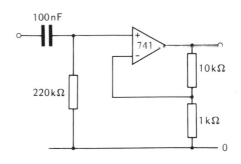

Figure 5.19

For example, consider the AC amplifier shown in figure 5.19. The half-life of the *RC* network at the input is $0.7 \times 220 \times 0.1 = 15$ ms. One cycle of the AC signal must be much shorter than this if it is to pass through to the amplifier unscathed. Therefore its frequency must be much greater than $1/15 = 0.065$ kHz or 65 Hz. So an AC signal whose frequency is much smaller than 65 Hz will not be amplified by the system because it will fail to get through the *RC* network at the input.

An *RC* network can be used to take out the DC component of a signal. It can also be used to add a DC signal to an AC one, i.e. it can change the average value of an AC signal. For example, look at figure 5.20. The voltage divider generates a DC signal of $+2.5$ V, and the coupling capacitor permits AC signals to be added to it. This sort of network is very widely used at the input end of transistor AC amplifiers, the topic of the next section.

THE COMMON-EMITTER AMPLIFIER

A simple one-transistor AC amplifier is shown in figure 5.21. Transistor amplifiers can look fairly complex, with many resistors and capacitors hooked onto the transistor. This particular sort of transistor amplifier is not very useful, but it illustrates how transistors can amplify. Transistor amplifiers which are more useful will be discussed in a later section.

The common-emitter amplifier contains four distinct parts. The signals enter from the left and are fed into a coupling capacitor. The voltage

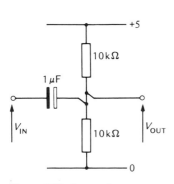

Figure 5.20 A coupling capacitor adding DC

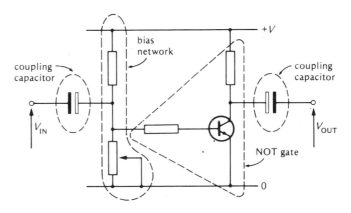

Figure 5.21 A common-emitter amplifier

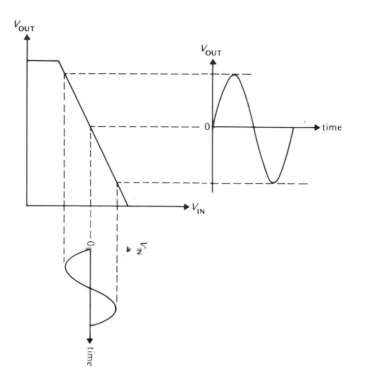

Figure 5.22 How a change of V_{IN} can cause a similar change of V_{OUT}

divider adds a DC signal to the AC one, and the combined signal is fed into a NOT gate. On the right is another coupling capacitor which allows the AC part of the NOT gate's output to leave the system.

The clue to understanding why the system amplifies is an appreciation of the role of the voltage divider. This generator of DC signals is called the **bias network**. Its function is to hold the input of the NOT gate at just the right voltage for its output to sit halfway between the supply rails. The NOT gate is biased so that it lies in the centre of its linear region (figure 5.22). Any AC signal fed into the input of the gate causes its output to wobble up and down in sympathy. If the slope of the V_{OUT}-V_{IN} graph is steep enough, the AC signal fed out of the NOT gate will be much larger than the AC signal fed into it.

The AC gain

This section will be devoted to getting an algebraic expression for the AC gain of a common-emitter amplifier. The contents of the formula will highlight the shortcomings of the system.

All of the symbols are defined in figure 5.23. We are going to obtain an expression for the slope of the linear part of the V_{OUT}-V_{IN} graph i.e. (change of V_{OUT})/(change of V_{IN}). To do this we shall first find the value of V_{IN} that will just saturate the transistor.

Consider the collector resistor. In general

$$R_C = \frac{V - V_{OUT}}{I_C}$$

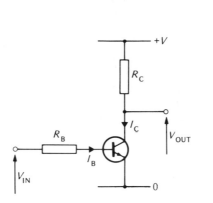

Figure 5.23 A NOT gate

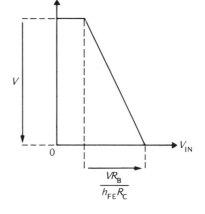

Figure 5.24 The input-output characteristics of a NOT gate

At saturation

$$V_{OUT} = 0$$

so

$$R_C = \frac{V}{I_C}$$

therefore

$$I_C = \frac{V}{R_C} \qquad (1)$$

The base and collector currents are connected by the current gain equation:

$$I_C = h_{FE}I_B \qquad (2)$$

Combining (1) and (2):

$$I_B = \frac{I_C}{h_{FE}} = \frac{V}{h_{FE}R_C} \qquad (3)$$

If we now consider the input side of the gate:

$$V_{IN} = I_B R_B + 0.7 \qquad (4)$$

Combining (4) and (3) to eliminate I_B, we get:

$$V_{IN} = \frac{VR_B}{h_{FE}R_C} + 0.7$$

Look at the graph of figure 5.24. V_{OUT} drops from V to 0 as V_{IN} goes from 0.7 to $0.7 + (VR_B)/(h_{FE}R_C)$. In other words when V_{OUT} changes by $-V$, V_{IN} changes by $+(VR_B)/(h_{FE}R_C)$.

$$\text{Therefore} \quad \frac{\text{change of } V_{OUT}}{\text{change of } V_{IN}} = \frac{-V}{(VR_B)/(h_{FE}R_C)}$$

$$= -h_{FE}(R_C/R_B)$$

So if the transistor is biased so that it lies in its linear region, any change of V_{IN} causes a change of V_{OUT} that is of the opposite sign and is $h_{FE}(R_C/R_B)$ bigger.

It is good practice to bias the transistor so that V_{OUT} can swing equal amounts in both directions. So the collector should sit halfway between the supply rails when no AC signal is fed into the amplifier. The DC signal fed in by the bias network will have to be halfway between 0.7 and 0.7 $+ (VR_B/h_{FE}R_C)$; i.e. $0.7 + \frac{1}{2}(VR_B/h_{FE}R_C)$.

Before we apply these two results to design a specific amplifier, note that both the AC gain and the DC bias are functions of h_{FE}. Since h_{FE} is a very variable quantity from transistor to transistor, each time that a common emitter amplifier is assembled it will have an unpredictable gain and bias.

Calculating the operating point

We are going to apply the results of the last section to complete the design of the circuit shown in figure 5.25. The values of the resistors R_1 and R_2 need to be worked out.

The bias network needs to hold point X at a voltage given by:

$$0.7 + \frac{1}{2}\frac{(VR_B)}{h_{FE}R_C} = 0.7 + \frac{0.5 \times 5 \times 10}{200 \times 1} = 0.85 \text{ V}$$

(There is no point in quoting the answer with a precision of more than 0.05 V; the value of V_{BE} can easily change by this amount as the transistor heats up.)

Using the voltage divider formula:

$$\frac{R_2 \times 5}{R_1 + R_2} = 0.85$$

$$\text{therefore} \quad \frac{R_2 + R_1}{R_2} = 6 \tag{1}$$

All we have to do now is choose a value for one of the resistors. It is important that the resistors be large so that the amplifier has a large input impedance. They must not be so large that the base current drawn by the transistor upsets the operation of the voltage divider. The collector must sit halfway between the supply rails, so the quiescent collector current is

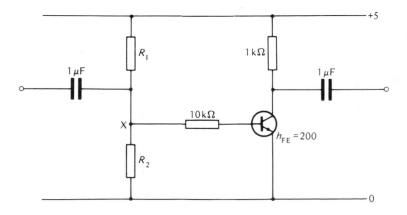

Figure 5.25 A common-emitter amplifier

2.5 mA. Therefore the quiescent base current will be $2.5/200 = 0.013$ mA. To be safe we must have at least 0.13 mA flowing through R_1 and R_2:

$$R_1 + R_2 = \frac{5}{0.13} = 38.5 \qquad (2)$$

Combining equations (1) and (2) we get the following results for the two resistors:

$$R_2 = 6.4\,\text{k}\Omega$$

$$R_1 = 32.1\,\text{k}\Omega$$

In practice, we would set R_1 equal to the nearest preferred value (47 kΩ probably) and use a potentiometer to set the value of R_2 (as in figure 5.21).

The AC gain of the amplifier will be

$$-h_{FE}(R_C/R_B) = -200 \times 1/10 = -20$$

Finally, it will only work properly for AC signals whose frequency is greater than $1/(0.7 \times 6.4 \times 1) = 0.22$ kHz.

EXPERIMENT
Investigating the properties of a common emitter amplifier

Assemble the circuit shown in figure 5.26. Start off with the DC bias voltage at 0 V i.e. point X at 0 V. Monitor the DC voltage of the collector (point Y) as you slowly increase the bias voltage. Note the value of bias voltage that just switches the transistor on. Then increase the bias voltage until the collector sits halfway between the supply rails.

The transistor is now biased at the centre of its linear region. The potentiometer should not be adjusted from now on.

Measure the voltage at point X. Use its value to calculate the current gain of the transistor. Then calculate the AC gain of the amplifier.

Feed a 1 kHz sine wave that is 100 mV peak-to-peak into the amplifier. Trigger a CRO on the waveform at Z and use it to look at the waveforms at points W, X, Y and Z. Draw graphs, one under the other and to the same time-scale, of those waveforms. State the actual gain of the amplifier.

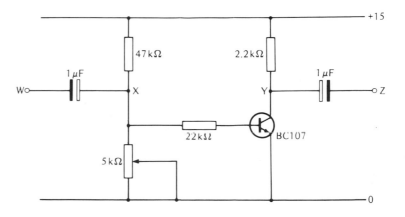

Figure 5.26

Stabilising the operating point

You have assembled a common emitter amplifier and sampled its short-comings for yourself. The bias voltage has to be carefully adjusted to get the collector sitting halfway between the supply rails. Once you have set it, this **operating point** is not very stable. For example, the value of V_{BE} will change by perhaps 0.1 V as the transistor warms up. This will be amplified and cause the collector voltage to change by much more. Thus, if we consider the circuit of figure 5.25, a rise in V_{BE} of 0.1 V will push the collector up by $20 \times 0.1 = 2$ V, leaving it just 0.5 V below the top supply rail. The system can now only feed out AC signals that are less than 0.5 V peak, rather than the 2.5 V peak it could manage previously.

We can stabilise the bias of the transistor by putting a resistor between the emitter and the ground supply rail. This also makes the AC gain independent of the current gain of the transistor and therefore predictable. Look at the circuit of figure 5.27. The same current will flow through both R_C and R_E. So we can safely predict that the voltage across R_C will always be ten times larger than the voltage across R_E. Now suppose that the transistor base is moved up a bit. The emitter follows the base, so the voltage across R_E also goes up. As the top end of R_C is held at +15 V, its other end (the collector) must fall ten times as far as the emitter rises.

So any AC signal fed in at the left through the coupling capacitor will result in an AC signal being fed out of the capacitor on the right. The signal fed out will have the opposite sign and will be ten times larger than the signal fed in. The AC gain of the amplifier is -10 or $-R_C/R_E$. Since the voltage gain is independent of the current gain of the transistor, we can accurately predict what its value will be.

The biasing of the transistor is easier too, and more or less independent of the transistor. The current flowing through the bias network is $15/(47 + 4.7) = 0.3$ mA. We shall assume for the moment that this is much larger than the base current. Using the voltage divider formula, the bias voltage will be 1.4 V. This is the voltage of the transistor base, so its emitter will sit 0.7 V lower, at 0.7 V. The voltage across R_C will be ten times this i.e. 7 V, so the DC voltage of the collector must be $15 - 7 = 8$ V. This is a good operating point, allowing the collector to swing up 7 V on either side of its quiescent value.

Since the emitter voltage is 0.7 V and the value of R_E is 0.22 kΩ, the

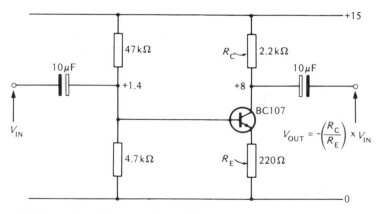

Figure 5.27 An emitter degenerated amplifier

emitter current must be $0.7/0.22 = 3.2\,mA$. If we assume a current gain of at least 100, the base current will be less than $3.2/100 = 0.032\,mA$. This is not going to be enough to upset our calculation of the bias voltage.

The amplifier has been improved because the calculated values of the resistors are now independent of the characteristics of the transistor used. The only requirement is that the current gain is not too small. This makes the manufacture of such AC amplifiers much easier; they do not need individual adjustment, the AC gain can be guaranteed and there is no possibility that the system will need regular servicing to tweak the bias voltage. However, the system is still far from perfect. It is not immune to thermal drifts in the operating point for example. Its gain at high frequencies is unnecessarily limited, and it has a relatively low gain. The elegant techniques of cascode transistors and bootstrapping are beyond the scope of this book, so we will simply assure you that the system can be improved upon.

EXPERIMENT

Assemble the amplifier shown in figure 5.27. Investigate its properties e.g. its AC gain, maximum signal that it can feed out, thermal stability of operating point, how its gain depends on the size of the signal and its frequency, etc.

PROBLEMS

1 This question refers to the circuit shown in figure 5.28. If the system is to have an AC gain of -20 for signals above 10 Hz, calculate the size of

   a) the collector resistor,

   b) the DC collector current,

   c) the DC emitter voltage,

   d) the bias network resistors

   e) the capacitors.

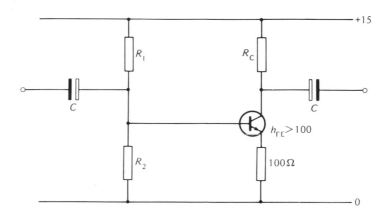

Figure 5.28

2 The circuit shown in figure 5.29 is called a phase splitter. A 2 V
peak-to-peak signal at 5 kHz is fed in at A. Draw graphs, one under the
other and to the same scale, to show what the waveforms at A, B, C, D,
E and F look like.

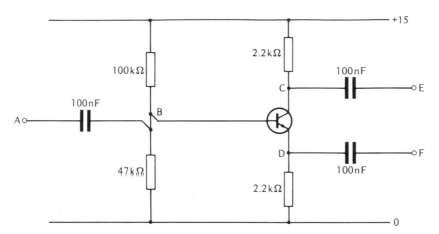

Figure 5.29 A phase splitter

TRISTATES

A **tristate** is a logic gate which, as its name implies, has three output states
instead of the normal two. Its symbol and truth table are shown in figure
5.30. As well as a digital input and a digital output, it has an enable input.
When the terminal marked E is held at logical 0, the state of Q is the same
as that of A. But when E is at logical 1 the output adopts its third state and
goes open-circuit. That is, the Q terminal is effectively disconnected from
the A terminal so that it can float up and down freely. The whole system
behaves like a logic gate with a built-in relay at the output; when the relay
contacts are closed the logic gate can feed out its signal and when they are
open it cannot.

Figure 5.31 shows one way in which a tristate can be put together. In a
moment we will show how the two transistors can be used to feed out a
logical 1, a logical 0 or behave like an open-circuit. First, a word or two
about the usefulness of tristates.

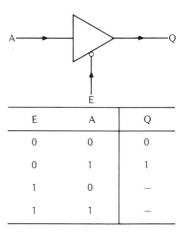

E	A	Q
0	0	0
0	1	1
1	0	–
1	1	–

Figure 5.30 A tristate

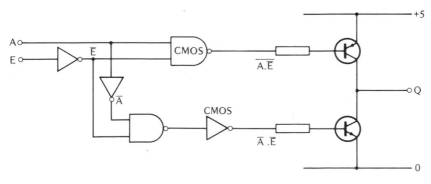

Figure 5.31 A possible tristate circuit

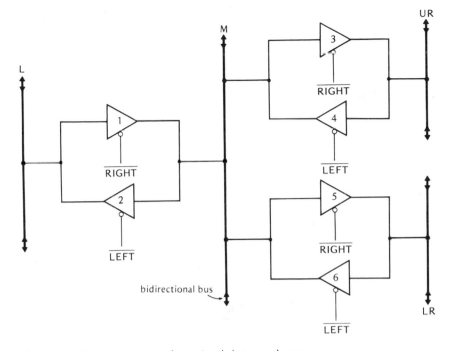

Figure 5.32 Using tristates to shunt signals between buses

Tristates allow the outputs and inputs of many logic systems to be connected to a common **data bus**. A data bus is just a piece of wire that can be used to send digital signals from one logic system to another. Figure 5.32 shows four data buses (L, M, UR and LR) which are interconnected by a series of tristates. The buses are bidirectional because they can be used for sending digital signals as well as receiving them (not at the same time though). The tristates can, if suitably enabled, transfer digital signals from any one of the buses to any other. For example, suppose that you wish to feed a signal from the L bus to the LR one. If you pull the enable pin of gate 1 down, signals can go from L to M. The signal can then be transferred to LR by pulling the enable pin of gate 5 low as well. Signals can be sent in the reverse direction by pulling the enable pins of gates 6 and 2 low and leaving all of the other gates in their open-circuit state.

Tristates are extensively used in microprocessor systems, and we will be showing how they are used to route signals to and from memories in Chapter Eleven.

Sinks and sources

Transistors can be used to make tristate output stages because they are current sources or sinks. Consider circuit (i) in figure 5.33 (overleaf). The npn transistor is switched off, so that no current can flow into its collector. On the other hand, the pnp transistor is switched on because current is being drawn out of its base. Since its base current is $(5 - 0.7)/10 = 0.43$ mA, its collector current will be 43 mA. If no load is connected to the output then that current will try to flow into the collector of the npn transistor. But that transistor will accept no current, so it behaves like a very large resistor and the output rises up to the top supply rail. If a load is connected, the pnp

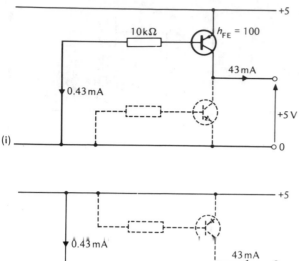

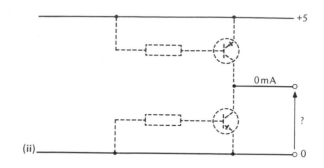

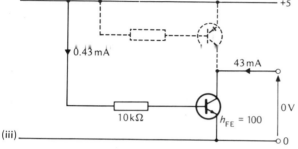

Figure 5.33

transistor will be able to source up to 43 mA into it and keep the output at +5 V.

Circuit (ii) of figure 5.33 has both of the transistors switched off. Neither transistor will allow current to flow in or out of its collector, so the output looks like an open-circuit to any device that is connected to it. The output is free to sit at any voltage between +5 V and 0 V, and will neither sink nor source current.

Circuit (iii) of figure 5.33 has its output pulled down to logical 0. The npn transistor has been switched on, so that it can sink up to 43 mA into its collector. As the pnp transistor is switched off it looks like a very large collector resistor for the npn transistor. That transistor is therefore saturated and the output sits at 0 V. Furthermore, the output can sink up to 43 mA from a load that is connected to it.

Returning to the circuit of figure 5.31, we are now in a position to explain how it works. To get Q to feed out a logical 0, we must have the npn transistor only on. Since $Q = 0$ when $\bar{E}.\bar{A} = 1$ (the first line of the truth table) we have to feed $\bar{E}.\bar{A}$ into the npn base resistor. Similarly, we have to switch on the pnp transistor only to get a logical 1 out of the system. This can be effected by feeding $\bar{E}.\bar{A}$ into the pnp base resistor (line two of the truth table). The other two lines of the truth table which have not been considered require both transistors to be off. So a logical 1 needs to be fed into the pnp base resistor and a logical 0 into the npn base resistor. If you study figure 5.31 you will see that this is the case.

Note that you have to use CMOS gates in figure 5.31. It is important that the outputs of the gates go all the way up to the top supply rail when they are logical 1. This is so that the pnp transistor gets firmly switched off. If TTL gates are used there is a danger that the outputs will not rise to within 0.7 V of the top supply rail.

EXPERIMENT
Designing and assembling a sawtooth waveform generator

Assemble the circuit shown in figure 5.34. Monitor the ramp output with a voltmeter and see what happens when the input (UP/$\overline{\text{DOWN}}$) is held at logical 1 and logical 0. You should find that a logical 1 makes the output ramp up and a logical 0 makes it ramp down; the output voltage should change at about 1 V/s.

Explain carefully why the system behaves as it does, including a calculation of the rate at which the output changes, assuming $h_{\text{FE}} = 200$ for both transistors.

Assemble the Schmitt trigger shown in figure 5.35. Then connect it to the ramp generator which you have already built in such a way that the system oscillates and feeds out a sawtooth waveform. Choose a capacitor value that will make the sawtooth waveform have a frequency of 100 kHz; it ought to look like the one shown in figure 5.36. You may have to adjust the base resistors to get a symmetrical waveform.

Draw a circuit diagram of the final system, showing all of the component values.

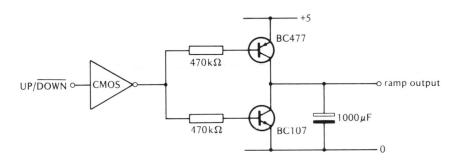

Figure 5.34 A ramp generator

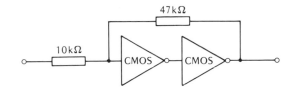

Figure 5.35 A Schmitt trigger

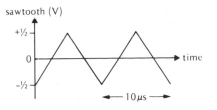

Figure 5.36 A ramp generator and a Schmitt trigger will generate a sawtooth waveform

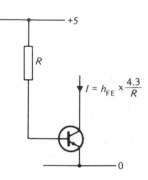

Figure 5.37 A constant current sink

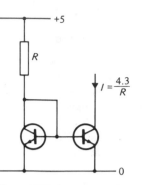

Figure 5.38 A current mirror

Current mirrors

A single npn transistor can act as a current sink, but the amount of current that it will sink depends on its current gain. This is illustrated in figure 5.37. The current flowing into the base will be 4.3/R, so a current that is h_{FE} larger will flow into the collector provided, of course, that it is above 0 V. The collector will sink a constant current regardless of its voltage, a property that was exploited in the ramp generator of figure 5.34. Unfortunately, the collector current is not predictable because it depends on h_{FE}.

If you use a pair of matched transistors, as shown in figure 5.38, this difficulty can be overcome. Both transistors need to be built on the same piece of silicon so that they are identical. Since the transistors have the same base-emitter voltage, they will have virtually the same collector current, regardless of the temperature or current gain. The collector current of one transistor is fixed by a resistor going up to the top supply rail; a little current is bled off to feed the bases, but this will be negligible if the current gain is large enough. The collector of the other transistor will have to sink exactly the same current (unless it saturates).

The circuit is called a **current mirror** because the current that is sourced into one collector is mirrored by the current that the other collector sinks. Because they rely on having matched transistors they are limited principally to being used within integrated circuits. As we shall see in the next section, they are central to the design of op-amps.

DIFFERENTIAL AMPLIFIERS

Now to find out what makes a 741 tick! In fact it is done with mirrors, and relies heavily on the close matching of the all of the transistors. It is therefore an example of a transistor circuit which works well only if all of the transistors are built at the same time on the same piece of silicon, i.e. is constructed as an IC.

A circuit which acts as a differential amplifier is shown in figure 5.39. It mimics the behaviour of the familiar 741 and illustrates the techniques used in its design. It is a differential amplifier, i.e. it only amplifies the voltage difference between its input terminals. The inputs can sit at any voltage

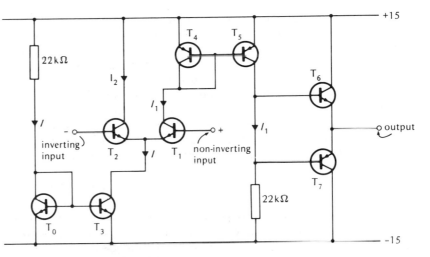

Figure 5.39 A differential amplifier

between the supply rails and pull in about $10\,\mu A$ from the signal source which drives them. The output can get within $0.7\,V$ of either supply rail. The DC gain is about 300. After we have explained how the system works we shall indicate how its performance could be boosted to 741 specifications.

The long-tailed pair

At the heart of the differential amplifier are two transistors known as a **long-tailed pair**. They are T_1 and T_2 in figure 5.39, and constitute the input stage of the amplifier. The tail of the pair is the lead which descends from their joined emitters; a constant current I must be sunk from the tail to get it to work. Look at figure 5.40. If a current I has to be sourced into the tail by the two transistors, then their collector currents must be linked by the equation:

$$I = I_1 + I_2$$

If the two transistors are identical, then $I_1 = I_2$ when $V_1 = V_2$. When the bases are at the same voltage (any voltage) a current of $\frac{1}{2}I$ flows into each collector. Now if V_1 is raised above V_2, T_1 will be switched on harder so

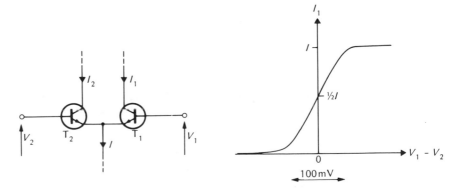

Figure 5.40 A long-tailed pair

that more current flows through it. Of course, the emitter of T_1 will try and sit a constant 0.7 V below its base, so it will rise with V_1. As the base of T_2 is held steady by an unchanged V_2, the rise in its emitter will tend to switch if off and decrease the current flowing into its collector. In fact, once V_1 is about 50 mV higher than V_2, I_1 is 0.9I and I_2 is 0.1I.

The graph of figure 5.40 summarises how I_1 depends on the value of $V_1 - V_2$. When V_2 is much larger than V_1, T_1 is completely switched off so that I_1 is 0. Conversely, if V_1 is more than 100 mV above V_2, T_2 is switched off and T_1 provides all of the current for the tail. Note how the device has **common-mode rejection**, i.e. it ignores the absolute value of the signals fed into it and concentrates on their difference.

Figure 5.39 shows how an npn current mirror can be used to sink a constant current from a long-tailed pair. T_3 sinks a current of $(30 - 0.7)/22 = 1.3$ mA from the tail provided that it remains above -15 V. A pnp current mirror (T_4 and T_5) is used to feed I_1 into a resistor and generate an output signal. T_6 and T_7 constitute a push-pull output stage, boosting the current output. (There is no need to compensate this stage for the V_{BE} drop as op-amps are designed to be used closed-loop; as you will find out in Chapter Twelve, negative feedback wipes out non-linear behaviour of the output stage.) A little thought will convince you that the output will sit at 0 V when the two inputs are held at the same voltage.

Improving the system

We will discuss three ways in which the circuit of figure 5.39 can be improved.

As it stands, each input will draw up to 10 μA from the signal source driving it. This can be reduced by a factor of at least 100 by using Darlington pairs in the long-tailed pair (figure 5.41).

The DC gain of the system is about 300. A change of 100 mV in the value of V_1 as it sweeps past V_2 will cause the output to go from one supply rail to the other, a swing of 30 V. The gain could be increased by another factor of 300 (i.e. to get a total gain of 90 000) by including a second long-tailed pair to compare I_1 and I_2. You would need a pnp current mirror to get I_2 out of the input stage long-tailed pair so that it could be fed through a 22 kΩ resistor in the way that I_1 is. The extra long-tailed pair would then compare the voltage across the two 22 kΩ resistors carrying I_1 and I_2.

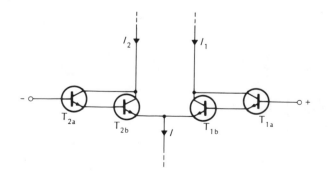

Figure 5.41 A Darlington input stage

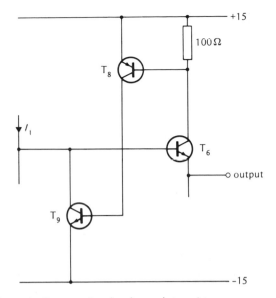

Figure 5.42 Short circuit protection for the push transistor

Finally, the system has no short-circuit output protection. If you pull too much current out of the output you will destroy the push-pull transistors. Any really useful IC has to have some means of protecting itself against this sort of abuse. One way in which this can be done is shown in figure 5.42; it shows the protection for the push transistor only, but a similar network can be used to protect the pull transistor too. When the current pulled out of the push transistor (T_6) is less than 7 mA, T_8 will be switched off as its base will not be 0.7 V below its emitter. If T_8 is switched off, there will be no base current flowing into T_9 so it is off too. Once T_8 is switched on i.e. more than 7 mA is pulled out of the emitter of T_6, T_9 will switch on too. T_9 gets its collector current from the base of T_6, so when T_9 is on it diverts the current that would have gone into T_6 down to the -15 V supply rail. As the base current of T_6 has been reduced, its collector current falls too. The whole arrangement stops the push transistor from sourcing more than 7 mA.

REVISION QUESTIONS

1 Draw a circuit diagram of a NOT gate built out of an npn transistor. If the current gain of the transistor is 100 and the collector resistor is 4.7 kΩ, what is a suitable value for the base resistor if logical 1 is any signal above $+2.5$ V?

2 Draw circuit diagrams of dual input NOR and NAND gates built out of bipolar transistors.

3 Draw the circuit diagram of a simple common emitter AC amplifier. Label its various parts i.e. NOT gate, bias network and coupling capacitors. Explain why the bias network is necessary. Sketch a series of graphs to show how the voltages at various places in the circuit vary with time when a small AC signal is being amplified.

4 What is the difference between a DC amplifier and an AC amplifier? Why are AC amplifiers useful?

5 Why is the common emitter AC amplifier generally unsatisfactory?

6 Draw the circuit diagram of an AC amplifier which has the following properties. Its voltage gain is a guaranteed -22 for an h_{FE} of more than 50, it can feed out AC signals which are up to 10 V peak-to-peak without distortion and has a 2.2 kΩ collector resistor. Show all component values, with reasons.

7 Draw the circuit symbol for a tristate, and explain what it can do. Show how one can be built out of a pair of bipolar transistors and some logic gates.

8 Show how a 1 mA constant current source can be constructed with a matched pair of pnp transistors and a resistor. Assuming that it runs off a 15 V supply, what should the value of the resistor be?

9 Show how a resistor, a pnp transistor and a capacitor can be put together to make a system whose output ramps steadily up at 1 V per second. Assume a 15 V power supply and that h_{FE} is 150.

10 Draw a long-tailed pair. Discuss how it can be used as a differential amplifier.

6
Field effect transistors

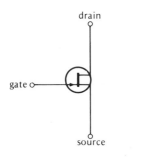

Figure 6.1 An n-channel JUGFET

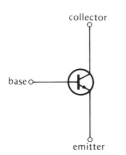

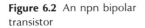

Figure 6.2 An npn bipolar transistor

At first glance, a **field effect transistor** (or FET) looks more or less the same as a bipolar one., Look at the symbol for an **n-channel junction gate FET** (or JUGFET) shown in figure 6.1; it is quite similar to the npn transistor shown in figure 6.2. Each of them has three terminals. Two of the terminals carry the major part of the current through the device and the third terminal is used to control the current. Then, each of the devices appears to contain a diode. The bipolar transistor base-emitter junction behaves like a diode. Indeed, it is the amount of current flowing through that junction which decides how much current will flow between the collector and the emitter. Bipolar transistors are, in effect, current-controlled current sources; feed a constant current into the base and a constant current will flow into the collector. The JUGFET gate-source junction also looks like a diode, but in normal operation that diode is always reverse biased. So no current will flow in or out of the **gate**. Instead, it is the voltage difference between the gate and the **source** which controls the properties of the FET. The FET, unlike a bipolar transistor, is a voltage-controlled device. Furthermore, an FET is two devices rolled into one. It can behave like a voltage-controlled current source (rather similar to a bipolar transistor) or like a voltage-controlled resistor.

FETs can be used to make virtually any circuit which can be made by a bipolar transistor. We shall be looking at AC amplifiers, power amps and current sources built out of FETs in the first part of this chapter. In some respects these systems behave better than their bipolar equivalents, but in other respects their performance is comparatively poor. Then, because of their property of voltage-controlled resistance, FETs can be used to make systems which have no bipolar equivalent. These systems will be studied in the middle of the chapter. The final sections of this chapter will be devoted to MOSFETs (**metal oxide semiconductor FETs**), the type of FET which is used in ICs.

First of all, we shall explain a simple model of FET behaviour. It is quite a precise model, i.e. quite a lot of it can be expressed as algebraic formulae, but it is not easy to apply it to circuit design. This is because there is a very wide spread of real FET characteristics. FETs of the same type are very variable in behaviour. Furthermore, different types of FET can have very different values for their parameters. This is similar to the situation which exists for transistors. The current gain of a transistor is an unknown quantity until you measure it. As it happens, there are ways of designing transistor circuits with predictable properties; the same sort of technique has to be applied to FETs, but it is not so easy as there is no FET equivalent of the constant 0.7 V base-emitter voltage.

JUGFET BEHAVIOUR

Our model of a JUGFET is fairly mathematical, so we will start off by explaining what some of the symbols mean. Figure 6.3 shows the normal arrangement of an n-channel JUGFET in a circuit (we shall look at its mirror-image, the p-channel JUGFET later on). The drain is held above the source so that the drain current (I_D) flows from drain to source. The names of the terminals imply that the current ought to flow the other way; do not worry about it! V_{DS} is the voltage between drain and source, and it will always be positive. V_{GS} is the voltage between gate and source, and it will always be negative or zero. This makes the gate-source junction into a reverse biased diode, so that no current (apart from leakage current) will flow out of the gate.

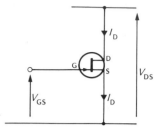

Figure 6.3

The graph of figure 6.4 shows how the drain current depends on the voltage between drain and source when the gate is held a fixed number of volts below the source. At low values of V_{DS}, the curve is a straight line through the origin. This means that V_{DS}/I_D is constant, and the FET behaves like a resistor. Then the curve flattens off and becomes horizontal, so that I_D is independent of V_{DS}. In this region, the FET behaves like a current source, with the drain current remaining constant regardless of changes in drain voltage. The characteristics of both regions depend on the value of V_{GS}. This is clear when you look at the expressions for their values as functions of the FET parameters:

$$R = R_S \bigg/ \left(1 + \frac{V_{GS}}{V_P}\right) \qquad (V_{DS} < V_P + V_{GS})$$

$$I_{DS} = I_{DSS}\left(1 + \frac{V_{GS}}{V_P}\right)^2 \qquad (V_{DS} > V_P + V_{GS})$$

R_S, V_P and I_{DSS} are the parameters of the FET, the numbers which fix its electrical behaviour.

The meanings of the FET parameters are made clearer in the graph of figure 6.5. This shows how I_D varies with V_{DS} when $V_{GS} = 0\,V$. For values of V_{DS} that are less than the **pinch-off voltage** (V_P), the FET has a resistance R_S between drain and source. When V_{DS} is greater than V_P the drain current

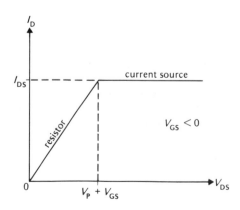

Figure 6.4

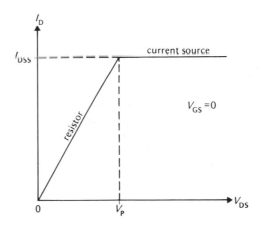

Figure 6.5

saturates to a constant value I_{DSS}. If you think about Ohm's Law, you will find that the three parameters are linked by the equation:

$$R_S = \frac{V_P}{I_{DSS}}$$

The two parameters which are usually quoted for an FET are I_{DSS} and V_P. Once you know two of them, you can work out the value of the third one.

Real JUGFETs

Real JUGFETs do not suddenly change from resistive to current source behaviour as V_{DS} is increased (as implied in figure 6.5). Instead, there is a region around V_P where the FET is neither a good resistor nor a good current source. Outside this region our model is good. Furthermore, JUGFETs are better current sources when V_{GS} is fairly low. The graph of figure 6.6 shows several curves for I_D as a function of V_{DS} for different values of V_{GS} for a real JUGFET.

For a signal JUGFET, i.e. a JUGFET used for processing signals rather than power amplification, V_P will be between 2 V and 8 V and I_{DSS} between 3 mA and 0.5 mA. The particular type which you will be using in practical work (the BF244) has V_P less than 2 V and I_{DSS} greater than 2 mA. A typical BF244 might have $V_P = 1.5$ V and $I_{DSS} = 3$ mA; we will be assuming these values in our examples from now on. A typical BF244 will have $R_S = 1.5/3 = 0.5$ kΩ. In practice, there is a large spread of parameters between FETs of a given type.

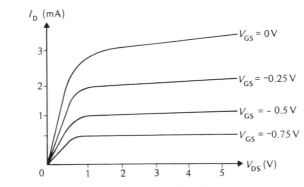

Figure 6.6 I_D as a function of V_{DS} for a typical real FET

CURRENT SOURCES

The simplest type of FET current source is shown in figure 6.7. The gate is anchored to the source, so that V_{GS} is 0 V. Provided that the drain is at least 1.5 V above the source, a constant current of 3 mA will be pulled through the load (we are assuming typical BF244 parameters).

If you refer to the graph of figure 6.6, you will see that FETs are rotten constant current devices when V_{GS} is 0 V. The curves are more horizontal when the gate is pulled below the source, so the drain current varies less as the drain voltage is changed. An improved current source is shown in figure 6.8. A voltage divider holds the gate 0.65 V below the source. Putting this into the drain current formula:

$$I_{DS} = I_{DSS}\left(1 + \frac{V_{GS}}{V_P}\right)^2$$

$$= 3 \times \left(1 - \frac{0.65}{1.5}\right)^2$$

$$= 1 \text{ mA (approximately)}$$

So the drain pulls a constant current of 1 mA through the load.

Note that the voltage divider uses very large resistors. You can do this because the gate current of a JUGFET is extremely low; it is about 5 nA for a BF244.

A neater arrangement for pulling the gate below the source is shown in figure 6.9. The voltage drop across the source resistor (1 kΩ) holds the source 0.75 V above the gate, hence the name **source biasing** for this technique. You can work out the size of drain current that you will get for a given source resistor by using a graph of I_{DS} as a function of V_{GS} (figure 6.10). The parabola is the drain current formula, and the straight line is Ohm's Law for the source resistor. Where the two cross is the actual operating point; I_{DS} will be 0.75 mA for a 1 kΩ source resistor.

In practice, since you do not usually know in advance the exact parameters of the FET, you make the source resistor adjustable. You then tweak its value until you get the desired current flowing into the drain.

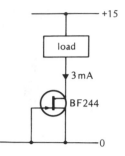

Figure 6.7 A constant current sink

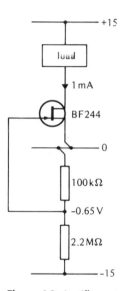

Figure 6.8 A stiff constant current sink

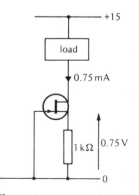

Figure 6.9 A source-biased constant current sink

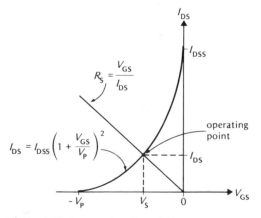

Figure 6.10 I_{DS} as a function of V_{GS}

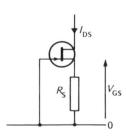

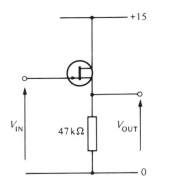

Figure 6.11 A source follower

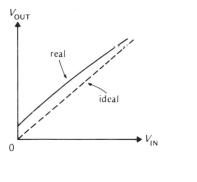

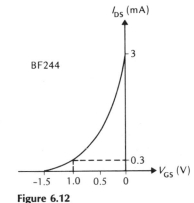

Figure 6.12

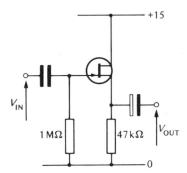

Figure 6.13 An AC source follower

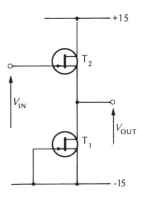

Figure 6.14 A zero offset follower

SOURCE FOLLOWERS

One of the advantages that FETs have over bipolar transistors is their comparatively huge input impedance. You always have to push a little current into the base of a transistor, but virtually no current flows out of the gate of an FET. This means that FETs can be used to extract signals from high impedance sources such as chemical and biological probes. The FET equivalent of the emitter follower is the **source follower** which is shown in figure 6.11. The FET is source biased, with the gate sitting between 1.0 and 1.5 V below the source. If you look at the graph of figure 6.12, you will see that if the current is less than 0.33 mA, then V_{GS} is between 1.0 and 1.5 V. The maximum drain current will occur when V_{IN} is at its largest value; the maximum drain current will be about $15/47 = 0.32$ mA. So when V_{IN} is large, we expect that $V_{OUT} = V_{IN} + 1$. As V_{IN} goes down, so does V_{OUT} and the drain current. When the drain current is very small, V_{GS} is a bit less than -1.5 V. So when V_{IN} is small, $V_{OUT} = V_{IN} + 1.5$.

The relatively large variation in V_{GS} as the size of V_{IN} is varied makes the source follower inferior to its bipolar equivalent, the emitter follower. Nevertheless, the high input impedance makes it valuable; the source of V_{IN} thinks that it is feeding a signal into several hundred megohms. On the other hand, the output impedance of the source follower itself is fairly high, as only a small current can flow out of the source. You can get the best of both worlds by using a source follower to buffer an emitter follower.

If you only want the follower to cope with AC signals the poor linearity of the source follower is not much of a drawback as V_{OUT} will not change by very much. Figure 6.13 shows what an AC source follower looks like. It uses the source biasing idea, with the source sitting 1.4 V above the gate. A large resistor allows the small gate current to get to ground, and the AC signals are fed into the system via a capacitor. Since the drain current is always small (about $1.4/47 = 0.0003$ mA) V_{GS} has got to stay fairly close to V_P (look at figure 6.12). So the AC signal at the source will be a reasonably faithful copy of the signal fed into the gate.

Zero offset followers

The use of matched pairs of FETs can overcome the problem of the uncertainty in V_{GS} for source followers. Like matched pairs of transistors, matched FETs have to be built onto the same slice of silicon. Figure 6.14 shows a pair of FETs put together to make a **zero offset follower** i.e. one where $V_{OUT} = V_{IN}$. T_1 has its gate anchored to its source so that V_{GS} is 0 V.

It acts as a current source for T_2 if there are a few volts between the drain and the source. The same drain current has to go through both T_1 and T_2. As both FETs are identical, they must have the same values of V_{GS} if they have the same values of I_D. So $V_{OUT} = V_{IN}$ over a wide range of inputs, as shown in the graph of figure 6.15.

A better follower is shown in figure 6.16. The two resistors have the same value, and serve to push the gate down below the source. You will remember that an FET acts as a better current source when V_{GS} is negative. So the drain current through T_4 remains more steady as its gate goes up and down, reducing any variations in V_{GS} to a negligible level. This type of zero offset follower is much used as a high impedance input stage for DC amplifiers, e.g. in oscilloscopes.

AC AMPLIFIERS

FET amplifiers have the advantage of very large input impedance, but the disadvantage of low gain. Typical input impedances are tens of megohms, but you are unlikely to get an AC gain of more than 10. For this reason, FET AC amplifiers are only used at the input end of an amplifying system, with bipolar transistors providing most of the gain in subsequent stages.

A simple FET AC amplifier is shown in figure 6.17. If R_S is large, then I_D will be small and the source will be V_P higher than the gate. It follows that the voltage across R_D will be larger than that across R_S by a factor of R_D/R_S. You need V_{DS} to be at least 2 V for the FET to be in its current source region, so the drain voltage must be at least a few volts. If we let R_D/R_S be 5, then the drain will sit $5 \times 1.5 = 7.5$ V below the top supply rail. This operating point allows the drain plenty of room to swing in either direction. A glance at the source follower of figure 6.13 should make it clear that an AC signal fed into the gate appears at the source. Furthermore, if R_S is large enough, the AC signal across it will be equal to the AC signal across R_G. Now, we have chosen R_D to be $5R_S$, so the AC signal across R_D must be five times larger than the one across R_S. So the AC gain of the whole system will be -5. The gain is negative because the drain must move down when the source moves up.

An example of such an amplifier is shown in figure 6.18. The gate resistor fixes the input impedance at $10\,M\Omega$, and the drain resistor fixes the output impedance at a few $k\Omega$. You need to have the output impedance as low as possible so that it can drive a transistor amplifier. The source resistor is

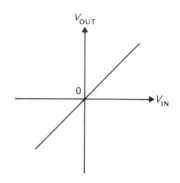

Figure 6.15 The input-output characteristics of a zero offset follower

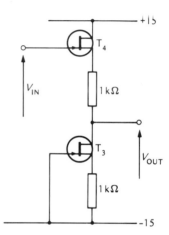

Figure 6.16 A better zero offset follower

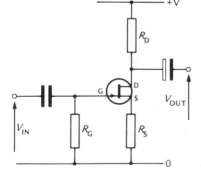

Figure 6.17 A source-biased AC amplifier

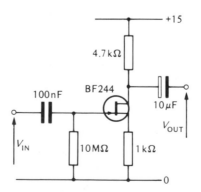

Figure 6.18 An AC amplifier with a gain of about -5

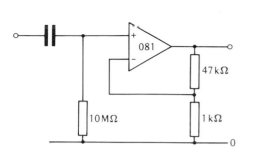

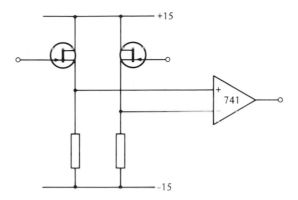

Figure 6.19 A high input impedance, high gain AC amplifier

Figure 6.20 A bipolar op-amp with an FET input stage

$1\,\text{k}\Omega$, so that we expect the AC gain of the system to be -4.7. The drain current is going to be fairly large (average value $0.75\,\text{mA}$), so we can expect that V_{GS} will vary quite a lot as the source wobbles up and down. Therefore the AC signal at the source is not going to be a very good copy of the signal fed in at the gate. The amplifier will distort the signal as well as amplify it.

This sort of distortion is no problem if you can use negative feedback to eliminate it (see Chapter Eight). Unfortunately, negative feedback lowers the input impedance of a circuit, so you cannot use it to reduce distortion in amplifiers which need a high input impedance.

By making the input stage of an op-amp out of FETs you can get the best of both worlds. The system will have the high input impedance typical of FET circuits, combined with the high DC gain and close matching of bipolar transistors. Furthermore, you can use negative feedback without losing the high input impedance (figure 6.19). For example, the 081 BiFET (i.e. bipolar mixed with FET) IC has a DC gain of 10^5 combined with an input impedance of $1\,\text{G}\Omega$.

Figure 6.20 shows how a pair of matched FETs can be used to increase the input impedance of a bipolar op-amp. You would never use this in practice as an 081 IC is a cheaper and simpler alternative, but it illustrates how it can be done. The FETs act as source followers, so that the voltages across the resistors will mirror the signals fed in at the gates. The op-amp then samples the difference in those voltages. A little thought will convince you that if the resistors and the FETs are matched, the variation of V_{GS} with I_D is of no importance when negative feedback is applied to the whole system as in figure 6.19.

p-CHANNEL JUGFETS

Just as the pnp transistor complements the npn one, the p-channel FET complements the n-channel one. As you can see from its circuit symbol (figure 6.21), the gate has to be pushed above the source (i.e. V_{GS} is positive) to control the drain current. An example of a p-channel JUGFET in operation is shown in figure 6.22 (overleaf). The system is a source follower which is designed to handle large AC signals without distorting them. Note the use of a pnp transistor to provide a constant drain current of $1.5\,\text{mA}$ for the FET so that V_{GS} remains constant as the gate is wobbled up and down by the AC signal. If V_P is a few volts, the source should be able to move at least $5\,\text{V}$ on either side of the operating point.

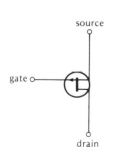

Figure 6.21 A p-channel JUGFET

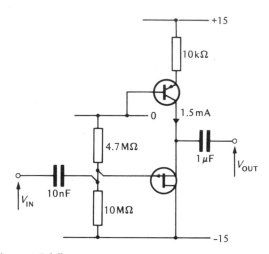

Figure 6.22 A linear AC follower

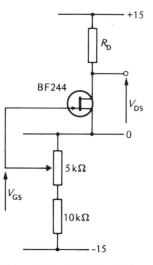

Figure 6.23 A circuit which can be used to investigate the properties of an FET

EXPERIMENT
Measuring the parameters of an n-channel JUGFET

The circuit is shown in figure 6.23. The source is held at 0 V, and the gate can be held at any voltage from 0 V to −5 V by the potentiometer. Make the drain resistor (R_D) 2.2 kΩ and adjust V_{GS} until the drain sits at +10 V. Since the voltage across R_D is $15 − 10 = 5$ V, you can calculate the drain current, i.e. $I_{DS} = 5/R_D$. Note the value of V_{GS} needed to get this value of I_D.

Repeat the exercise with different values of R_D e.g. 4.7 kΩ, 10 kΩ, 22 kΩ, ..., 470 kΩ. Have V_{DS} equal to 10 V every time.

Use your results to draw a graph of I_{DS} as a function of V_{GS}. It should look a bit like the one in figure 6.12.

Our model of a JUGFET says that:

$$I_{DS} = I_{DSS} \left(1 + \frac{V_{GS}}{V_P}\right)^2$$

If we take the square root of both sides and do a bit of shuffling we get:

$$\sqrt{I_{DS}} = V_{GS} \frac{\sqrt{I_{DSS}}}{V_P} + \sqrt{I_{DSS}}$$

So if you plot $\sqrt{I_{DS}}$ against V_{GS} with your results you should get a straight line (figure 6.24). Draw the best straight line through your points and use it to read off V_P and I_{DSS} from the graph.

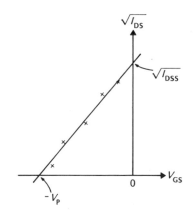

Figure 6.24

PROBLEMS

1 A particular n-channel JUGFET has $V_P = 3$ V and $I_{DSS} = 2$ mA. Draw a graph of I_{DS} as a function of V_{GS}. The same FET is set up as shown in figure 6.25 with $R_S = 1$ kΩ, 10 kΩ and 100 kΩ in turn. Work out the voltage of the source each time, using the graph that you have drawn or by some other method.

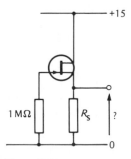

Figure 6.25

2 Draw the circuit diagram of an AC source follower similar to the one shown in figure 6.22, but using an n-channel JUGFET. Given that the system is to operate off a 15 V split supply and that the FET has $V_P = 6\,V$ and $I_{DSS} = 0.5\,mA$, choose suitable values for the components. State and explain the maximum size of AC signal that your system can cope with.

3 Draw the circuit diagram of an AC amplifier which uses a p-channel JUGFET.

4 Carefully explain why the performance of an FET AC amplifier improves as the quiescent drain current is lowered. Illustrate your answer by considering a system with a gain of -2 built around an n-channel JUGFET which has $V_P = 4\,V$ and $I_{DSS} = 1\,mA$, and runs off a 10 V supply.

5 This question refers to the three circuits in figure 6.26. The same FET is used in all three. Use the data provided by the two circuits on the left to calculate, with reasons, the values of V_P and I_{DSS} for the FET. Then use those parameters to work out the quiescent voltage of the source and the drain in the third circuit. Finally draw graphs, one under the other and to the same scale, to show the waveforms at the gate, the source and the drain when a 500 mV peak-to-peak signal at 2 kHz is fed into the amplifier.

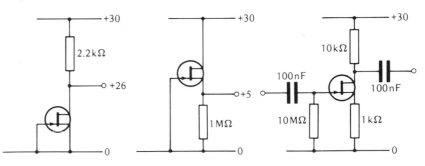

Figure 6.26

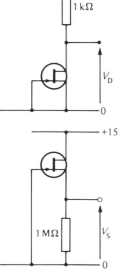

Figure 6.27

EXPERIMENT
*Designing and assembling an FET AC amplifier
which has a gain of -10*

You will first need to get some idea of the values of V_P and I_{DSS} for your FET. Insert it into the two circuits of figure 6.27 and measure the values of V_D and V_S; they should allow you to state the FET's parameters.

Then design the AC amplifier. Not only must the gain be -10, the output impedance must be of the order of a few kΩ. Assemble your design and test its performance by feeding in small signals (e.g. 100 mV peak-to-peak at 1 kHz) and looking at waveforms within the circuit with a CRO.

FETs AS VARIABLE RESISTORS

According to our FET model, if V_{DS} is less than $V_P + V_{GS}$ then the FET behaves like a resistor. The resistance between drain and source is given by:

$$R = R_S \left/ \left(1 + \frac{V_{GS}}{V_P}\right)\right.$$

This relationship is plotted on the graph of figure 6.28. Notice that the resistance can be varied from R_S (typically a few hundred ohms) to infinity. The FET makes a very good resistor providing the voltage across it does not get too big, especially when V_{GS} is 0 V. As the graph of figure 6.29 shows, you can swing V_{DS} from -0.2 V to $+0.2$ V and keep a linear relationship between I_D and V_{DS}, i.e. $V_{DS}/I_D = $ constant. At the other extreme, where V_{GS} is lower than $-V_P$, no current flows through the FET, so it behaves like an infinite resistor i.e. an open-circuit.

Analogue switches

An **analogue switch** is a solid state relay. It has all the properties of a relay, but no moving parts. Figure 6.30 shows what an analogue switch looks like in a circuit diagram. The control terminal is usually TTL compatible i.e. is designed to be held at $+5$ V or 0 V. When the control terminal is held at logical 1 the two I/O terminals behave as if they are joined by a resistor of about 0.5 kΩ. A logical 0 at the control terminal places an open-circuit between the I/O terminals. As we shall see, an analogue switch does for analogue signals what a tristate does for digital ones.

The simplest way of generating an analogue switch is with a p-channel JUGFET, as shown in figure 6.31. The FET needs to have a low value of V_P, in the region of 1 V. Let us assume that an analogue signal between $+1$ and $+3$ V is fed in at L. When C is a logical 0, $V_{GS} = 0$ V for the FET; the 1 MΩ resistor pulls the gate up so that it follows the source. So L and R are joined by a resistance of about 0.5 kΩ, and the signal at L is fed through to R. If we now bring C up to logical 1, the diode is forward biased and the gate is pulled up to about $+4.3$ V. The source is held lower than this by L, so V_{GS} is positive and the FET behaves like an open-circuit. R is pulled down to 0 V by the 100 kΩ resistor. The behaviour of the system is summarised in the graphs of figure 6.31.

The analogue switch which we have been discussing is a bidirectional device. This means that we can feed analogue signals either way through it. When an FET is used as a small resistor you can swap the drain and the source around; when you use an analogue switch V_{DS} will always be small, so that V_{GS} is more or less the same as V_{GD}. The only drawback is that the system can only cope with analogue signals which lie between -0.7 V and $+3.3$ V (assuming $V_P = 1$ V). By using MOSFETs, which we will be discussing later, you can build analogue switches that will cope with signals which lie anywhere between logical 1 and logical 0. For example, the CMOS IC 4066

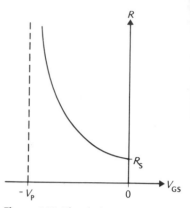

Figure 6.28 The drain-source resistance as a function of V_{GS}

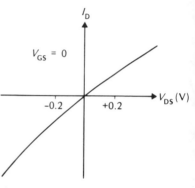

Figure 6.29

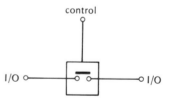

Figure 6.30 An analogue switch

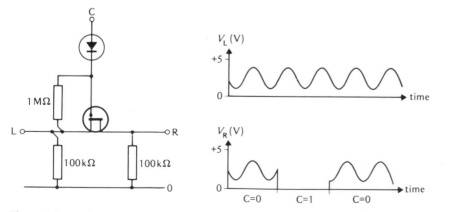

Figure 6.31 Implementing an analogue switch with a p-channel FET

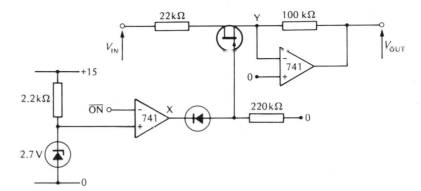

Figure 6.32 An FET controlling an amplifier

contains four analogue switches in a single 14 pin package; it is a very convenient component for marrying digital systems to analogue ones.

The resistive nature of FETs can be used to good effect when they are introduced into op-amp circuits. Figure 6.32 shows an <u>inverting</u> amplifier which can be switched on and off by a signal. Suppose that $\overline{\text{ON}}$ is a logical 0. X will be at $+13\,\text{V}$, the diode will be reverse biased and the $220\,\text{k}\Omega$ resistor will pull the gate down to $0\,\text{V}$. Since the op-amp on the right has negative feedback applied to it, point Y will be a virtual earth i.e. also at $0\,\text{V}$. Therefore $V_{\text{GS}} = 0\,\text{V}$ and the FET behaves like a small resistor. Since the $22\,\text{k}\Omega$ resistor is at least 50 times larger than the FET, V_{DS} is unlikely to get outside the range $+100\,\text{mV}$ to $-100\,\text{mV}$ as V_{IN} swings from $+3\,\text{V}$ to $-3\,\text{V}$. The op-amp behaves like an amplifier, with $V_{\text{OUT}} = -4.5V_{\text{IN}}$. But when $\overline{\text{ON}}$ is raised to logical 1 the amplifier is switched off and $V_{\text{OUT}} = 0\,\text{V}$. This is because the gate will now be at $-12\,\text{V}$ and the FET will look like an open-circuit. The op-amp on the right now behaves like a follower, so its output sits at the same voltage as its non-inverting terminal, i.e. $0\,\text{V}$.

If you use p-channel JUGFETs with pinch-off voltages of less than 5 V, you

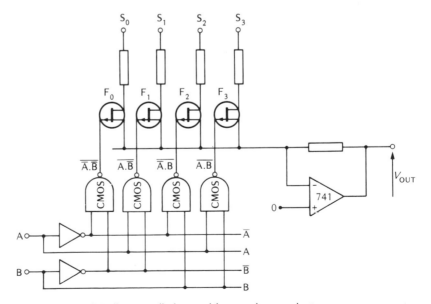

Figure 6.33 A digitally controlled one-of-four analogue selector

can control the op-amp from CMOS logic systems run off a 5 V supply. For instance, figure 6.33 shows a system which allows external logic signals to feed one of four analogue signals (S_0, S_1, S_2 and S_3) through to the output. The two inputs A and B take in an address, BA, which can range from 0 to 3. The logic system pulls the gate of one of the FETs down to logical 0, permitting one of the signals through to the output. For instance, if A is 0 and B is 1, the address is 10_2 or 2_{10}. The gate of F_2 is pulled down to logical 0, the others being held at logical 1. So $V_{OUT} = -S_2$.

EXPERIMENT

*Assembling and evaluating a voltage controlled amplifier
which exploits the variable resistance of an FET.*

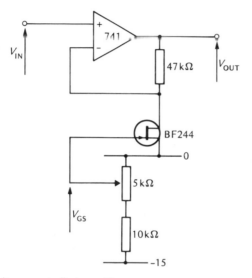

Figure 6.34 A voltage controlled amplifier

The circuit is shown in figure 6.34. Feed in a 100 mV peak-to-peak signal at 1 kHz into the amplifier. Look at the output waveform with a CRO, and measure its peak-to-peak value for values of V_{GS} from 0 V to -5 V. Use your results to plot a graph of the gain of the system as a function of V_{GS}. Explain how the system works and obtain an expression for its gain as a function of V_{GS}. Does the system behave as predicted by the model of FET behaviour?

Finally, explore the range of input signals which the system can process without distortion.

EXPERIMENT

*Designing, assembling and evaluating a two tone oscillator.
The system will produce a square wave which continuously alternates
between two frequencies, somewhat like a police siren.*

The heart of the system is the circuit shown in figure 6.35. You should be able to design the whole system if you add a Schmitt trigger, a relaxation oscillator and an inverter; use 741 based circuits throughout.

When you have got the system to work, draw the circuit diagram and carefully explain how the system works.

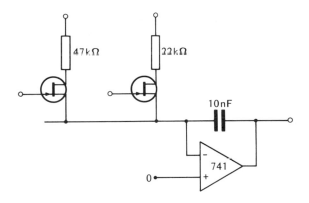

Figure 6.35 An FET controlled ramp generator

PROBLEMS

1 Draw a circuit diagram to show how a bidirectional 2 input multiplexer can be made out of analogue switches and logic gates.

2 Figure 6.36 shows a system for sending four analogue signals down a single pair of wires (the signal and ground lines). Signals from D_n are routed onto the signal line and then steered out to O_n, where n is the address fed into the address lines. The 7442 IC is a one-of-ten selector, with one of its outputs going to logical 0 when an address is fed in at D, C, B and A.

a) Explain how analogue switches of the type shown could be assembled from those on a 4066 IC and logic gates.

b) How many wires would you need in order to send the four signals without the multiplexing scheme of figure 6.36?

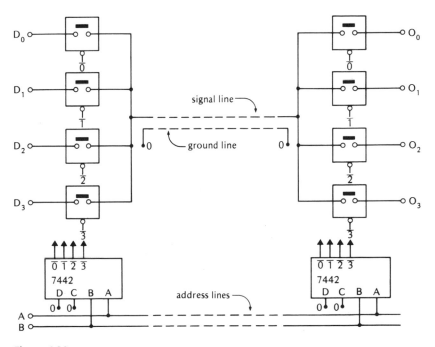

Figure 6.36

c) How many wires would you need to send 64 multiplexed signals?

d) The system of figure 6.36 can be built without using the 7442 ICs, but with rather more analogue switches and some NOT gates. Draw a circuit diagram to show how this could be done. (Hint: think of Russian dolls.)

3 Look at the relaxation oscillator of figure 6.37. Work out the relationship between the frequency of its square wave and the binary number CBA fed into the analogue switches. The Schmitt trigger is the system (using CMOS gates) shown in figure 6.38.

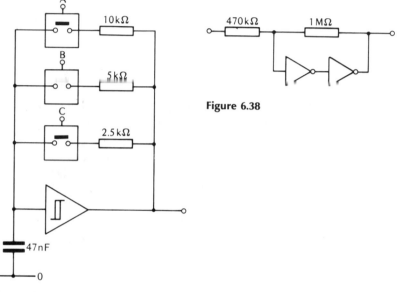

Figure 6.38

Figure 6.37

4 Figure 6.39 shows a ramp generator.

a) If V_{IN} is held at some non-zero voltage, state and explain what happens to the voltage at V_{OUT} when $V_{CONTROL}$ is held at i) $-13\,V$, and ii) $+13\,V$.

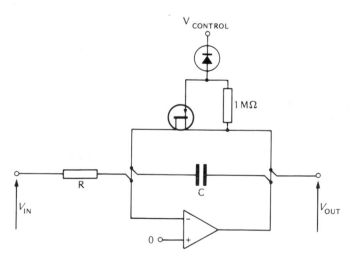

Figure 6.39

b) Draw a circuit diagram to show how the system of figure 6.39 can be used to generate a continuous sawtooth waveform which rises slowly, before rapidly returning to 0 V.

5 Figure 6.40 shows a sample-and-hold system. The circuit has three terminals. An analogue signal is fed in at one and out at the other when the third terminal is held at $+13$ V. A -13 V at the third terminal freezes the analogue output. Describe, and explain how the circuit functions.

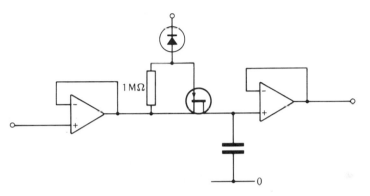

Figure 6.40

MOSFETs

We are going to finish this chapter by briefly describing the properties of metal oxide semiconductor FETs (or MOSFETs). This type of transistor is used extensively in complex ICs, principally because they can be packed close together on a chip of silicon. Most of the discussion will be centred on how the two types of MOSFET are used to make simple logic gates.

Figure 6.41 illustrates how an n-channel MOSFET (an NMOS FET) can be used as a power amplifier. When the gate is held at the same voltage as the source, the FET has an infinite resistance between drain and source. This resistance becomes very small when the gate is raised a few volts above the source. MOSFETs are available which can handle a few amps with a drain-source resistance of only a few ohms. They are very useful for allowing digital signals to control large currents as no current flows in or out of the gate at all. The gate has a layer of dielectric between it and the rest of the FET.

Note that the type of MOSFET used in figure 6.41 has no JUGFET equivalent. When V_{GS} is 0 V, a MOSFET is in its insulating state, i.e. no current can get through it. The MOSFET only conducts when V_{GS} is positive. This is exactly opposite to the behaviour of a p-channel JUGFET.

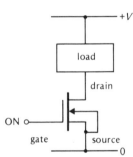

Figure 6.41 An n-channel MOSFET as a power amp

Logic gates

The behaviour of an NMOS FET will become clearer when you see how it can be used to make a NOT gate (figure 6.42). The circuit looks very similar to that of a NOT gate built around an npn transistor. When A is a logical 0, the FET has a drain current of 0; the pull-up resistor holds Q at logical 1. If A is held at logical 1, the FET conducts, its drain current generates 5 V across the resistor, and Q is pushed down to logical 0.

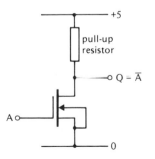

Figure 6.42 An NMOS NOT gate

Figure 6.43 illustrates an NMOS NOR gate. Once again, it is very similar to the npn version (figure 5.4), except that the pull-up resistor has been replaced with an FET. Resistors take up a lot of room on an IC, so it pays to simulate them with MOSFETs wherever possible. Q_3 will have a non-zero drain current when either or both of Q_1 and Q_2 are switched on. All of the MOSFETs will be identical as they are built at the same time, so they will only carry the same drain current if they have the same value of V_{GS}. If A is at logical 1 then V_{GS} for Q_2 is $+5\,V$. V_{GS} for Q_3 must therefore also be $+5\,V$. Since the gate of Q_3 is anchored to the $+5\,V$ supply, its source must be pushed down to $0\,V$, taking Q with it. On the other hand, if there is no drain current, V_{GS} is zero for all three MOSFETs and Q will be pulled up to logical 1. (The gate is designed to feed its signal into similar gates, none of which will ever draw any current from it. So we do not need to worry about current flowing in or out of Q.)

It should be clear that when an NMOS output is at logical 0, some current is flowing from the top supply rail to the bottom one. For instance in figure 6.43, Q_3 and Q_1 will both be in their conducting states when $A = 0$ and $B = 1$. On the other hand, when the output is logical 1 no current flows at all. Using the same example as before, when Q is logical 1 all three MOSFETs are insulators, with $V_{GS} = 0\,V$. So a logic system made out of NMOS logic gates alone will consume quite a lot of energy, although it will have high input impedances. However, by making logic gates out of mixtures of NMOS and PMOS FETs you can assemble logic systems which hardly consume any power at all.

PMOS FETs are mirror images of NMOS ones. Figure 6.44 shows one in action as part of a NOT gate. When A is a logical 1, $V_{GS} = 0\,V$, the FET behaves like an open-circuit and Q is pulled down to logical 0 by the resistor. If we now pull A down to logical 0, V_{GS} is negative, the FET conducts, a voltage appears across the resistor due to the drain current flowing through it and Q is pushed up to logical 1. A PMOS FET (p-channel MOSFET) behaves in exactly the opposite way to an n-channel JUGFET.

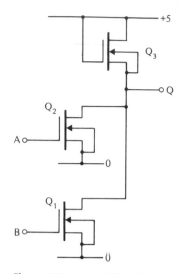

Figure 6.43 An NMOS NOR gate

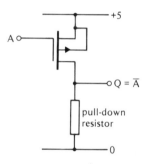

Figure 6.44 A PMOS NOT gate

CMOS logic gates

Now look at figure 6.45. It shows a NOT gate which uses a mixture of NMOS and PMOS FETs; an IC which uses this mix is known as a CMOS one, as it uses complementary MOSFETs. Here is how it works. When A is at logical 1, M_1 is an insulator and M_2 is a current source. Q will therefore be pushed down to logical 0, as a finite current will be trying to get through an infinite resistor. When Q gets low enough M_2 will stop behaving like a current source and start to look like a small resistor instead. M_1 and M_2 now look like a pair of resistors in series between the supply rails, one very large and the other very small. The net result is that Q sits at $0\,V$ and no current flows through M_1 or M_2 (remember that Q will be feeding its signal to other CMOS gates and that the gate current of a MOSFET is zero). So when A is a logical 1, Q is a logical 0. It should be obvious from the symmetry of the circuit that when A is a logical 0, Q must be a logical 1.

You may have realised that no current flows through M_1 or M_2 when Q is at logical 1 as well as when it is at logical 0. In fact, the only time that an appreciable amount of current flows is the instant when Q is on its way from one state to the other. When Q is not jammed against one of the supply rails, some current must be flowing but, as we have seen, this state is self-extinguishing. So if you have a logic system which is static, uses only

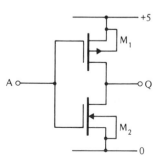

Figure 6.45 A CMOS NOT gate

CMOS gates and is not driving any output transducers, it will consume a negligible amount of current.

Even in the fastest computers, the logic gates spend most of their time sitting at logical 0 or logical 1, so CMOS gates can consume far less electrical energy than their NMOS or PMOS counterparts. In fact, CMOS systems are often used as active switches, systems which stay switched on all of the time and whose sole function is to activate other systems when they are required. The CMOS part consumes no power until it is triggered; it can then switch on the power supply to the rest of the system with a power MOSFET. Calculators with touch keyboards frequently use this sort of idea.

PROBLEMS

1 Draw the circuit diagram of an all-PMOS NOT gate i.e. with no resistors. Explain how the system works.

2 Look at the CMOS NAND gate of figure 1.58. Explain how it works.

3 Using figures 1.58 and 6.43, draw the circuit diagram of a CMOS NOR gate.

4 Work out how a CMOS tristate can be constructed.

REVISION QUESTIONS

1 Draw the circuit symbol of an n-channel JUGFET. Draw a graph to show how the drain current changes as the drain-source voltage changes for a fixed gate-source voltage. State how the graph changes as the value of the gate-source voltage is changed.

2 Explain, with a circuit diagram, how an FET can be used to make a constant current source.

3 Explain what the parameters I_{DSS}, V_P and R_S mean.

4 Draw the circuit diagram of a source follower using a single FET. Discuss its shortcomings, and show how a pair of matched FETs can make a better source follower.

5 Draw a circuit diagram of an AC amplifier employing an FET. If V_P is 4 V and I_{DSS} is 1 mA, work out the component values necessary for an AC gain of -10. In what way is an FET AC amplifier better than a bipolar AC amplifier?

6 Discuss how an FET can be used as a voltage controlled resistor. Illustrate your answer by considering a voltage controlled amplifier which contains an FET.

7 Show how an n-channel FET can be made into an analogue switch.

8 Draw the circuit diagram of a CMOS NOT gate. Explain how it works and why it draws very little current in use.

7

Transistor multivibrators

This chapter is about just three circuits, the three types of multivibrator. You have already met two of them in Chapter Two (the monostable and astable circuits), so only the third circuit, the **bistable**, will be completely new. You are going to find out how to assemble all three circuits from transistors.

Before the advent of integrated circuit logic gates, transistor multivibrators had an important place in the scheme of electronics. In practice, multivibrators constructed from logic gates work far better than their transistor equivalents. However, the transistor multivibrators can be amended to make two useful systems namely the voltage-to-frequency converter and the voltage-to-pulse-width converter. Furthermore, the transistor version of the bistable is perhaps easier to understand than its logic gate equivalent. So transistor multivibrators are worth studying, even though they are unlikely to be used much in circuits that you build.

BISTABLE SYSTEMS

As its name implies, a bistable circuit has two stable states. The simplest bistable consists of a pair of NOT gates feeding their outputs into each other, as shown in figure 7.1. In order to analyse the behaviour of the system we need to guess the state of one of the two outputs. We are forced to do this because the circuit has no inputs, the traditional starting point of circuit analysis. Having made a guess, we will then chase its consequences round the circuit until we get back to where we started.

So let $\bar{Q} = 1$. $\bar{Q}$ is the input to gate 2, so its output must have the opposite state i.e. $Q = 0$. But Q is the input to gate 1, so the output of that gate must be $\bar{0} = 1$ i.e. $\bar{Q} = 1$. We have come full circle and got back to our starting point, i.e. a value for $\bar{Q}$. Furthermore, we appear to have established that our initial guess was correct. It should be obvious that the circuit will sit with $\bar{Q} = 1$ indefinitely, that it is a stable state. So once $\bar{Q} = 1$, it will remain at this value until the power supply is switched off.

Just because we have found one stable state does not mean that it is the only one. Suppose we had guessed that $\bar{Q}$ was 0 instead of 1. If you chase

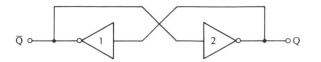

Figure 7.1 A bistable system

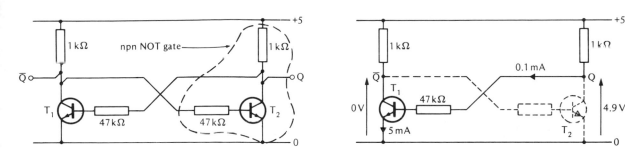

Figure 7.2 A transistor bistable

Figure 7.3 A transistor bistable with $\bar{Q}$ at logical 0

the consequences of this value around the circuit you will end up by agreeing that $\bar{Q}$ does indeed equal 0. (Try it.) So the bistable of figure 7.1 has two stable states. One state has $Q = 1$, $\bar{Q} = 0$. The other state has $Q = 0$, $\bar{Q} = 1$. The circuit is blessed with the property of having a memory; it "remembers" a particular value of Q until its power supply is disconnected. Unfortunately, there is no way of telling which state the system will adopt when it is connected to its power supply. At the instant of connection both outputs will be at 0 V. So both inputs will also be logical 0, forcing both outputs up to logical 1. The first output to get there pushes the other one back down to logical 0. Thereafter the system is stable.

TRANSISTOR BISTABLES

A transistor bistable is shown in figure 7.2. As you can see, it is simply the circuit of figure 7.1 with npn NOT gates. It is important to be aware of the criteria used for selecting the resistors in the circuit, so we shall take a detailed look at the circuit. Suppose that $\bar{Q} = 0$ and $Q = 1$ as shown in figure 7.3. To get $\bar{Q}$ down to 0 V, T_1 must be saturated. So the collector current of T_1 has to be 5 mA, enough to push its collector down to 0 V. Moving across to T_2, its base-emitter voltage will be 0 V. So T_2 must be switched off and have zero collector current. This will allow the collector resistor to pull T_1's base resistor to within 0.1 V of the top supply rail. The base current of T_1 will be $5/48 = 0.1$ mA, ample to ensure a collector current of 5 mA if the current gain is more than 50. In all transistor multivibrators you have to make sure that enough current can flow into the base of a transistor to make it saturate. In other words, the base resistor has to be smaller than the current gain times the collector resistor. (Think about it.)

Transistor flip-flops

A flip-flop is a bistable which can have its state set by a signal fed in from outside it. In other words, it is a bistable circuit which has inputs. The circuit of figure 7.2 is perfectly useless; after all, who wants to remember a random logic state? The addition of a couple of transistors converts that bistable into a flip-flop, the fundamental building block of electronic computers.

Figure 7.4 shows how a bistable is converted into a flip-flop. T_0 and T_3 are the extra transistors, otherwise the circuit is the same as the one of figure 7.2. T_1 and T_2 form the basic bistable and T_0 and T_3 are used to put the system into one or other of its stable states. The four graphs of figure

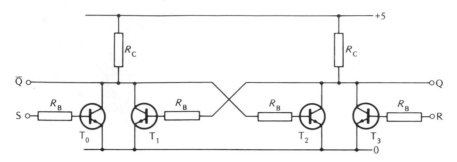

Figure 7.4 A transistor SR flip-flop

7.5 describe the behaviour of the system. They show how the states of the inputs change the outputs as time progresses. The quartet of graphs is called a **timing diagram**; we shall now use it to describe and explain the operation of the transistor flip-flop.

Initially both R and S are held at logical 0 (the far left of the timing diagram). Since we have no way of knowing the initial values of Q and $\bar{Q}$ they are drawn as being at 1 and 0 simultaneously. The convention is that when the line goes up, the state goes from 0 to 1. So up is 1, down is 0.

Then S is raised to logical 1. Referring to figure 7.4, this will saturate T_0. Its collector ($\bar{Q}$) will therefore be pushed down to 0 V. This will switch off T_2 (it will have no base current). T_3 is off anyway (R = 0) so Q rises to logical 1. Finally, T_1 will switch on and saturate because the +5 V at Q will push current into its base. The system is now stable, with T_1 on and T_2 off (as in figure 7.3).

The timing diagram shows that at the instant that S rises to 1, Q goes to 1 and $\bar{Q}$ goes to 0. It then shows that Q and $\bar{Q}$ remain in these states when S is lowered back down to 0. This is because T_1 remains switched on when T_0 is switched off. In fact, the sole function of T_0 is to pull $\bar{Q}$ down to logical 0. Once it has done this T_1 is switched on so that $\bar{Q}$ will stay at logical 0 regardless of what happens to the state of S.

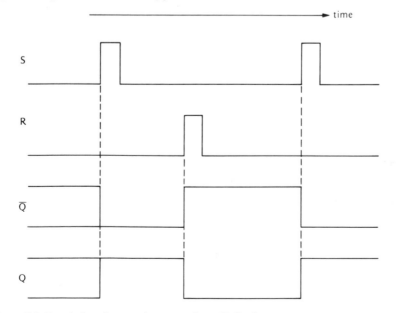

Figure 7.5 The timing diagram for a transistor SR flip-flop

Q and Q̄ remain stable until R is raised to logical 1. The timing diagram shows that Q and Q̄ quickly swap states when this happens, and stay stuck in those states. A logical 1 at R will turn T_3 on. The collector current of T_3 will pull Q down to 0V. In its turn, this switches off T_1 (it has no base current) so that Q̄ rises to +5V. Finally, this rise of Q̄ switches on T_2. This will hold Q at its new state of logical 0 when R is no longer at logical 1.

It should be clear that a logical 1 at S or R triggers the flip-flop into adopting one or other of its stable states. Once it is triggered into a state it stays there. You can think of the circuit as remembering which of the S or R terminals was last raised to logical 1.

THE SR FLIP-FLOP

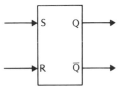

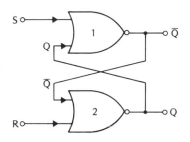

Figure 7.6 An SR flip-flop

Figure 7.7 A NOR gate SR flip-flop

An SR flip-flop is any device which has the timing diagram of figure 7.5. Its circuit symbol is shown in figure 7.6. The two input terminals are given special names. The S or SET input sets the flip-flop, i.e. makes Q = 1. The flip-flop is reset (Q = 0) with the R or RESET input. Of course, the states of Q and Q̄ are different at all times. The standard method of building an SR flip-flop is shown in figure 7.7. The outputs of a pair of NOR gates are fed back into each others inputs; the two spare inputs are the S and R terminals. The circuit is the same as that of the transistor flip-flop. Look at figure 7.4. T_0, T_1, their base resistors and the collector resistor make up a transistor NOR gate (figure 5.4). So figure 7.4 is just a pair of connected NOR gates.

What do the methods of Boolean algebra tell us about the behaviour of the SR flip-flop? Consider gate 1. We can use the standard NOR relationship to get Q̄ as a function of S and Q:

$$\bar{Q} = \overline{S + Q}$$

This tells us that Q̄ will be 0 whenever S is 1. Furthermore, Q̄ will stay at 0 if Q stays at 1. That is, if the output of gate 2 is a logical 1 the output of gate 1 must be a logical 0.

Now for the second gate. The relationship between its inputs and the output is given by:

$$Q = \overline{R + \bar{Q}}$$

So Q is 0 whenever R is 1, otherwise it will be 0 provided that Q̄ is 1.

The behaviour of the system can be summarised in the form of a **state diagram** (figure 7.8). The system has two states which can be specified by their value for Q. Each circle in the state diagram contains a value of Q. It therefore represents one of the states. The diagram shows that to push the system from Q = 1 to Q = 0 you have to arrange for R . S̄ = 1 (i.e. R = 1 and S = 0). Similarly, to flip the system back to its original state you need S . R̄ = 1. When S̄ . R̄ = 1 (both S and R = 0) the system is stable and the two states feed themselves. (The fourth input state, R . S = 1, is one to be

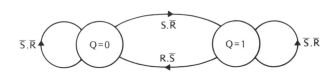

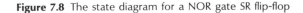

Figure 7.8 The state diagram for a NOR gate SR flip-flop

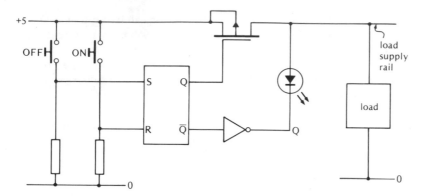

Figure 7.9 An SR flip-flop in action

avoided in flip-flop applications. It results in both outputs being at logical 0, not a very useful output state.)

USING FLIP-FLOPS

If you combine several SR flip-flops you can make a device called the D-type flip-flop. The whole of Chapters Nine, Ten and Eleven are concerned with systems based on D flip-flops. So for the present we shall only consider how an SR flip-flop can be used in its own right, rather than as a sub-unit of a more complex flip-flop.

Flip-flops are used for remembering short-lived digital signals. Look at the circuit of figure 7.9. Imagine that the ON switch is pressed and released. This pulls R up to logical 1 for an instant and resets the flip-flop. Q goes to logical 0 and stays there. This pulls the gate of the MOSFET below its source so that it becomes a short-circuit. The drain of the MOSFET acts as the top supply rail for the load, which could be some electronic system. So power is switched through to the load when ON is pressed and released. The LED also comes on to show that the power has been switched on. The MOSFET will stay on until OFF is pressed. This pulls S high, sets the flip-flop so that Q goes back up to 1 and puts the MOSFET into its insulating state.

The circuit is interesting for several reasons. The two switches only have to carry very small currents so they can be light and cheap; the MOSFET does all the high current switching. The flip-flop and switch/resistor networks have to be connected to the supply all of the time. If they are run off a mains-derived power supply this is no problem, but it might make it difficult for battery operated systems. You will recall, however, that CMOS NOR gates will consume no current if their outputs are static. So you could leave a CMOS flip-flop connected to its supply all the time without worrying about draining the battery. Finally, you could use digital pulses to trigger the flip-flop instead of push switches. Thus a brief logical 1 into R would establish a permanent logical 0 at Q.

EXPERIMENT

Verifying the properties of an SR flip-flop

Assemble the circuit shown in figure 7.10, using TTL NOR gates. Verify that the system behaves as described in the last few sections. Remember that a lit LED will indicate a logical 0.

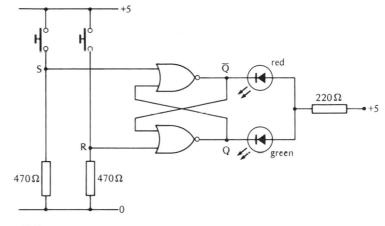

Figure 7.10

EXPERIMENT

Investigating the circuit of figure 7.11

Look at figure 7.11. There are two circuits, labelled A and B. B runs a pair of LEDs, one called heads, the other called tails. When the input to B (ENABLE) is held at logical 0, the LEDs will come on and off rapidly, taking it in turns to be lit. When ENABLE is raised to logical 1 the state of the LEDs is frozen so that one of them will be permanently lit. Circuit A must be made out of CMOS gates. Its output (OUT) controls circuit B. Its two inputs (T and U) can be pulled down to logical 0 by touching the contacts at X and Y with a finger (skin resistance is less than 100 kΩ). When X is bridged by a finger the LEDs flash on and off. The instant that Y is bridged the LEDs stop flashing.

The whole circuit acts like a coin being tossed in the air. You briefly touch X to throw the coin. When you think it has spun for long enough you press Y momentarily to "catch" the coin. The LEDs display whether it landed heads or tails up.

Design and assemble block A. Check that it works. Then assemble block B and get the whole system to function as described. Draw the complete circuit diagram and explain how it works.

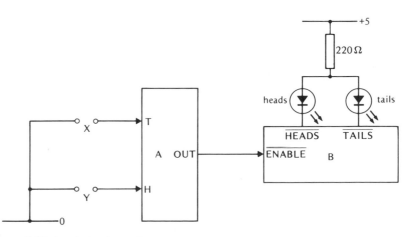

Figure 7.11 A coin tossing system

PROBLEMS

1 Draw the circuit diagram of a bistable based upon

   a) relays

   b) pnp transistors

   c) op-amps

2 Draw the circuit diagram of a flip-flop based upon pnp transistors. State and explain how the system can be set and reset. Draw a timing diagram for the flip-flop.

3 Figure 7.12 shows a flip-flop, a partial truth table and an unfinished timing diagram. The truth table shows how $\overline{S}$ and $\overline{R}$ are varied as time progresses. The first line shows the initial state of the system; work out the subsequent values of Q and $\overline{Q}$. Finally, copy and complete the timing diagram.

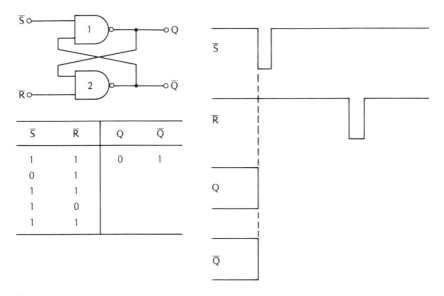

$\overline{S}$	$\overline{R}$	Q	$\overline{Q}$
1	1	0	1
0	1		
1	1		
1	0		
1	1		

Figure 7.12

4 A pair of simple push switches can be used to set and reset the bistable of figure 7.2. Draw a circuit diagram to show where the switches would have to be placed. Explain their action on the flip-flop. (There are two ways in which it can be done!)

5 An LDR is used to discourage the theft of a valuable painting. It is placed on the wall behind the painting so that it is kept in the dark. The LDR is connected to an electronic circuit which contains a loudspeaker. Whenever the LDR is exposed to light the loudspeaker emits a noise until a concealed switch is pressed. The alarm must continue to sound even if the LDR is covered up again. Devise a suitable electronic circuit and explain how it works.

MONOSTABLES

We are going to assume that you are already familiar with the monostable systems discussed in Chapter Two. For most practical purposes monostables based on logic gates are easier to build than their transistor counterparts but, as we shall see, their pulse widths are not so predictable. A transistor monostable has eight discrete components, whereas an integrated circuit one has only three. Nevertheless, a transistor monostable is useful for several reasons. It is not fussy about its power supply rails, it can be adapted so that the pulse width can be controlled by an external signal, and the trigger threshold is adjustable.

TRANSISTOR MONOSTABLES

Figure 7.13 shows a monostable built out of a NOR gate and a NOT gate. Its behaviour is summarised in the timing diagram. When T is raised to logical 1 the system is triggered into holding Q at logical 1 for a fixed time. If CMOS logic gates are used, the width of the pulse will be about $0.7RC$. This will only be approximate because there is quite a wide spread of 1/0 thresholds in the manufacture of CMOS ICs.

We are going to replace each of the gates with transistor circuits, so that we end up with an all-transistor monostable. The transistor logic gates we are going to use are not quite the same as the ones that you met in Chapter Five, so we shall discuss their properties first.

The transistor NOR gate is shown in figure 7.14. It is going to be used instead of the standard NOR gate (figure 5.4) because it only uses one transistor. If the circuit is a NOR gate, C_1 has to be a logical 0 when either, or both, the inputs (T and Q) are held at logical 1. Now, $+5V$ at either T or Q will saturate T_1 if its current gain is more than 50. C_1 only goes up to logical 1 when both T and Q are held at 0 V to switch T_1 off. So the system does indeed behave like a NOR gate.

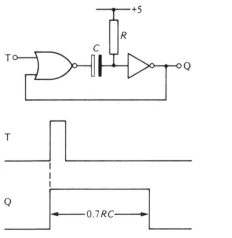

Figure 7.13 A rising edge triggered monostable

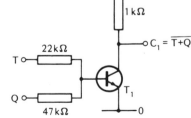

Figure 7.14 A transistor NOR gate

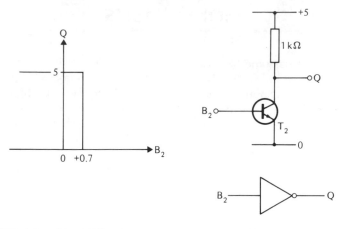

Figure 7.15 A transistor NOT gate

The transistor NOT gate is a bit special too. Look at figure 7.15. Any signal at B_2 which is less than $+0.7\,V$ will switch T_2 off. This allows the $1\,k\Omega$ resistor to pull Q up to $+5\,V$. A signal which attempts to push B_2 above $+0.7\,V$ will switch on T_2 and saturate it. Q will be pushed down to $0\,V$, and B_2 will be clamped at $+0.7\,V$. (Remember that the base-emitter junction behaves like a diode.) So the circuit is a NOT gate which recognises $+0.7\,V$ as logical 1, and any signal below that as logical 0.

When you insert the circuits of figure 7.14 and 7.15 into that of figure 7.13 you get an all-transistor monostable. It is shown in figure 7.16. Although we know what Q will do when T is raised to logical 1 (see the timing diagram of figure 7.13), we need to take a detailed look at the waveforms at C_1 and B_2 in order to calculate the pulse width.

How a transistor monostable works

The four graphs of figure 7.17 summarise the behaviour of a transistor monostable when it is triggered.

Before it is triggered, Q will sit at $0\,V$. For this to happen, R must allow enough base current to flow into T_2 to allow it to saturate. Of course, if T_2 is saturated, B_2 must sit at a steady $+0.7\,V$. Going across to T_1, it must be switched off as both Q and T are at $0\,V$. So C_2 will be pulled up to $+5\,V$ by the $1\,k\Omega$ resistor. The left hand plate of C will be $4.3\,V$ higher than the right hand one.

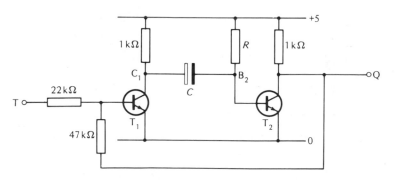

Figure 7.16 A transistor monostable

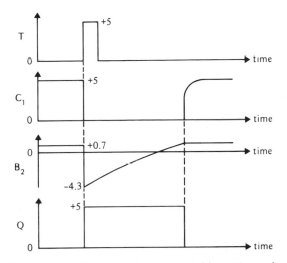

Figure 7.17 What happens when a transistor monostable is triggered

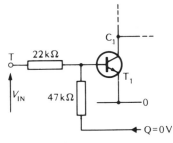

Figure 7.18

To trigger the circuit, T_1 has to be switched on. Look at figure 7.18; it shows T_1 and its two base resistors just before it switches on. V_{IN} is the signal fed in at T, and it has pulled the base of T_1 up to $+0.7$ V. If T_1 has not switched on yet, we do not need to worry about its base current. So we can use the voltage divider formula to find the value of V_{IN}:

$$0.7 = \frac{V_{IN} \times 47}{22 + 47}$$

therefore $V_{IN} = \dfrac{0.7 \times 69}{47} = 1.0$

So T_1 will not switch on until T has been raised above $+1.0$ V.

Once the circuit has been triggered a number of voltages change rapidly. The train of events is set into motion by switching on T_1. Its collector current pushes C_1 down towards 0 V. The capacitor plates stay 4.3 V apart because there is no time for the charge on them to change. So as C_1 goes down it pushes B_2 down below $+0.7$ V. This switches off T_2 immediately; Q rises straight up to $+5$ V, pulled up by the 1 kΩ resistor. The signal at Q is fed back to T_1 via the 47 kΩ resistor. As Q rises, it feeds more base current into T_1 so that it definitely saturates. So C_1 is inexorably pushed down to 0 V, and B_2 goes all the way down to -4.3 V.

Note the way in which the feedback from T_2 to T_1 ensures that the latter is saturated rapidly once it has been switched on. This is why we say that a 1.0 V signal at T triggers the circuit into action.

Now we will consider what subsequently happens to the voltage at B_2. C_1 remains at a constant 0 V, so the voltage of the left-hand plate of the capacitor does not change. The resistor R slowly pulls the right-hand plate (and B_2) up towards the top supply rail. While B_2 is less than $+0.7$ V there will be no base current into T_2, so we can use the capacitor discharge formula:

$$t = RC \log_e (V_0/V)$$

The initial voltage across R is $5 + 4.3 = 9.3\,$V. Just before T_2 is switched on again, the voltage across R will be $5 - 0.7 = 4.3\,$V. So we can calculate how long T_2 is switched off:

$$t = RC\log_e(9.3/4.3) = 0.77RC$$

When B_2 gets back up to $+0.7\,$V it sets off another series of changes around the circuit. First of all, T_2 is switched on. So Q falls towards $0\,$V. This is fed back to T_1 via the $47\,$kΩ resistor. If T is held at $0\,$V, a drop in Q is going to eventually start to turn off T_1. So C_1 will start to rise, pulling B_2 up with it. This switches T_2 firmly on so that Q is firmly pushed down to $0\,$V.

Note, once again, how the feedback quickly forces Q towards its final value once it has started to change.

Finally, once T_2 has switched on and T_1 has been switched off, C_1 is pulled up to $+5\,$V by the $1\,$kΩ resistor. Since B_2 is clamped at $+0.7\,$V by the base of T_2, charge is going to have to flow onto the capacitor as C_1 rises. So there is a delay of about $3 \times 1 \times C$ before C_1 gets up to $+5\,$V. The circuit is now stable, and will sit quietly until the next trigger pulse is fed in at T.

EXPERIMENT

Investigating a monostable

Assemble a monostable shown in figure 7.19. You are going to choose the values of R and C such that the pulse fed out at Q is about 10 seconds long. Bear in mind that R must be small enough to ensure that T_2 is saturated.

Use the potentiometer to trigger the circuit; check that the pulse width and trigger threshold agree with their theoretical values. Use a voltmeter to look at the waveforms at C_1 and B_2 after the system has been triggered.

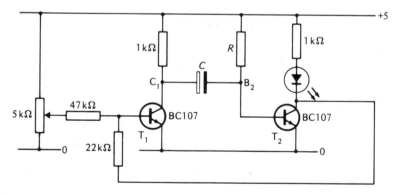

Figure 7.19

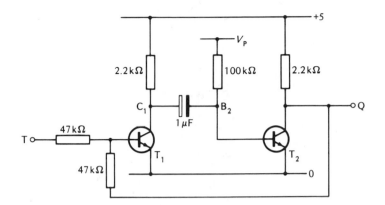

Figure 7.20 A variable pulse width monostable

Altering the pulse width

Look at figure 7.20. It is a standard transistor monostable, run off a 5 V supply, with the base resistor of T_2 held at V_P instead of the supply voltage. V_P is a signal which controls the width of the pulse fed out at Q.

Suppose that $V_P = +5$ V i.e. the system is a standard monostable of the type shown in figure 7.16. A little thought will convince you that the pulse width is given by:

$$t = RC \log_e \left(\frac{V_0}{V}\right)$$

$$= 100 \times 1 \times \log_e \left(\frac{5 + 4.3}{5 - 0.7}\right)$$

$$= 100 \times 1 \times 0.77$$

$$= 77$$

So with V_P held at $+5$ V, the pulse width will be 77 ms.

If we raise V_P up to $+10$ V, the pulse width will get smaller. More precisely, it will be given by:

$$t = 100 \times 1 \times \log_e \left(\frac{10 + 4.3}{10 - 0.7}\right)$$

$$= 100 \times \log_e(1.5)$$

$$= 43$$

The pulse width is now 43 ms.

Figure 7.21 shows how the pulse width depends on the value of V_P. There is no upper limit to the value of V_P, but there is a lower limit. V_P must be big enough to allow T_2 to saturate. If we assume a current gain of 150 for T_2, it needs a base current of at least 0.015 mA (work it out). So V_P must be greater than $0.015 \times 100 + 0.7 = 2.2$ V.

In subsequent chapters you will find out how this sort of monostable can be used, particularly for communication purposes.

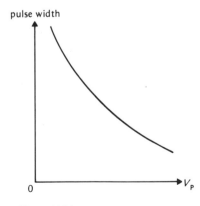

pulse width

Figure 7.21

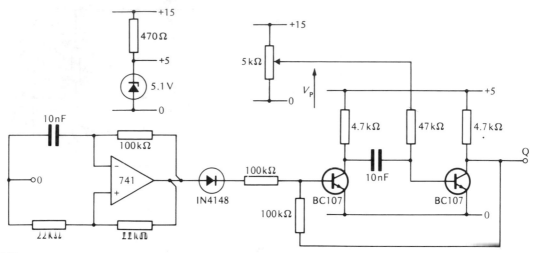

Figure 7.22

EXPERIMENT

*Assembling and evaluating a monostable whose pulse width
can be controlled with an external signal*

The circuit is shown in figure 7.22. On the left is a relaxation oscillator
which triggers the monostable 500 times a second. A potentiometer can be
used to alter the value of V_P. Note the power supply arrangements.

Trigger a CRO on the rising edge of the square wave coming out of the
oscillator. Look at the waveform at Q, and see what happens when V_P is
varied.

Measure the width of the pulse at Q for several values of V_P. Use them
to draw a graph of pulse width as a function of V_P.

Finally, use the op-amp follower and loudspeaker shown in figure 7.23 to
listen to the waveform at Q.

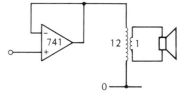

Figure 7.23

PROBLEMS

1 This question refers to the monostable circuit shown in figure 7.24.

   a) What is the minimum voltage at T that will trigger the system?

   b) Calculate the pulse width when V_P is held at i) $+5\,V$, ii) $+10\,V$, and
   iii) $+15\,V$.

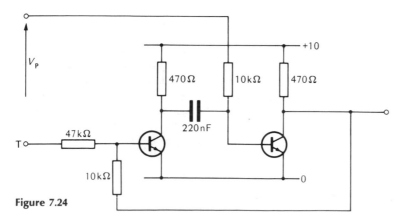

Figure 7.24

2 Draw the circuit diagram of a transistor monostable which has the following characteristics. It is run off a 12 V supply, has a pulse width of 5 ms, is triggered by any voltage above +6 V and has transistors which cannot have a collector current of more than 100 mA, and which have an h_{FE} of 150.

3 Draw the circuit diagram of a pnp transistor monostable. Draw a series of graphs, one under the other and to the same scale, to show what happens to the voltages in the circuit when it is triggered. (It may help to look at figures 7.16 and 7.17 while standing on your head!)

ASTABLES

Once again, we are going to assume that you are familiar with the astables built out of logic gates which were dealt with in Chapter Two. So you will recognise the system shown in figure 7.25, and appreciate that it can oscillate. The behaviour of the astable is summarised in the timing diagram. Note that when gate 2 falls from 1 to 0 this forces gate 1 up to 1 for $0.7R_1C_1$ milliseconds. Similarly, whenever gate 1 falls from 1 to 0 gate 2 is pushed high for $0.7R_2C_2$ milliseconds.

TRANSISTOR ASTABLES

A transistor astable can be created by replacing the two NOT gates of figure 7.25 with their transistor equivalents. We have to use the NOT gate

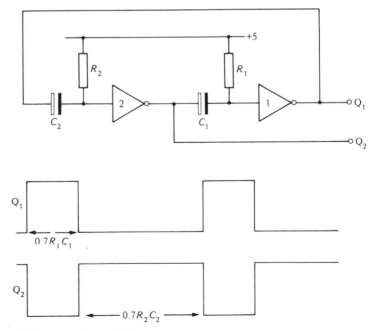

Figure 7.25 An astable multivibrator

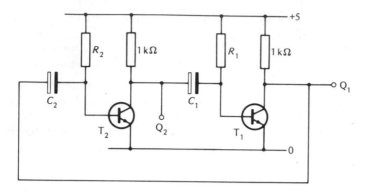

Figure 7.26 A transistor astable

of figure 7.15 as we need a gate with an input which is clamped. The result is shown in figure 7.26.

We are not going to describe how the circuit works in great detail because it is nearly the same as the logic gate astable. Instead we will point out a few important points about its operation.

To start with, the system does not have to oscillate. When both transistors are saturated ($Q_1 = Q_2 = 0$), the system is stable. To start it oscillating you have to switch one of the transistors off momentarily; this can be done by shorting one of the bases to ground. For example, suppose that the base of T_2 is pulled down to $0\,V$ for an instant. Its collector rises to $+5\,V$ and then falls again when the base is released. That rapid fall of the collector of T_2 pushes the base of T_1 down to $-4.3\,V$. So T_1 is switched off, and remains off for $0.77R_1C_1$ milliseconds. (Exactly the same system is used for timing the pulse in a transistor monostable; convince yourself by comparing figures 7.26 and 7.16.) When T_1 switches on again, the drop in its collector voltage is fed back to the base of T_2 etc.

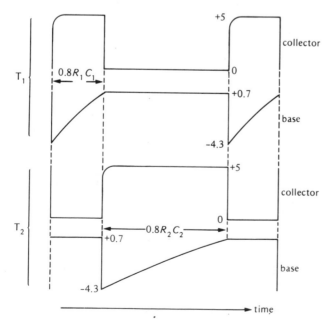

Figure 7.27 How a transistor astable works

The four graphs of figure 7.27 show how the voltages at the base and collector of each transistor change with time once the astable has started oscillating. Initially, T_2 has just been switched on, so that its collector falls rapidly from $+5\,V$ to $0\,V$. Work out the consequences of this around the circuit, and satisfy yourself that the graphs do indeed show all of the subsequent changes of voltage.

The astable has no inputs, so there is no reason why the circuit has to be drawn with signals travelling from left to right. The signals within an astable travel in a continuous loop, with outputs taken off at diametrically opposite points. It is customary to reflect this symmetry in the circuit diagram by having one output on the left and the other on the right. This is shown in figure 7.28. The arrangement of the transistors is similar to that used for the bistable of figure 7.2. Indeed, the only difference is that DC signals can travel around the bistable and only AC ones can go around the astable.

Voltage controlled frequency

Since the timing within a transistor astable is done the same way as in a transistor monostable, we can carry over the method of altering the timing with an external signal. Look at figure 7.29. The astable shown there will generate a square wave going from $0\,V$ to $+5\,V$.

The frequency of the square wave will depend on the voltage V_F. Suppose that $V_F = 5\,V$; i.e. the system looks like the standard astable of figure 7.28, with the base resistors going up to the top supply rail. Then each transistor will be switched off for a time t given by:

$$t = RC\log_e\left(\frac{V_0}{V}\right)$$

$$= 22 \times 0.1 \times \log_e\left(\frac{5 + 4.3}{5 - 0.7}\right)$$

$$= 2.2 \times \log_e 2.2$$

$$= 1.7$$

One cycle of the square wave will therefore take $2 \times 1.7 = 3.4\,ms$. The frequency must be $1/3.4 = 0.29\,kHz$.

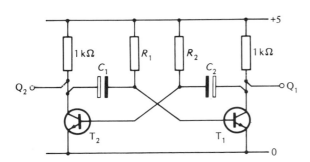

Figure 7.28 Another way of drawing a transistor astable

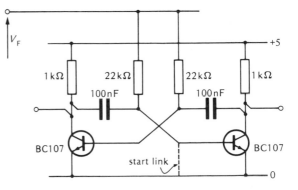

Figure 7.29 A voltage-to-frequency converter

Now haul V_F up to $+15\,V$. Running through the calculation again:

$$t = 22 \times 0.1 \times \log_e \left(\frac{15 + 4.3}{15 - 0.7} \right)$$

$$= 2.2 \times \log_e 1.35$$

$$= 0.66$$

The frequency will now be $1/(2 \times 0.66) = 0.76\,kHz$.

The system is sometimes described as a voltage-to-frequency converter. It takes in an analogue signal (V_F) and pumps out an AC signal whose frequency is a function of the analogue signal. Transmitting and storing DC analogue signals directly is very difficult. Doing the same for AC signals is easy, as you will find out later. A voltage-to-frequency converter takes an awkward DC signal and makes it into an easy AC one. For example, suppose that a thermistor is being used to monitor the temperature in a building. If it is made into part of a voltage divider, it will feed out a slowly varying DC signal. That signal can be buffered and used to control an astable like the one shown in figure 7.29. The result is an AC signal whose frequency depends on the temperature. That signal could be recorded on magnetic tape (i.e. stored), or it could be transmitted by radio to some central monitoring station. We shall be looking at the uses of voltage-to-frequency converters in more detail in Chapter Thirteen.

EXPERIMENT
Investigating an astable

Assemble the astable of figure 7.29. Run it off a $0\,V$ and $5\,V$ supply; use the zener diode/resistor network of figure 7.22 to get the $+5\,V$ from $+15\,V$. Use a $5\,k\Omega$ potentiometer to generate values of V_F betwen $0\,V$ and $15\,V$.

Short a transistor base to ground to set the system oscillating. Use a CRO to look at the waveforms at the base and collector of each transistor when $V_F = +5\,V$. Verify that they agree with the waveforms shown in figure 7.27.

Measure the frequency of the output waveform for values of V_F between $+3\,V$ and $+15\,V$. Draw a graph of the frequency as a function of V_F.

Finally, use the arrangement of 7.23 to listen to the output waveform. Design a triangle waveform generator based on 741s which will wobble V_F up and down between $+3\,V$ and $+12\,V$; when you have got it to work you ought to get an interesting noise out of the speaker! Draw the circuit diagram of your final system. Say what it does and explain how it works.

PROBLEMS

1 This question refers to the astable shown in figure 7.30. If $V_\Gamma = V_S -$ $+5\,V$, draw a graph to show how the voltage at Q changes with time. Label the time and the voltage axes.

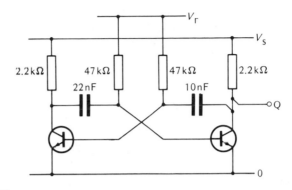

Figure 7.30

2 Once again, this one concerns the astable of figure 7.30. V_F has the same value as V_S, but V_S can vary from $+5\,V$ to $+15\,V$. Calculate the frequency of the output waveform for $V_S = 5\,V$, $10\,V$ and $15\,V$ in turn.

3 This is the final question concerned with figure 7.30. V_S is fixed at $+10\,V$. V_F can go up to $+30\,V$.

a) How far down can V_F go before the astable stops oscillating? ($h_{FE} = 150$ for both transistors.)

b) Calculate the frequency of the output waveform for several values of V_F and use them to draw a graph of frequency against V_F.

4 Draw the circuit diagram of an astable which has the following characteristics. The transistors have a current gain of 100 and can take a maximum collector current of 50 mA. The output must be a 2 V square wave with a frequency of 10 kHz.

5 Draw the circuit diagram of an astable made from pnp transistors. Draw four graphs, one under the other and to the same scale, to show how the voltages at the collector and base of each transistor change with time.

6 If you add a third npn transistor to the circuit of figure 7.28 you can feed in a signal which will start, or stop, the system oscillating. Draw a circuit diagram to show how to do this, and explain how it works. (Hint: look at figure 7.4.)

REVISION QUESTIONS

1 Draw a bistable constructed from a pair of NOR gates. Describe its properties, and draw some timing diagrams to illustrate them.

2 Show how an SR flip-flop can be built out of npn transistors.

3 Draw the circuit diagram of a transistor monostable. Sketch graphs to show how the voltages at various places in the circuit change when the circuit is triggered by a rising edge. Use them to explain how the system works and why the pulse width is roughly 0.77RC when it is run off a 5 V supply.

4 A monostable has to be triggered when its input goes above +2.5 V. The output pulse must be 50 ms long. If the system is run off a 5 V supply and uses transistors with a minimum h_{FE} of 100 and a maximum collector current of 200 mA, draw its circuit diagram showing all of the component values.

5 Draw a transistor astable multivibrator. Sketch graphs to show how the voltages at various places change with time. Use them to explain why the system oscillates.

6 Show how an astable's frequency can be controlled by an external signal.

7 Draw the circuit diagram of an astable which feeds out 100 μs pulses that are 5 V high at intervals of 1 ms. Show all component values, with reasons.

8

Audio systems

An **audio system** is a circuit which processes AC signals that have frequencies between 16 Hz and 16 kHz. It handles signals which, when converted to sound by a loudspeaker, can be heard by the human ear. Alternatively, you can think of an audio system as a device which is designed to handle the signals generated by a microphone.

For example, consider the circuit shown in figure 8.1. The audio system shown there consists of three main parts. On the left is the microphone, a transducer which converts sound waves into AC signals. The transistor amplifier in the centre amplifies the audio signal, making it about 150 times larger. Finally, on the right, there is a loudspeaker which converts the audio signal into a sound wave.

The function of an audio amplifier is to make an enlarged, but faithful, copy of the signal which is fed into it. The signals which emerge from the microphone contain all of the information that is needed to recreate the original sound wave. If the loudspeaker is matched to the microphone, it will behave like the microphone run backwards. Then the sound wave which emerges from the speaker should sound exactly the same as the one which was originally picked up by the microphone. Take a look at figure 8.2 (overleaf). The microphone picks up the sound wave, sends an AC signal down the line and causes the speaker to generate an identical sound wave. The role of an audio amplifier is to boost the signal on its way from microphone to speaker so that an adequate volume of sound emerges from the speaker.

There are several different types of microphone, each with different characteristics. The simplest is just a loudspeaker. Its only advantage is its low

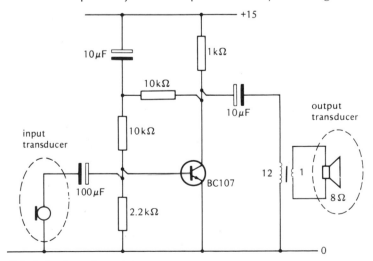

Figure 8.1 A simple audio system

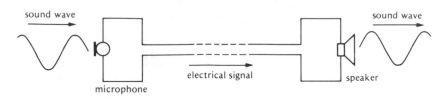

Figure 8.2 A microphone driving a loudspeaker

output impedance; indeed, because of this it is suitable for driving the amplifier of figure 8.1. (It has a low input impedance, about $0.5\,k\Omega$.) Crystal microphones have a higher output signal, but this goes with a higher output impedance. Of course, each type of microphone reacts differently to the sound waves which impinge on it. Capacitor microphones give the best performance, but they have enormous output impedances.

The situation is even worse when it comes to loudspeakers. The majority of speakers have the same construction, i.e. are moving coil types, but their characteristics are heavily dependent on how they are mounted and the room that they are placed in. Often, more than one speaker is used to cover the audio range, with one speaker handling the low frequency signals and a smaller one dealing with the high frequencies.

In this chapter we are going to concentrate on the electronic aspects of audio systems. We will have to assume that the microphones and loud-speakers which feed signals in and out of the electronics are ideal.

IDEAL AUDIO AMPLIFIERS

An ideal audio amplifier must have three properties. It must not **distort** the signals which it processes. It must not add any **noise** to the signal. Finally, it must not alter the **frequency balance** of the signal.

In order to appreciate what these three requirements mean, it is instructive to look at a non-ideal amplifier in detail e.g. the amplifier of figure 8.1. We shall deal with distortion first. Look at the graphs of figure 8.3. They

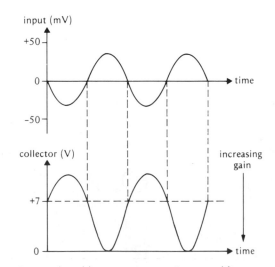

Figure 8.3 Distortion produced by a common-emitter amplifier

compare the voltage at the collector with the signal fed out of the microphone. Notice that the two waveforms have a different shape, so that the output of the amplifier is not an exact copy of the signal fed into it. The amplifier has distorted the signal.

In order to understand the cause of the distortion we need to look at the amplifier in detail. The signal from the microphone is fed directly into the base of the transistor. (Note the bias arrangements; they force the collector to sit at $10 \times 0.7 = +7\,V$.) As the base is wobbled up and down, the collector current wobbles up and down in sympathy. It can be shown, using a sophisticated model of the transistor, that the AC gain of the amplifier is about -150, proportional to the collector current. This means that the positive peaks of the signal are amplified more than the negative ones.

The term noise is used to describe the signal which comes out of the amplifier when no signal is fed into it. If the microphone is removed, the bias resistors and the coupling capacitor are the only signal sources for the transistor. Each of those components will produce small voltage fluctuations which will be amplified and fed into the speaker; the resultant sound is a mixture of hisses, crackles and random swishes. All electronic components generate noise, but some components are much noisier than others.

The amplifier should be able to handle signals in the region of 16 Hz to 16 kHz. It should amplify signals in this range by an equal amount, so that the frequency balance of the signal is preserved. This is certainly not the case for the amplifier which we are considering. The transformer will not be equally efficient at all of the frequencies, and signals below 40 Hz will be attenuated by the coupling capacitor at the input. So the low frequency signals will not be amplified as much as the others.

AUDIO SIGNALS

Before we proceed to show how the defects of audio amplifiers can be eliminated, we need to explain some of the properties of audio signals and how the human ear responds to sound waves. In general, the signals that are processed by an audio system are very complex. They have waveforms which have intricate shapes and which change rapidly with time. Despite this complexity it is possible to describe the important features (as far as the listener is concerned) of an audio signal with a graph known as a **frequency spectrum**. The secret of generating the frequency spectrum of a signal is an appreciation of what happens when sine waves are added to each other.

Sine waves

The simplest audio signal is a sine wave. One of these is shown in figure 8.4 (overleaf). As you can see, it consists of cycles which are repeated over and over again. The wave can be represented by some algebra:

$$V = A \cos(2\pi ft + \phi)$$

(Do not worry that a cosine function is used to describe a sine wave!) The amplitude of the wave is A, sometimes called its peak value. The frequency of the wave (f) is the number of cycles the wave goes through in one second. Time is denoted by the symbol t. The last symbol (ϕ) is the **phase** of the wave. It tells you what the size of the wave was at $t = 0$; more precisely, when $t = 0$, $V = A \cos(\phi)$. The phase, of course, is measured in

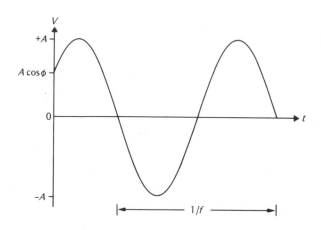

Figure 8.4 A sine wave

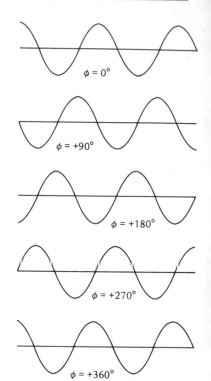

Figure 8.5 A sine wave with different phase shifts

degrees. By changing the phase of the signal you shift backwards or forwards along the time axis. Figure 8.5 shows the effect of increasing the phase in steps of 90°. Note that a change of 180° changes the sign of the wave, and that a change of 360° has no effect at all.

Signal synthesis

Sine waves are very special because any AC signal can be made by adding enough of them together. Any signal which varies with time and has an average value of zero can be represented by a sum of sine waves with appropriate amplitudes and phases. For example, a square wave of amplitude A and frequency f is equal to:

$$A \times \frac{4}{\pi} \times \left[\cos(2\pi ft) - \frac{\cos(2\pi 3ft)}{3} + \frac{\cos(2\pi 5ft)}{5} - \frac{\cos(2\pi 7ft)}{7} + \cdots \right]$$

Figure 8.6 shows what happens when the first three components of the series are added together. In order to get a true square wave you need an infinite number of sine waves.

A convenient way of showing which sine waves go to make a complex waveform is with a frequency spectrum. Figure 8.7 shows the frequency spectrum of a 1 kHz square wave. As you can see, there are components at 1 kHz, 3 kHz, 5 kHz etc., and that the amplitude of the components gets progressively smaller as their frequency gets higher.

Figure 8.6 Adding three sine waves to make a rough square wave

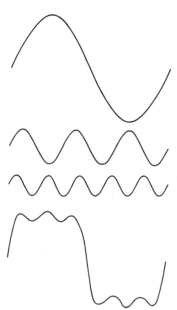

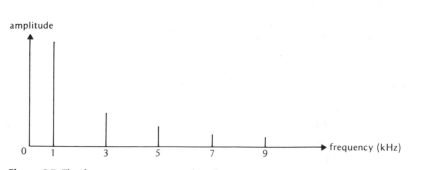

Figure 8.7 The frequency spectrum of a 1 kHz square wave

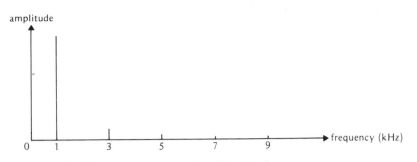

Figure 8.8 The frequency spectrum of a 1 kHz triangle wave

Figure 8.8 shows the frequency spectrum of a triangle wave. It looks very similar to that of a square wave, with components at 3, 5, 7 etc. times the fundamental frequency, but they drop much more rapidly as the frequency goes up.

Detecting sound

The frequency spectrum of a signal is particularly useful because it shows what the human ear detects when it hears a sound wave. Our ears are built to detect sine waves. For example, suppose that a 3 kHz square wave is fed into a perfect loudspeaker. When the resulting sound wave sets our eardrums vibrating, the receptors in our inner ear detect three sine waves. There will be a fundamental sine wave at 3 kHz, and two harmonics at 9 kHz and 15 kHz; they will have relative amplitudes of $1:\frac{1}{3}:\frac{1}{5}$. (All the other components lie outside the audible range.) This combination of frequencies and amplitudes is recognised as a 3 kHz square wave by the brain. Indeed, if a signal is synthesised from three sine waves with the right frequencies and amplitudes, the result will sound exactly the same as a 3 kHz square wave.

You may have noticed that a frequency spectrum contains no information about the phases of the component sine waves. As far as an audio signal is concerned the phase is not important. This is because the human ear cannot distinguish the relative phase of two sine waves, only their relative amplitudes. The following experiment demonstrates this. A pair of signal generators feed out two signals at 1 kHz and 2 kHz respectively. The signals are added together and fed into a loudspeaker; a CRO looks at the signal fed into the loudspeaker. If the 2 kHz signal is not exactly twice the frequency of the 1 kHz one then the relative phase of the signals will slowly change with time. Figure 8.9 shows some of the complex waveforms that will be fed into the loudspeaker. Although the waveforms are all different, they sound identical. The listener cannot respond to the change of phase; all he or she detects is the two frequencies and their amplitudes.

The inability of the human ear to measure the phase of a sine wave is very fortunate as far as amplifiers are concerned. Some of the remedies applied to amplifiers to cure the defects of audio systems alter the phases of the frequency components. If human beings were able to detect phase, some of the ailments of audio systems would be virtually incurable.

Response of the ear

An audio signal is generally made up of components which can be sorted into two classes. The bass components have frequencies of less than

Figure 8.9 Some waveforms generated by adding sine waves at 1 kHz and 2 kHz

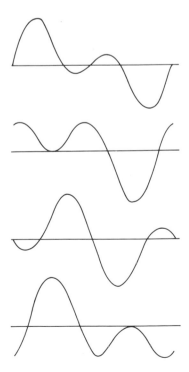

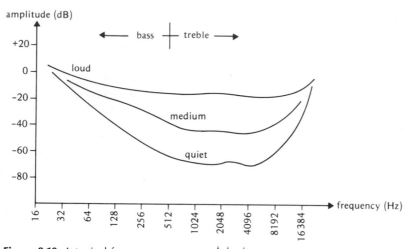

Figure 8.10 A typical frequency response of the human ear

0.5 kHz, the treble components have frequencies greater than 0.5 kHz. This classification into two classes is useful because the ear is more efficient at detecting treble signals than bass ones. That is, if you measure the amplitude of the signals picked up by a microphone of a bass sound and a treble sound which have the same apparent loudness, the treble signal will be much smaller than the bass one.

The graph of figure 8.10 should make this clearer. It illustrates how the amplitude of the detected signal varies with frequency for a sine wave that is kept at the same loudness all the time. Before we discuss the significance of the three curves we ought to explain the units used to mark off the axes.

The vertical axis is marked off in decibels (or dB). This is a very convenient unit for measuring signals which vary by several factors of ten. To convert a signal into dB you first have to express it as a ratio. Then you take the logarithm of the ratio and multiply that by 20. For example, suppose that a signal is 100 mV peak-to-peak. If we take the unit signal to be 1 mV peak-to-peak, then the value of the signal in dB will be given by:

$$20 \times \log_{10}(100/1) = +40\,\text{dB}$$

Of course, the value of the signal expressed in dB depends on the size of the unit signal. Suppose that we had chosen 1 V peak-to-peak rather than 1 mV peak-to-peak. Then the value of the signal in dB would have been:

$$20 \times \log_{10}(0.1/1) = -20\,\text{dB}$$

Anyway, a little thought should convince you that when a signal increases by a factor of ten, its value in dB has 20 added to it.

The horizontal axis is marked off in octaves. An octave is a unit borrowed from the world of music. Whenever the frequency of a sine wave is doubled, it goes up by one octave. So there are two octaves between 4 kHz and 1 kHz (think about it). Signals which are a whole number of octaves apart are perceived as similar by the listener. In musical terms, 4096 Hz is as far from 2048 Hz as 256 Hz is from 128 Hz.

Now take another look at figure 8.10. Start off by considering the middle curve, the one labelled "medium". All of the signals on that curve sounded equally loud to the person who was used to plot the curves (the exact response varies from one person to another). As you can see, the bass notes are, on average about 20 dB higher than the treble ones i.e. they are about ten times larger. Now look at the curve labelled "quiet". All of the points on the curve have the same loudness, but sounded much quieter than those on the previous curve. The shape of the curve has changed, as well as being lower down. Finally, when the sound is very loud there is not much difference in the amplitudes of the signals over the whole frequency range.

So the frequency response of the human ear depends on the loudness of the sound. This creates real difficulties for the faithful reproduction of recorded signals. Suppose that you record a piece of live music at a concert on magnetic tape. If the electronic components and tape are ideal then when the music is played back at the original volume it will sound just like the real thing. But what if you play it back at a lower volume? The sound will be wrong. The bass parts of the frequency spectrum will not appear to be as loud as before, so the music will appear to have too much treble. On the other hand, if you play the music back at a higher volume than the original then it will appear to have too much bass!

In order to get around this difficulty, quality audio amplifiers are provided with **tone controls**. These allow the listener to adjust the frequency response of the amplifier to compensate for any perceived imbalance in the frequency spectrum of the music. As we shall see later on this introduces phase shifts into the components that are enhanced or suppressed, but this does not matter as we cannot hear them.

DISTORTION

When an amplifier distorts, its output signal is not a faithful copy of its input. Ideally, of course, the output should be strictly proportional to the input. This requires a linear relationship between input and output, so ideal amplifiers are called **linear amplifiers**. A **non-linear amplifier** will distort because its output is not proportional to its input.

For example, consider the circuit of figure 8.11. A sine wave is fed into it, but the output is not a sine wave; the clamping action of the diodes flattens the top and bottom of the output waveform. Because the output is not a sine wave it has to have more than one component in its frequency spectrum (figure 8.12). In fact, extra components will appear at 3, 5, 7 etc. times the original frequency. (The distortion goes a little way to converting the output into a square wave, so we expect the harmonics to look like

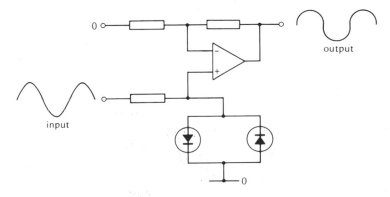

Figure 8.11 A non-linear amplifier

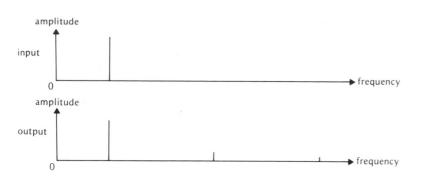

Figure 8.12 The effect of a non-linear amplifier on the frequency spectrum of a signal

those of a square wave). The extra frequencies introduced by the amplifier are readily detected by the listener, but are not always annoying. This is because the harmonics are integral multiples of the original frequency. The sound from musical instruments usually consists of a similar mix of fundamental and harmonics, so the distortion does not sound unnatural.

Mixing signals

As we explained above, non-linear amplifiers distort a sine wave by adding higher frequencies into the frequency spectrum of the signal. This harmonic distortion is a minor nuisance when compared with the effect that a non-linear amplifier has on a mixture of sine waves! The amplifier mixes the signals together to generate extra frequencies which do not occur naturally in music. This is called **inter-modulation distortion**, and it sounds terrible.

Mixing is quite a complicated phenomenon. In order to explain it properly we shall have to use some algebra. If the gain of the amplifier is G, then $V_{OUT} = G \times V_{IN}$. Here V_{OUT} and V_{IN} are the instantaneous values of the AC signals at the amplifier output and input terminals. For a linear amplifier, G is just a number. For a non-linear amplifier, the value of G depends on the value of V_{IN}, and we have to express G as a polynomial expansion in powers of V_{IN}. So, for a non-linear amplifier:

$$V_{OUT} = G_1 V_{IN} + G_2 V_{IN}^2 + G_3 V_{IN}^3 + G_4 V_{IN}^4 + \cdots$$

For an exact description of the amplifier we will need to use an infinite number of terms in the expansion. This is not a problem, because the exact properties of the amplifier are not important when negative feedback is applied to it. Furthermore, the coefficients for the higher powers of V_{IN} get progressively smaller. So $G_1 \gg G_2 \gg G_3$ for a well designed amplifier.

Now consider the transistor amplifier of figure 8.1. It has a V_{OUT}-V_{IN} graph like the one shown in figure 8.13. That curve can be quite well represented by the equation:

$$V_{OUT} = G_1 V_{IN} + G_2 V_{IN}^2$$

G_1 is about -150, twice the size of G_2. Suppose that we feed in a pair of sine waves simultaneously. That is:

$$V_{IN} = A \cos(2\pi f_1 t) + B \cos(2\pi f_2 t)$$

therefore $V_{OUT} = G_1 [A \cos(2\pi f_1 t) + B \cos(2\pi f_2 t)]$

$$+ G_2 [A \cos(2\pi f_1 t) + B \cos(2\pi f_2 t)]^2$$

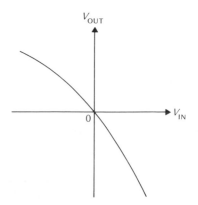

Figure 8.13 The output of a common emitter amplifier as a function of its input

The first term of the expression for V_{OUT} describes the output signal for a linear amplifier. That signal is just a large number (G_1) times the input signal. The second term in the expression describes the distortion; it tells us where the extra frequencies are put in the frequency spectrum, and how big those extra components are. Multiplying out the squared term:

$$A^2 \cos^2(2\pi f_1 t) + 2AB\cos(2\pi f_1 t)\cos(2\pi f_2 t) + B^2\cos^2(2\pi f_2 t)$$

Whenever you have a pair of sine waves multiplied together you can replace them with a sum of two other sine waves. You use the identity:

$$\cos(x)\cos(y) = \tfrac{1}{2}\cos(x+y) + \tfrac{1}{2}\cos(x-y)$$

So, using this rule:

$$\cos^2(2\pi f_1 t) = \tfrac{1}{2}[1 + \cos(2\pi 2 f_1 t)]$$

$$\cos^2(2\pi f_2 t) = \tfrac{1}{2}[1 + \cos(2\pi 2 f_2 t)]$$

$$\text{and} \quad \cos(2\pi f_1 t)\cos(2\pi f_2 t) = \tfrac{1}{2}\cos[2\pi(f_1 + f_2)t]$$
$$+ \tfrac{1}{2}\cos[2\pi(f_1 - f_2)t]$$

Thus the signal contains several distortion components. There is a signal at $2f_1$ with an amplitude of $\tfrac{1}{2}G_2 A^2$, and another at $2f_2$ with an amplitude of $\tfrac{1}{2}G_2 B^2$; these are the **harmonic distortion** components. The other two signals are the results of inter-modulation distortion. One is at $f_1 + f_2$, the other at $f_1 - f_2$; both have an amplitude of $G_2 AB$. The frequency spectrum of the total output is shown in figure 8.14.

At first sight, the fact that a non-linear amplifier introduces a number of extra components into the frequency spectrum of a signal implies that its output will sound terrible. In practice, one only has to worry about the non-linearity of the final amplification stages. This is because the percentage distortion of a signal depends on its size.

Consider a signal of amplitude A fed into our amplifier. The undistorted output will have an amplitude $G_1 A$, and the harmonic distortion will have an amplitude of $\tfrac{1}{2}G_2 A^2$. So the percentage distortion will be given by:

$$\text{percentage distortion} = \frac{\tfrac{1}{2}G_2 A^2}{G_1 A} \times 100$$

$$= \frac{50\,G_2 A}{G_1}$$

$$= 25\,A$$

(We have assumed that G_1 is twice the size of G_2.) Now, suppose that A is small, 0.1 mV. Then the percentage distortion is also very small, $25 \times 10^{-4} = 0.0025\%$. So if the amplifier is only used to amplify small

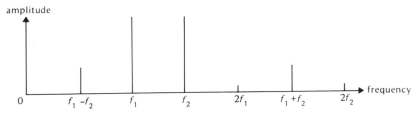

amplitude

0     $f_1 - f_2$ f_1 f_2 $2f_1$ $f_1 + f_2$ $2f_2$ frequency

Figure 8.14

signals, the distortion is negligible. On the other hand, as the amplitude of the signal rises, so does the distortion. A 50 mV signal has a percentage distortion of 1.25%, according to our approximations. (In practice it will be somewhat larger.)

EXPERIMENT
Listening to the distortion produced by a non-linear amplifier

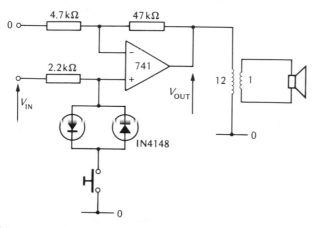

Figure 8.15

Assemble the amplifier shown in figure 8.15. Feed in a 1 V peak signal at 1 kHz, and look at the output with a CRO as well as listening. By pressing the switch you will make the amplifier non-linear; you should be able to hear and see the difference quite clearly.

Try different frequencies. Explain why you cannot hear any difference between the distorted and undistorted signals when it has a frequency of 10 kHz.

NEGATIVE FEEDBACK

The secret of building a linear amplifier is to use negative feedback. You organise the amplifier so that its gain is as large as you can make it, without worrying about its linearity. Then you put a feedback network around the amplifier which feeds a fraction of the output back to the input. The end result is an amplifier which is linear.

Look at figure 8.16. The triangular block is an amplifier which has a gain of $-A$ for AC signals. The quantity A is known as the **open-loop gain** of the system; it is usually quite large. The signal to be amplified (V_{IN}) is fed into one end of a voltage divider and the output (V_{OUT}) is fed into the other end. The amplifier gets its input signal from the centre of the voltage divider. That signal is given by:

$$V_{IN} + (V_{OUT} - V_{IN})\frac{BR}{R + BR}$$

So

$$V_{OUT} = -A\left[V_{IN} + (V_{OUT} - V_{IN})\frac{BR}{R + BR}\right]$$

therefore

$$\frac{V_{OUT}}{A} = -V_{OUT}\frac{B}{1 + B} - V_{IN} + V_{IN}\frac{B}{1 + B}$$

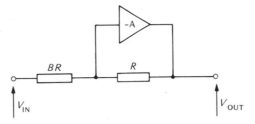

Figure 8.16 An amplifier with negative feedback

therefore $\quad V_{\text{OUT}}\left(B + \dfrac{1 + B}{A}\right) = -V_{\text{IN}}(1 + B - B)$

therefore $\quad V_{\text{OUT}}(AB + 1 + B) = -V_{\text{IN}}\,A$

therefore $\qquad\qquad G = \dfrac{V_{\text{OUT}}}{V_{\text{IN}}}$

$$= -\frac{A}{1 + (A + 1)B}$$

$$= -\frac{A}{1 + AB} \quad \text{if} \quad A \gg 1$$

G is the **closed-loop gain** of the system, its actual gain when the feedback network is in place. As you can see, its size only depends on the relative sizes of B and A. The size of B, the **feedback fraction**, is dictated by the two resistors.

Suppose that A is 300 and that B is 0.1. Then the **loop gain** (AB) will be $300 \times 0.1 = 30$. The closed-loop gain, G will be equal to $300/31 \simeq 10$. Note that this is equal to $1/B$. In fact, provided that the loop gain is much larger than 1, the closed-loop gain only depends on the feedback fraction.

$$G = -\frac{A}{1 + AB} = -\frac{A}{AB} = -\frac{1}{B} \quad \text{if} \quad AB \gg 1$$

Now if the closed-loop gain only depends on the size of two resistors (which are linear) and the precise value of the open-loop gain is unimportant, then the whole system must be linear. In practice, as you raise the value of the loop gain (AB) the behaviour of the system depends less on the properties of the amplifying block. So a poor amplifier can be made as linear as you like, provided that its open-loop gain is large enough and you settle for a small enough closed-loop gain.

EXPERIMENT
Verifying that the closed-loop gain of an amplifier with
negative feedback is equal to $-A/(1 + AB)$

The basic amplifier is shown in figure 8.17 (overleaf). It is a CMOS NOT gate, with some feedback to bias the input at $+2.5\,\text{V}$. The capacitor ensures that the feedback only takes place at frequencies of less than a couple of Hz. No DC current worth talking about flows through the feedback resistors, so they impose the condition $V_{\text{OUT}} = V_{\text{IN}}$. The graph of figure 8.17 shows that the only stable DC value of V_{OUT} is $+2.5\,\text{V}$.

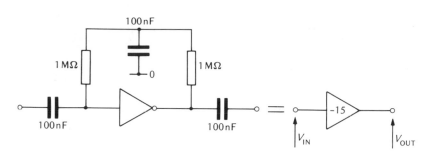

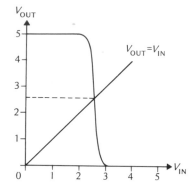

Figure 8.17 An AC amplifier made out of a CMOS NOT gate

Feed in a 50 mV peak sine wave at 1 kHz and look at the AC component of the output of the amplifier. Measure its peak value, and use this to calculate the AC open-loop gain of the amplifier; it should be about 15.

Now add negative feedback, as shown in figure 8.18. Let the resistor R be 2.2, 4.7, 10, 22, 47 and 100 kΩ in turn. Measure the closed-loop gain each time. Check that it agrees with its theoretical value.

Power-amp distortion

The ultimate destination of most audio signals is to be converted into a sound wave by a loudspeaker. This requires the use of a power amplifier to drive the low resistance of the loudspeaker. Figure 8.19 shows a simple push-pull output stage which is suitable for driving a small speaker. The speaker is rated at 500 mW at 8 ohms, so it will need a fairly large current at full blast. The two 47 ohm resistors prevent the transistors from overheating when the output is accidentally short-circuited to ground. Note their power ratings!

The graph of figure 8.20 shows what happens when a 2 V peak sine wave is fed into the power amp. The transistors will not turn on until the input is greater than 0.7 V; the result is a severely distorted waveform at the output. Obviously, you can reduce this **crossover distortion** by increasing the size of the signal. Unfortunately the speaker can only take a 2.8 V peak sine wave before you exceed its power rating, so something else has to be done to reduce the distortion.

One solution is to use negative feedback, as shown in figure 8.21. The open-loop gain of the 741 is 100 000, and all of the output is fed back to the input. As a consequence, the closed-loop gain will be 100 000/(1 + 100 000) = 1. The behaviour of the system is illustrated in figure 8.22; the negative feedback has reduced the crossover distortion to a negligible level. What distortion there is, is due to the finite speed with which the 741 can move its output from +0.7 V to −0.7 V (it takes about 3 μs); even this can be removed by using a faster op-amp such as the 081.

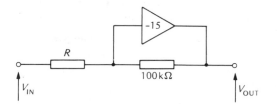

Figure 8.18

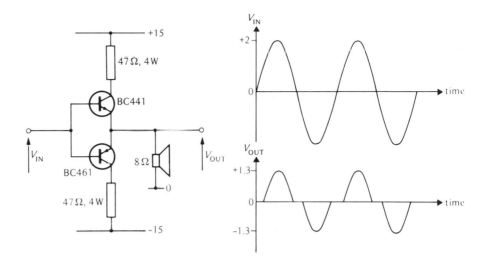

Figure 8.19 A push-pull output stage driving a loudspeaker

Figure 8.20 Crossover distortion

There is one form of output stage distortion which cannot be cured by negative feedback. It occurs when the output saturates, and is known as **clipping distortion**. For example, if the signal fed into the system of figure 8.21 exceeds 2.2 V its top and bottom are flattened. The signal is partly converted to a square wave by the clipping, so components at 3, 5, 7 etc. times the signal frequency are introduced by it.

You can avoid clipping distortion by designing the power amp stage so that it is capable of handling signals which are several times larger than the average. This inevitably means that the amplifier is capable of delivering much more power to the speaker than it normally does. High quality audio amplifiers have power ratings that far exceed the average power delivered

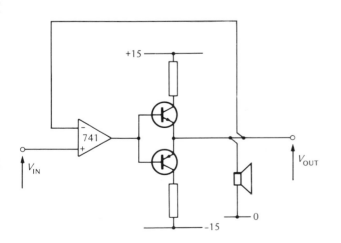

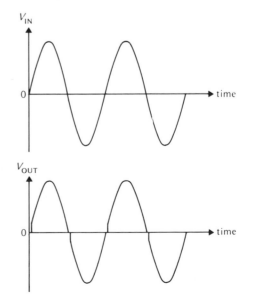

Figure 8.21 Using negative feedback to reduce crossover distortion

Figure 8.22

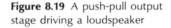

in practice. This allows plenty of room for transient loud signals to be amplified without them banging against the supply rails.

EXPERIMENT
Listening to crossover and clipping distortion produced by a push-pull output stage

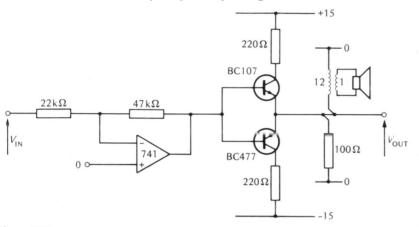

Figure 8.23

Assemble the circuit shown in figure 8.23. The 741 amplifies the input signal by a factor of 2 and feeds into the push-pull output stage. The output is fed into a $100\,\Omega$ load in parallel with a speaker/transformer system.

Look at the signal across the $100\,\Omega$ with a CRO. Feed in a 1 V peak sine wave at 1 kHz. Sketch the output waveform and explain its shape. Note the difference in sound when the speaker is driven by the input signal directly.

Incorporate the output stage into the feedback loop. (Look at figure 8.21.) Feed in the same signal as before. Sketch the output waveform and explain why the distortion has virtually disappeared; comment on any residual distortion that you notice.

Now listen to clipping distortion by increasing the amplitude of the input signal. You should be able to hear clearly the extra harmonics introduced when the output flattens off at 4.7 V. Try listening to a 10 kHz signal as it is clipped; why can you not hear the distortion?

NOISE

As we stated earlier on, noise in electronic circuits is inevitable. Every component in a circuit generates small electrical fluctuations. These may arise from variations in the physical properties of the component; its resistance might change for example. Most noise arises from the grainy nature of electric current; current is a flow of particles, and is therefore not smooth on a microscopic scale. But whatever its cause, noise is essentially random. This means that it is a constantly changing mix of signals at all sorts of frequencies. We therefore represent its frequency spectrum by a band rather than with a series of spikes. Figure 8.24 compares the frequency spectra of a typical noise signal and a square wave. Noise which has equal amplitudes at all frequencies is called **white noise**; it generally arises from fluctuations in currents and thermal noise in resistors. Noise from unstable components tends to have larger amplitudes at lower frequencies.

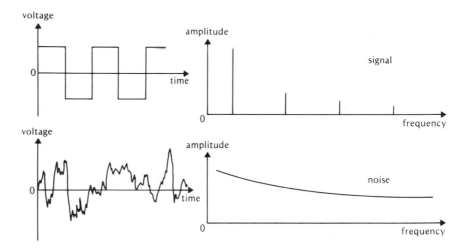

Figure 8.24 The frequency spectra of signal and noise compared

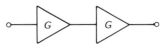

Figure 8.25

If we consider the noise generated in an amplifier, it can easily be shown that most of it originates in the first amplifying stage. Suppose that we have two identical amplifying stages as shown in figure 8.25. Each generates its own noise signal of amplitude N_1 and N_2 respectively, but they have the same gain, G. If a signal of amplitude S is fed in at the left, then the output at the right is given by:

$$(S \times G + N_1) \times G + N_2 = S \times G^2 + N_1 \times G + N_2$$

You can see that, since G is much larger than one, most of the noise is due to amplified noise from the first stage.

So it pays to make the first stage of an audio amplifier as quiet as possible. Then, provided that each subsequent stage has a lot of gain, you need not worry too much about noise generated elsewhere in the system.

Noise can be minimised by several strategies. Firstly, use components which are physically stable. For example, carbon composition resistors are notoriously noisy; they creak and crackle as they heat up and their environment changes. Metal-oxide resistors are much more stable and hence quieter. Nevertheless, even the stablest resistor will generate noise due to thermal fluctuations of the free electrons within it. This **Johnson noise** cannot be eliminated, but it can be reduced by using small value resistors and keeping their temperature down. The Johnson noise from a $10\,k\Omega$ resistor at room temperature has an amplitude of about $1\,\mu V$ in the audio region. Capacitors can creak too; avoid electrolytic ones.

Secondly, avoid small currents. As electric current consists of a flow of particles, charge is transferred from one part of a circuit to another in discrete lumps. The percentage fluctuation in the amount of charge transferred goes up as the number of particles involved goes down. That is, the amount of noise generated by a steady current goes up as the size of the current goes down. Large currents flow quietly.

Finally, choose a quiet amplifying element. JUGFETs are generally quieter than bipolar transistors, but some types of transistor are considerably quieter than others. Low noise op-amps have become cheaply available in the last few years; the 071 is a low noise version of the 081, which is hardly more expensive than the 741.

EXPERIMENT

Assemble a noise source and listen to its output.
After adding the noise to a sine wave, find out how much noise
you can add before you notice it.

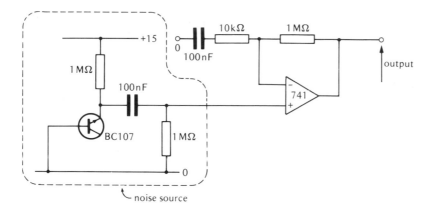

Figure 8.26 A noise generator

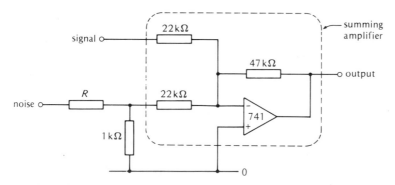

Figure 8.27 A circuit for adding noise to a signal

Assemble the circuit of figure 8.26. The noise is generated by the random zener breakdown of the reverse biased base-emitter junction of the transistor. This is amplified by a non-inverting amplifier; note that the noise source has a high output impedance (1 MΩ). Listen to the noise with a loudspeaker and matching transformer. Have a look at it with a CRO.

Use the summing amplifier of figure 8.27 to add a 2.5 V peak sine wave at 1 kHz to the noise. Listen to the output and look at it with a CRO. Increase the value of R from 1 kΩ until you cannot hear the noise above the 1 kHz signal. Use the CRO to measure the size of the noise that is just noticeable. (Measure the peak value of the noise entering the summing amplifier.) Then calculate the signal-to-noise ratio; express your answer in decibels.

INTERFERENCE

Interference describes the small signals which an audio system can pick up from other electrical systems. Whereas noise is generated from within a system, interference comes from outside it. Furthermore, interference may contain information whereas noise does not; it is therefore more annoying to the listener.

Although there are many sources of interference, they can be classified into two types. Magnetic sources inject signals into a circuit by setting up oscillating magnetic fields; when such fields pass through a conducting loop they set up AC currents in it. Electrical sources inject signals via oscillating electric fields; in most cases, the source and the audio system behave like the two plates of a capacitor so that signals may pass from one to the other.

Magnetic interference

The classic example of magnetic interference is mains hum. The 50 Hz currents flowing through the mains supply cables in a building flood it with AC magnetic fields. Any conducting loop placed in that field will have an AC current induced in it. If the loop has a large impedance, then a large signal at 50 Hz will appear in it. For example, if you touch the input terminal of a CRO with your finger you will see a large 50 Hz signal on the screen. The mains cable of the CRO, yourself and the ground under you form the conducting loop. The current induced in that loop has to flow through a large impedance (yourself), so the signal is large. If you make the loop smaller by touching the earth terminal of the CRO with your other finger the induced current will be much smaller.

Mains cables are not the only source of magnetic interference. Even a well designed mains transformer gives off a fairly hefty magnetic field. Large electric motors can also induce a lot of mains hum. The best cure for both of these sources is to get as far away from them as possible. Magnetic shields can be used, but they do leak.

In general, the cure for mains hum is to keep the area of conducting loops as small as possible. A signal has to be sent along a pair of wires so that the current can flow in a loop. If the wires are twisted together (to make a **twisted pair**) then the area of the loop is minimised. It also helps to keep the impedance of the loop as low as possible so that any currents which do flow cannot set up an appreciable voltage. We shall return to this point later when we consider the problems of transducers at the ends of long cables.

Electrical interference

Any rapid change of current in an electrical circuit will cause voltage spikes to appear in a circuit placed near to it. The circuits behave like two plates of a capacitor; when the voltage on one plate changes rapidly the other plate has to follow it. In this fashion, one part of an electronic system can "talk" to another part. So if you consider a stereo audio amplifier, signals from one channel will inevitably leak into the other. This effect is known as **crosstalk**.

Crosstalk can be reduced by several strategies. The most obvious is to physically separate the two circuits as much as possible. Then each circuit is made as small as possible, with short wires between components. Finally, a grounded conducting sheet is placed between them. This electrical screening makes each circuit think that its other capacitor plate is the ground, so that the crosstalk signals get fed to earth.

Other sources of electrical interference are radio transmitters, heavy-duty mains switches, high voltage overhead cables and electric motors. These emit radio waves which induce currents in conductors. The answer is electrical screening; surround the circuit with an aluminium box that is grounded.

Signal cables

Interference is a particular problem with audio transducers such as microphones, tape-heads and gramophone pickups. These are necessarily connected to an audio system by long cables. Any signals picked up by those cables will be amplified a great deal when they get into the amplifier; they could easily drown out the small signal from the transducer.

The first step towards eliminating interference in the cable is to screen it. The signal is sent down **coaxial cable**. A single wire down the centre of the cable carries the signal. Copper braid around the outside of the cable provides the return path for the signal current. If that braid is earthed at one point it acts as an electrical screen for the centre wire, sheltering it from electrical interference.

The second step is to boost the signal before it goes into the cable. You put a small amplifier right next to the transducer to amplify the signal to a level at which interference becomes negligible. A simple amplifier will suffice as the signals are very small; the linearity of the amplifier is not important. Figure 8.28 shows how an FET can be used to amplify a microphone signal by -22 before feeding it down a coaxial cable. An FET amplifier is particularly suitable for this purpose as it can be very quiet and will not worry about the output impedance of the transducer. Note how the centre wire of the coaxial cable carries the DC supply for the amplifier as well as the AC signal.

Using an amplifier next to a transducer allows you to lower the output impedance in the cable loop from that of the transducer to that of the amplifier. This serves to reduce the voltage generated by currents induced in the loop.

Figure 8.29 illustrates what you have to do if interference is a big problem. You might have to resort to this in a very noisy environment, such as a factory full of electrical machinery. The signal has to be buffered, so that it is fed out of a low impedance source. It is then fed into a twisted pair which runs down the centre of a coaxial cable; the braid on the outside of the cable is grounded so that it provides screening for the wires inside it. At the receiving end, the twisted pair feeds its signal into a **difference amplifier**. This has the property of common-mode rejection, so that it only detects the voltage difference between the two wires. Both wires have followed the same path, so they will have picked up identical interference; this signal will be ignored by the difference amplifier.

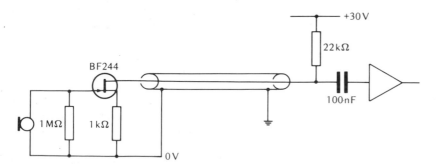

Figure 8.28 Amplifying a signal before feeding it into a coaxial cable

PROBLEMS

1 Show that the difference amplifier in figure 8.29 does indeed have an output which is equal to the difference in the signals at its inputs.

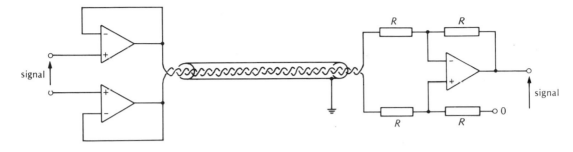

Figure 8.29 Sending a differential signal down a screened twisted pair

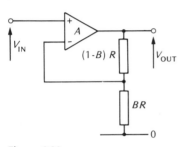

Figure 8.30

2 Show how the amplifier of figure 8.1 can be adapted so that it can feed its signal down the same coaxial cable that provides its supply. In other words, work out how to build the bipolar equivalent of the circuit shown in figure 8.28. (Hint: if one plate of a capacitor has to be held at a fixed voltage, the size of the voltage is immaterial.)

3 A quality audio amplifier will have a grounded aluminium sheet between the output stage and the other amplifying stages. Explain why.

4 Figure 8.30 shows an op-amp arranged as a non-inverting amplifier. The voltage divider feeds a fraction of the output back to the input. That fraction is B. The open-loop gain of the op-amp is A. Show that the closed-loop gain is given by $G = A/(1 + AB)$.

FILTERING

So far we have considered how to prevent an amplifier from adding extra components into the frequency spectrum of a signal. We are now going to find out how to enhance and suppress parts of a signal's frequency spectrum; this is known as **filtering**. Technically, this is a form of distortion, but there are sound reasons for doing it. Filtering can be used to compensate for the imperfect frequency response of a transducer, it can be used to eliminate some noise and it can restore the frequency balance of a signal which is played back too softly or too loudly.

Filtering square waves

The simple circuit of figure 8.31 is a **treble cut filter**. It suppresses high frequencies and lets low frequencies through. We shall be seeing what it does to sine waves in a minute, but it is instructive to see what it does to square waves first.

Figure 8.32 shows the three types of waveform which emerge from the filter when a square wave is fed into it. The behaviour of the filter depends

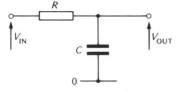

Figure 8.31 A simple treble cut filter

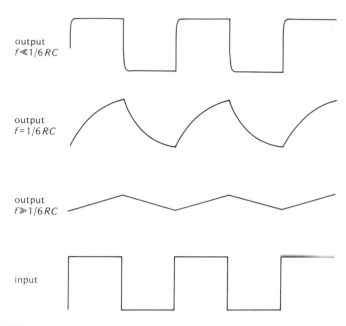

output
$f \ll 1/6RC$

output
$f = 1/6RC$

output
$f \gg 1/6RC$

input

Figure 8.32

on the value of the frequency (f) compared with the value of $1/(6RC)$. At low frequencies, (i.e. $f \ll 1/(6RC)$) virtually all of the signal gets through. This is because V_{OUT} will follow V_{IN} with a delay of only $3RC$ ms; if the period of V_{IN} is much larger than $6RC$, then V_{OUT} will be a good copy of V_{IN}. At high frequencies though, V_{OUT} will be much smaller than V_{IN}. This is because the capacitor cannot charge up quickly enough to follow the changes of V_{IN}. The cross-over point between the two regions of high and low frequencies occurs when $f = 1/(6RC)$. At this frequency the output is neither a square wave nor a triangle one.

Reactance

How do we analyse the behaviour of a simple treble cut filter when a sine wave is fed into it? By introducing the idea of **reactance** we can use the voltage divider formula to calculate the approximate behaviour of the filter.

Appendix A shows that when an alternating voltage (V) of frequency f is applied across a capacitor (C), the current which flows through it (I) is given by:

$$\frac{V}{I} = X = \frac{1}{2\pi f C}$$

The quantity X is the reactance of the capacitor. It describes, in a very loose sense, the resistance of the capacitor for AC signals. (Remember that $V/I = R$ for a true resistor.) If f and C are measured in kHz and μF respectively, X will be in kΩ. When a sine wave is fed into a network which contains capacitors, we can replace each capacitor with its reactance and deal with the system as if it only contained resistors.

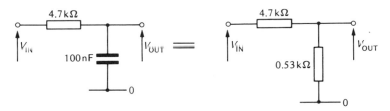

Figure 8.33

For example, suppose that a 5 V peak sine wave at 3 kHz is fed into the filter of figure 8.33. What is the size of the signal fed out?

The first step is to calculate the reactance of the capacitor.

$$X = \frac{1}{2 \pi f C}$$

$$= \frac{1}{2\pi \times 3 \times 0.1}$$

$$= 0.53 \, k\Omega$$

We then replace the capacitor with a resistor of this size and use the voltage divider formula to calculate the output:

$$V_{OUT} = \frac{V_{IN} R_2}{R_1 + R_2}$$

$$\text{therefore} \quad V_{OUT} = \frac{5 \times 0.53}{4.7 + 0.53}$$

$$= 0.5 \, V$$

So the output waveform is a 0.5 V peak sine wave at 3 kHz. A glance at the three waveforms of figure 8.32 should convince you that the output will also have suffered a phase shift of 90°.

The phase shift that is introduced by the capacitor means that it does not behave exactly like a resistor for AC signals. You can safely replace the capacitor with a resistor in a filter network if the reactance is much larger or smaller than the true resistor in the circuit. On the other hand, when the reactance is roughly the same as the resistance in a network then the voltage divider formula will not give the correct answer. (At worst, it will be wrong by a factor of $\sqrt{2}$.)

EXPERIMENT

Measuring the reactance of a 10 nF capacitor at several frequencies

Assemble the network shown in figure 8.34. Feed in a 2 V peak value sine wave at the frequencies shown in the table; note that each frequency has its own value of R. Use a CRO to measure the peak value of the output each time; it ought to be about 160 mV peak each time.

Use the voltage divider formula to calculate the reactance of the capacitor at each of the six frequencies. Plot your results on log-log graph paper; figure 8.35 (overleaf) shows what it ought to look like. Calculate the theoretical value of X at each frequency and put those values on the same graph.

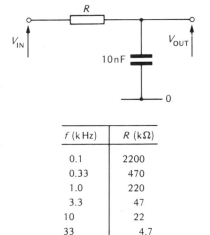

f (kHz)	R (kΩ)
0.1	2200
0.33	470
1.0	220
3.3	47
10	22
33	4.7

Figure 8.34

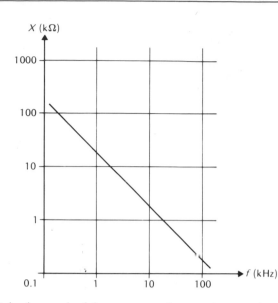

Figure 8.35 A log-log graph of the reactance of a capacitor as a function of frequency

PASSIVE FILTERS

You have already met the simple treble cut filter shown in figure 8.36. The behaviour of the filter is approximately given by:

$$G = \frac{V_{OUT}}{V_{IN}} = \frac{X}{R + X}$$

$$\text{where} \quad X = \frac{1}{2\pi f C}$$

We define the **break frequency** of the filter as follows:

$$f_0 = \frac{1}{2\pi RC}$$

The filter behaves differently above and below the break frequency. Below the break frequency, X will be much larger than R so that $V_{OUT} = V_{IN}$ and $G = 1$. Well above the break frequency X will be much smaller

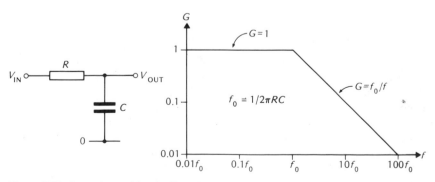

Figure 8.36 A passive treble cut filter

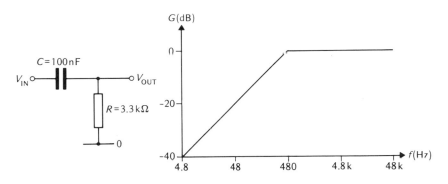

Figure 8.37 A passive bass cut filter

than R. This means that G will be less than 1. More precisely:

$$G = \frac{X}{R + X} \simeq \frac{X}{R} = \frac{1}{2\pi f R C} = \frac{f_0}{f} \quad \text{when } R \gg X$$

At the break frequency itself, $X = R$; we cannot use the voltage divider formula to get an exact value of G under these circumstances. In fact, at the break frequency $G = 1/\sqrt{2} = 0.7$.

It is customary to display the characteristics of a filter on a log-log plot of gain as a function of frequency. Look at the graph of figure 8.36. It has been drawn using the "two line approximation". Below the break frequency the graph is horizontal; above it, the line drops at an angle of 45°. In practice, there will be a smooth change of slope at the point where the two lines meet, but this makes little difference on a log-log plot.

A similar technique can be used to build a bass cut filter. Look at the circuit of figure 8.37. The break frequency will be given by:

$$f_0 = \frac{1}{2\pi R C} = \frac{1}{2\pi \times 3.3 \times 0.1} = 0.48 \, \text{kHz}$$

Below this frequency X is larger than R and $G = f/f_0$. Above the break frequency X is smaller than R so that $G = 1$.

Drawbacks of passive filters

The two filter circuits that we have just looked at suffer from a major defect; their behaviour is very dependent on the current drawn from them. They only behave as described if the output signal is fed into a very large impedance. So in practice they need to be buffered by some sort of amplifier. This will degrade the performance of the filter in other ways. The active element used to buffer it will introduce noise and distortion; the passive filter is linear and can be made of quiet components.

Anyway, you can see that even the most complicated network of resistors and capacitors can only attenuate. If you want to assemble filters which will boost selected frequencies you need to include an active element to provide the amplification.

ACTIVE FILTERS

We are going to introduce four active filters. They are based on op-amps which have resistors and capacitors providing negative feedback. The

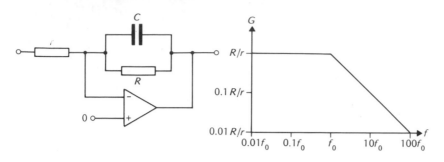

Figure 8.38 An active treble cut filter

closed-loop gain of the filters is usually small so that they are linear and well-behaved. They by no means represent the ultimate in active filters, but more complex systems cannot be treated properly with the approximations used in this book.

Treble filters

Figure 8.38 illustrates a treble cut filter and a log-log graph to show its behaviour. As usual, the break frequency is given by $f_0 = 1/(2\pi RC)$. Below that frequency the gain of the circuit is R/r; above it the gain drops by a factor of ten for every tenfold rise in frequency.

In order to get a better insight into the behaviour of the circuit, we shall resort to some algebra. Figure 8.39 shows an op-amp set up as an inverting amplifier. Its gain, of course, is equal to $-R_F/R_{IN}$. If we ignore the sign of the gain (it corresponds to a phase shift of 180°) we can use this to find the gain of the filter in figure 8.38.

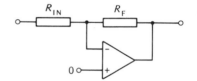

Figure 8.39

First of all, at low frequencies X is very much larger than R. Since the two are in parallel we can ignore the larger of the two when calculating the size of R_F. So $R_F = R$, and $G = R/r$. The gain is independent of the frequency.

At high frequencies X will be much smaller than R. So $R_F = X$, as the capacitor effectively short-circuits the resistor. The gain will be given by:

$$G = \frac{R_F}{R_{IN}} = \frac{X}{r} = \frac{1}{2\pi f C \times r} = \left(\frac{R}{r}\right) \times \left(\frac{1}{2\pi RC}\right) \times \frac{1}{f}$$

therefore $G = \left(\frac{R}{r}\right) \times \left(\frac{f_0}{f}\right)$

So as f increases by a factor of ten, the gain drops by a factor of ten.

If the components are swapped around you create a treble boost filter as shown in figure 8.40. As before, we consider two frequency ranges, with a break frequency given by $f_0 = 1/(2\pi RC)$. Below this frequency X is larger than R so we can ignore the capacitor. The gain will therefore be r/R. Above the break frequency X is smaller than R so we can ignore the resistor. The gain will be given by:

$$G = \frac{R_F}{R_{IN}} = \frac{r}{X} = 2\pi f C \times r = \left(\frac{r}{R}\right) \times 2\pi RC \times f$$

therefore $G = \left(\frac{r}{R}\right) \times \left(\frac{f}{f_0}\right)$

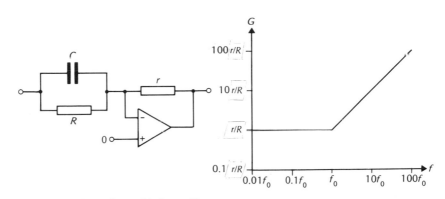

Figure 8.40 An active treble boost filter

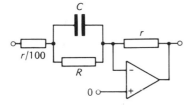

Figure 8.41 A realistic treble boost filter

So above the break frequency, the gain rises by a factor of ten for each tenfold increase of frequency.

In practice, the gain will not increase for ever. There will come a point where the amplifier will start to pack in, flattening off the frequency response and eventually making it drop. This is a good thing, as a system whose gain rises indefinitely with frequency is liable to be unstable and oscillate. It is a good idea to limit the gain of the filter at high frequencies by inserting a small resistor in series with the parallel *RC* network. For example, the circuit of figure 8.41 has a maximum gain of 100.

Bass filters

Bass filters exploit the frequency dependent behaviour of a network consisting of a resistor in series with a capacitor. At low frequencies X is much larger than R so that the effective resistance of the network is given by:

$$R + X = X = \frac{1}{2\pi f C} = R \times \left(\frac{f_0}{f}\right)$$

The resistance of the network rises as the frequency falls. Above the break frequency X is much smaller than R, so that the resistance of the network is just R, independent of the frequency.

Figure 8.42 shows a series RC network used to convert an op-amp into a bass cut filter. Below the break frequency the gain will be given by:

$$G = \frac{r}{R + X} = \frac{r}{X} = \left(\frac{r}{R}\right) \times \left(\frac{f}{f_0}\right)$$

Above the break frequency $G = r/R$.

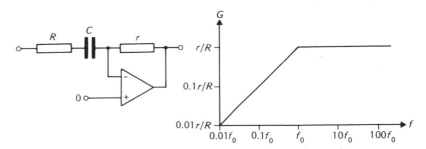

Figure 8.42 An active bass cut filter

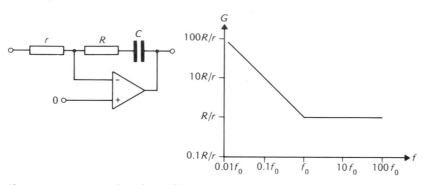

Figure 8.43 An active bass boost filter

A bass boost filter is shown in figure 8.43. You should have no difficulty convincing yourself that its gain below the break frequency is given by:

$$G = \left(\frac{R}{r}\right) \times \left(\frac{f_0}{f}\right)$$

A real bass boost filter must have a large resistor in parallel with the series RC network. This stops the gain from rising indefinitely as the frequency goes down. Without it, the system would have an infinite gain for DC signals, and the output would inevitably saturate. The amended bass boost circuit of figure 8.44 has a maximum gain of 100.

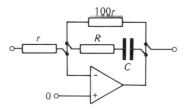

Figure 8.44 A realistic bass boost filter

EXPERIMENT

Investigating the behaviour of four active filters

Assemble each of the circuits shown in figure 8.45 in turn. Feed in a 1 V peak-to-peak sine wave at 0.1, 0.33, 1.0, 3.3, 10 and 33 kHz in turn. Use a CRO to measure the peak-to-peak value of the output waveform. Plot a graph of gain as a function of frequency on log-log graph paper. Use the two line approximation to join up your points; read off the break frequency and check that it agrees with its theoretical value.

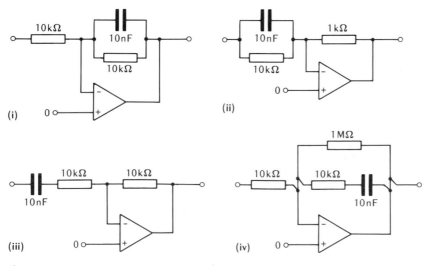

Figure 8.45

PROBLEMS

1 For each of the circuits shown in figure 8.46, calculate its break frequency. Then plot a log log graph of its gain as a function of frequency.

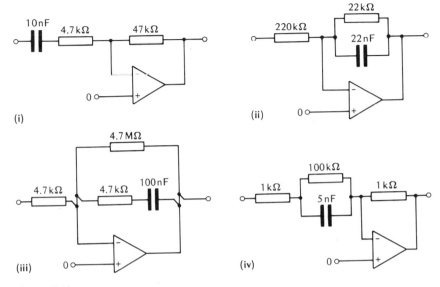

(i)

(ii)

(iii)

(iv)

Figure 8.46

2 Draw circuit diagrams of filters which have the characteristics shown in the graphs of figure 8.47.

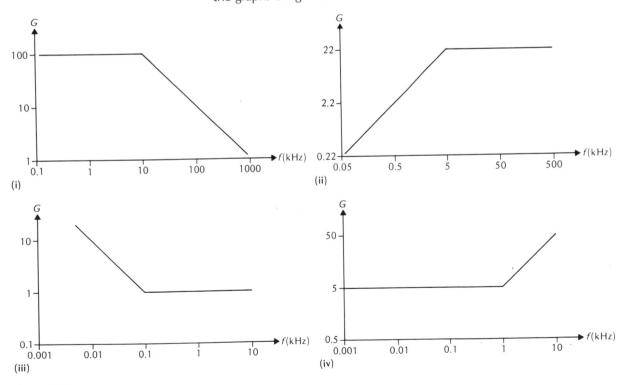

(i)

(ii)

(iii)

(iv)

Figure 8.47

3 The filter shown in figure 8.48 is a combined bass cut and treble cut filter. Calculate the two break frequencies and hence draw a graph of the gain as a function of frequency.

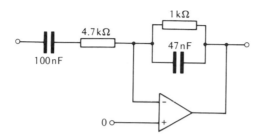

Figure 8.48

4 Work out the characteristics of the filter system shown in figure 8.49. Calculate the break frequencies, the maximum gains in the bass and treble regions and hence draw a log-log graph of the gain as a function of frequency.

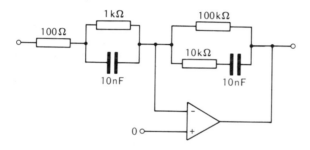

Figure 8.49

FILTERS IN RECORDING SYSTEMS

The two principal methods of storing sound grossly distort the frequency balance of the signals which they store. Magnetic tape and gramophone records tend to increase the amplitude of treble signals while cutting down the amplitude of bass ones. Filters are placed in the early stages of audio amplifiers to compensate for this, so that the final signal going into the speaker has its original frequency balance.

For example, consider sound which is stored on a gramophone record. The groove which is etched into the plastic of the disc is effectively a CRO trace of the audio signal which has been wound into a spiral. The side to side movement of the groove as you travel down it mirrors the changing voltage of the audio signal as time progresses. Before it is put onto the disc, the audio signal is fed through a filter to suppress the bass and boost the

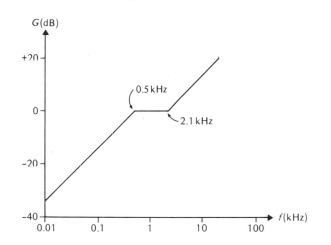

Figure 8.50 The frequency response of the standard recording filter

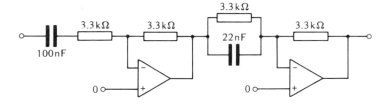

Figure 8.51 A possible circuit for the recording filter

treble. The characteristics of the filter are shown in figure 8.50. Signals below 0.5 kHz are attenuated by 20 dB per decade, and those above 2.1 kHz are boosted by 20 dB per decade.

Why is the frequency balance mutilated in this way? The reasons are sound ones. Music which sounds balanced tends to have equal volumes of bass and treble. But bass signals tend to have much larger amplitudes than treble ones at the same medium volume. So a system which records audio signals directly has to deal with large bass signals and small treble ones simultaneously. Large amplitude signals are a bad thing as far as gramophone records are concerned because they take up a lot of room on the surface of the record. Equally importantly, gramophone pickups, like transistor amplifiers, work best when they are dealing with small amplitude signals. If the needle has to swing a large distance from side to side as it follows the groove the voltage given out by the pickup ceases to be proportional to the displacement of the needle. (The pickup becomes nonlinear.)

The circuit of figure 8.51 shows how a filter with the recording characteristic can be assembled. It is just two active filters in series. The first filter is a bass cut with a break frequency of 0.5 kHz, the second a treble boost with a break frequency of 2.1 kHz.

In order to restore the correct frequency balance to the signal coming out of the pickup, the signal has to be fed through a filter with the characteristics shown in figure 8.52. This is the **equalisation filter**; it boosts the bass and cuts the treble.

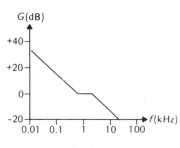

Figure 8.52 The frequency response of the equalising filter

PROBLEMS

1 Devise a filter system which has the characteristics shown in figure 8.52.

2 One method of reducing the noise present in a tape recording is to boost the treble signals before they are recorded on the tape. The playback amplifier incorporates an equalising filter which cuts the treble so that the signal has its original frequency balance restored.

a) The treble boost filter has the characteristics shown in figure 8.53. Draw the circuit diagram of a suitable filter, showing all of the component values.

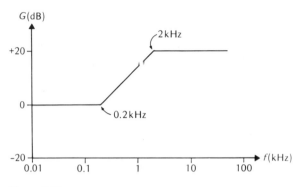

Figure 8.53

b) Draw a log-log graph of the characteristics of the equalising filter. Draw a circuit diagram for it.

c) The tape adds white noise to the signal. Draw the frequency spectrum of the noise signal picked up by the tape-head before and after it has gone through the equalisation filter.

d) Explain why the presence of the two filters reduces the background hiss heard when the tape is played.

EXPERIMENT

You are going to have some fun distorting your own voice!

The circuit shown in figure 8.54 allows the signal at IN to appear at OUT in bursts. On the left is a relaxation oscillator. Its output is a square wave which controls the FET. The FET alternates between being a closed-circuit to being an open-circuit at a rate decided by the setting of the 5 kΩ potentiometer.

Assemble the circuit and feed a 3 V peak sine wave at 1 kHz into IN. Monitor OUT with a CRO and listen to it with a loudspeaker/transformer. Note how the frequency spectrum of the sine wave is enriched by the chopping. Try feeding different frequencies into the system.

Finally, try feeding your own speech through the system. A microphone typically gives out a few mV, so you will have to amplify the signal quite a bit before putting it into IN. The AC amplifier shown in figure 8.55 is a suitable amplifying unit; it has a gain of 45, with a high input impedance so that it can deal with crystal or moving-coil microphones. You will probably have to use two in series to get a decent signal.

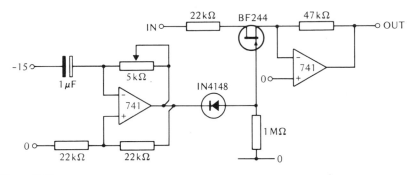

Figure 8.54

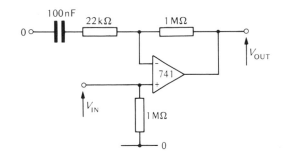

Figure 8.55

REVISION QUESTIONS

1 Explain what the term "audio system" means.

2 Outline the properties of an ideal audio amplifier.

3 Draw the frequency spectrum of

   a) a 2 kHz sine wave

   b) a 2 kHz square wave.

4 Explain why quality audio amplifiers are provided with tone controls.

5 Discuss the distortion imposed on a signal by a non-linear audio amplifier.

6 Use algebra to explain why negative feedback can improve the linearity of a non-linear amplifier. In other words, show that the closed-loop gain of the system is $-A/(1 + AB)$

7 Explain what clipping distortion and cross-over distortion are and where they come from. Describe how negative feedback can be used to eliminate the latter distortion.

8 Discuss sources of noise and interference in an audio system. Explain how noise and interference can be minimised.

9 Write down the formula for the reactance of a capacitor as a function of frequency. Explain what it means.

10 Draw the circuit diagram of each of the following filters:

   a) bass cut

   b) treble cut

   c) bass boost

   d) treble boost.

11 Draw circuit diagrams and log-log plots of gain as a function of frequency for the following filters. Show all the component values, with reasons.

   a) bass cut with break frequency of 200 Hz and a high frequency gain of 10,

   b) treble boost, break frequency 2 kHz, low frequency gain of 0.1,

   c) bass boost, break frequency 330 Hz, high frequency gain of 5

   d) treble cut, break frequency 16 kHz, low frequency gain of 22.

12 How is an audio signal filtered before it is recorded on a gramophone disc? Why is the filtering done? Explain why audio amplifiers include an equaliser filter for gramophone signals.

9

Sequential systems

A **sequential system** is a logic system which is capable of processing digital signals in a sequence of steps. That is, the signals are read in at one instant, combined some time later and eventually fed out again. In order to build such a system you need a circuit that is capable of storing a logic signal, i.e. a device which has a memory. You will remember from Chapter Seven that the SR flip-flop has precisely this property; it remembers which of its input terminals was last raised to logical 1. So this chapter describes some of the systems which can be made out of flip-flops. We shall show how they can be arranged to make memories which can store binary numbers, and how binary counters can be constructed.

THE D FLIP-FLOP

We shall be describing sequential systems solely in terms of D flip-flops. Other types of flip-flop exist (such as JK flip-flops and SR flip-flops) but they are not so versatile. (The D flip-flop is actually made out of a couple of SR flip-flops.) We will, however, ignore the interior of the D flip-flop for the moment and concentrate on its behaviour. Since it is a fundamental building block of all of the systems you will meet in this chapter it is important that you appreciate what it can do.

The circuit symbol of a D flip-flop is shown in figure 9.1. The inputs (DATA and CLOCK) are on the left, the outputs (Q and $\bar{Q}$) on the right. At the top and bottom are the set and reset terminals ($\bar{S}$ and $\bar{R}$). You get two of these flip-flops on a 7474 IC.

Figure 9.2 is a timing diagram which illustrates the essential properties of the D flip-flop. Starting at the left hand side of the diagram, DATA = 1 and CLOCK = 0; these two signals are fed into the flip-flop from some external

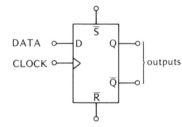

Figure 9.1 A D flip-flop

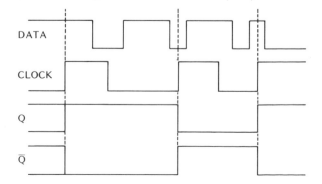

Figure 9.2 The timing diagram of a D flip-flop

system. Initially we have no way of predicting what the values of Q and $\bar{Q}$ will be so we have drawn them as simultaneously 1 and 0. Both $\bar{S}$ and $\bar{R}$ are left to float up to 1 (the 7474 is a TTL IC).

At the instant that CLOCK goes from 0 to 1 the signal fed into D (DATA) is placed on the Q terminal and left there. The $\bar{Q}$ terminal feeds out $\overline{DATA}$ at the same instant. So immediately after the rising edge of CLOCK, Q = 1 and $\bar{Q}$ = 0. Subsequent changes in the value of DATA have no effect on the states of Q and $\bar{Q}$. Furthermore, CLOCK can be lowered back down to logical 0 without altering the flip-flop outputs.

As you can see from the timing diagram, all changes of Q and $\bar{Q}$ coincide with the rising edges of CLOCK. The flip-flop responds to the 0 to 1 transitions of CLOCK by reading in the instantaneous value of DATA and storing it on Q; $\bar{Q}$ always has the opposite state to that of Q.

The set and reset terminals can be used to force the flip-flop into one of its two states. Both are **active-low** inputs, i.e. you have to feed in a logical 0 to get it to do its stuff. When $\bar{S}$ is held at 0, Q is 1 (regardless of the states of DATA and CLOCK) and the flip-flop is **set**. To **reset** the flip-flop, so that Q = 0, you hold $\bar{R}$ at 0. You try to avoid holding both $\bar{R}$ and $\bar{S}$ at 0 at the same time!

The $\overline{SR}$ flip-flop

At this stage it is instructive to contrast the edge-triggered D flip-flop with the more primitive $\overline{SR}$ flip-flop. The latter is shown in figure 9.3. The labelling of its inputs and outputs are the same as those of the equivalent terminals of a D flip-flop. Thus the inputs ($\bar{S}$ and $\bar{R}$) are active-low. When $\bar{S}$ is pulled low the flip-flop is set (i.e. Q = 1) and when $\bar{R}$ is pulled low the flip-flop is reset (i.e. Q = 0). As with the D flip-flop, it is usual practice to avoid pulling both input terminals down to 0 at the same time.

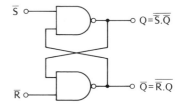

Figure 9.3 An $\overline{SR}$ flip-flop

You can easily work out what the flip-flop does by considering the algebraic expressions for its outputs. That of the Q terminal is given by:

$$Q = \overline{\bar{S} . \bar{Q}}$$

So Q is going to be 1 whenever $\bar{S}$ is 0. It will remain at 1 when $\bar{S}$ goes back up to 1 if $\bar{Q}$ is 0. To see if this happens we need to look at the expression for $\bar{Q}$:

$$\bar{Q} = \overline{\bar{R} . Q}$$

It should be obvious that if both $\bar{R}$ and Q are 1 then $\bar{Q}$ must be 0. So once Q has been pushed up to 1 by pulling $\bar{S}$ low it will stay there provided that $\bar{R}$ is left high.

The symmetry of the flip-flop means that exactly the opposite happens when $\bar{R}$ is briefly pulled down to logical 0. $\bar{Q}$ becomes 1 and, if $\bar{S}$ is left high, Q goes to 0.

Later on we shall show how the addition of extra gates to control the input signals can convert a $\overline{SR}$ flip-flop into a D flip-flop.

EXPERIMENT
Exploring the properties of a D flip-flop

Assemble the circuit shown in figure 9.4. The two input signals are generated by a pair of push switches, and two LEDs monitor the states of the Q and $\bar{Q}$ outputs. Let both $\bar{S}$ and $\bar{R}$ terminals float up to logical 1.

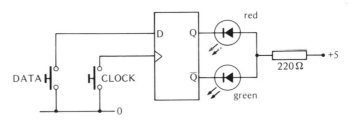

Figure 9.4

Set up the state of DATA by pressing or releasing the DATA switch (close the switch to make D = 0). When the CLOCK switch is pressed and released, the state of D should be fed through to Q (a lit LED indicates 0). Furthermore, any subsequent change of D should be ignored by the flip-flop if the CLOCK line is left high.

Find out what happens when the $\overline{S}$ and $\overline{R}$ terminals are pulled low (first separately and then together).

Summarise the properties of the D flip-flop with a timing diagram. You should show how the states of Q and $\overline{Q}$ depend on the states of $\overline{S}$, $\overline{R}$, DATA and CLOCK.

MEMORIES

A memory is a system which stores information in the form of digital signals. In Chapter One you met read-only memories, the sort of memory whose contents are established during its manufacture. Such memories are known as ROMs. You can read information from them, but you cannot write information into them. D flip-flops allow you to build memories which can take in signals as well as push them out. Such memories are known as **RAMs**, short for **random access memory**. It is possible to **read-and-write** to RAMs.

Before showing how a RAM can be constructed, we need to say a few words about how information can be stored as a series of digital signals.

Binary words

A single digital signal has only two possible states. Each of those states can be used to represent a piece of information. For example, let A represent the presence or absence of one pupil in a class, Anne. When Anne is there, A = 1 and when she is not, A = 0. If the class contains four pupils (David, Charles, Beatrice and Anne) then we need four digital signals to hold all of the information about the presence of the pupils. That information could be held in a four bit **binary word** DCBA. So when DCBA = 1101, all of the pupils are present except for Beatrice.

Four bit words are common in digital systems because they can represent any number from 0 to 9. Throughout this book we assume that numbers are coded according to the conventions of **binary coded decimal** (**BCD** for short), unless otherwise stated. According to this convention, the binary DCBA holds a number whose value is given by:

$$D \times 8 + C \times 4 + B \times 2 + A \times 1$$

where D, C, B and A are 1 or 0. Thus DCBA = 1001 holds the number 9 (the largest one allowed in BCD) and DCBA = 0111 holds 7.

It is clear that if you wish to store information in digital form, you must be able to store binary words. The standard binary word (the **byte**) is eight bits long (HGFEDCBA), but it can be dealt with as two four bit words, or **nibbles**. Nibbles are good for storing numbers, but bytes are usually needed for storing anything else.

The four bit latch

The simplest form of memory that can store a binary word is shown in figure 9.5. It is known as a four bit **latch**. The four bits of a nibble are placed on D_0, D_1, D_2 and D_3 at the same time. When a brief logical 1 is fed into ENABLE, the nibble is transferred to Q_0, Q_1, Q_2 and Q_3 and held there. Provided that no further pulses are fed into ENABLE, the binary word held at the outputs will remain constant regardless of any subsequent changes at the inputs. Of course, every time that a pulse is fed into ENABLE a fresh nibble is stored at the outputs and the old nibble is lost.

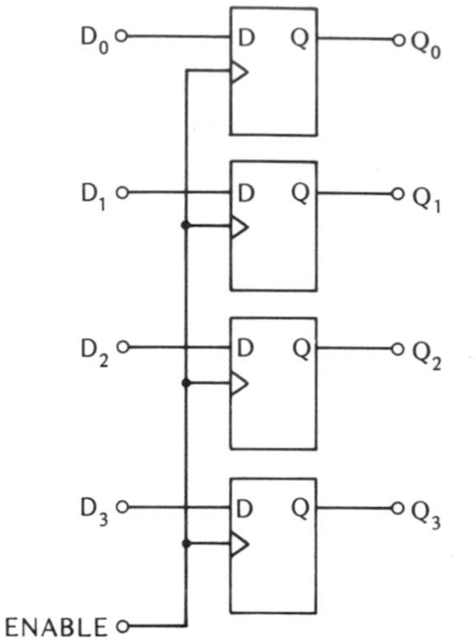

1 × 4 bit RAM

Useful memory systems need to be able to read in and feed out digital signals along the same line. The latch of figure 9.5 does not do this; the four bit word arrives along four lines at the left and is fed out along four lines to the right. It makes economic sense to use the same four lines to both send the word and receive it, especially as you do not want to store a word and retrieve it at the same instant!

Groups of lines that are used to get words in and out of memories are called **data buses**. A four bit data bus is shown in figure 9.6. The circuit attached to the bus is a 1 × 4 bit RAM. It can put a nibble onto the bus when the $\overline{\text{READ}}$ terminal is pulled low, and will store the contents of the bus when WRITE is pulsed up to logical 1. (It is usual to use the words read and write from the standpoint of the device that is using the memory. We therefore talk of reading words out of memory and writing words into it.)

You should have no difficulty in understanding how the RAM stores data. Consider the flip-flop labelled 0. When WRITE goes from 0 to 1, the state of the D_0 line is stored on its Q terminal. The read operation is not so obvious. The output of each flip-flop is connected to a line on the bus via a tristate. (Notice that the tristate has a circle at its enable input; this means that a logical 0 fed into that terminal allows the input of the tristate to get through to the output. A logical 1 at the enable input makes the tristate

Figure 9.5 A four bit latch

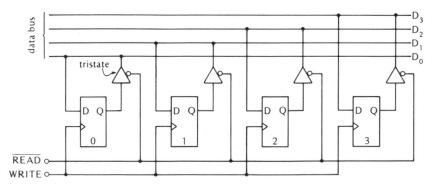

Figure 9.6 A 1 × 4 bit RAM attached to a 4 bit data bus

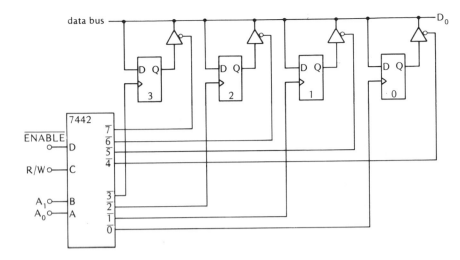

Figure 9.7 A 4 × 1 bit RAM

output behave like an open-circuit.) So when the $\overline{\text{READ}}$ line goes low, the flip-flop outputs are fed out to the four lines of the data bus. Of course, you have to make sure that nothing else is trying to put signals onto the data bus when you take $\overline{\text{READ}}$ low!

4 × 1 bit RAM

The circuit shown in figure 9.6 is called a 1 × 4 bit RAM. It stores one word that is four bits long. Because it can only store a single nibble, its use is limited. Useful memories have to be able to store more than one word. This means that the memory needs some address inputs as well as a pair of control inputs. The control inputs tell the RAM to feed out a word or to read it in; the address inputs tell the RAM which word to feed out, or where to store the word that it is reading in.

Figure 9.7 shows an example of a 4 × 1 bit RAM. It can hold four one bit words, one bit per flip-flop. The flip-flops are connected to a common data bus which carries data to and from the RAM. The RAM is controlled by four terminals, $\overline{\text{ENABLE}}$, R/$\overline{\text{W}}$, A_1 and A_0. Before we explain how you get the RAM to read and write, you need to know a bit more about the 7442 IC.

The official title of the 7442 IC is **BCD-to-decimal converter**. It is in fact a one-of-ten selector. The value of the BCD word fed into the four input terminals (D, C, B and A) selects the output ($\overline{0}$ to $\overline{9}$) which gets pulled low (figure 9.8). For example, if DCBA = 0110, then $\overline{6}$ goes low, with all the other outputs left high. The 7442 is a very useful, cheap and versatile building block.

Figure 9.9 shows you how its insides work. Each NAND gate will only feed out a logical 0 when all four of its inputs are at logical 1. If you consider the gate which provides $\overline{3}$, you will see that its inputs are all 1 when DCBA = 0011.

Returning to figure 9.7, let us suppose that we wish to store a bit in the flip-flop labelled 2. First set $A_1 = 1$ and $A_0 = 0$; this establishes the address of the flip-flop, as A_1A_0 is now the binary equivalent of 2. Next pull R/$\overline{\text{W}}$ down to 0. This informs the RAM that it is going to be written into (R/$\overline{\text{W}}$ is short for READ/$\overline{\text{WRITE}}$). The normal state of $\overline{\text{ENABLE}}$ is 1, so at this stage

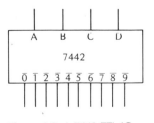

Figure 9.8 A 7442 TTL IC

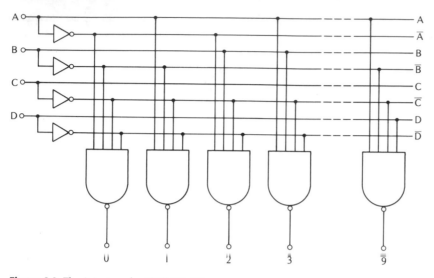

Figure 9.9 The interior of a 7442 TTL IC

DCBA = 1010. The decimal value of this is 10, a number which is not allowed in the BCD system. So none of the outputs of the 7442 will be low. Indeed, a little thought should convince you that provided D = 1 all of the outputs from $\bar{0}$ to $\bar{7}$ will stay at 1 regardless of the values of C, B and A. Finally, when the bit to be stored is waiting on the data bus, the $\overline{\text{ENABLE}}$ input is pulled low for an instant. DCBA is 0010 for that instant, so $\bar{2}$ is pulsed low. As $\bar{2}$ goes back up to 1, a rising edge is fed into the clock terminal of flip-flop 2 and the bit on the data bus is stored by it.

When you want to read data from the RAM you follow exactly the same procedure, but set R/$\overline{\text{W}}$ to 1. You select the address of the bit that you want to look at by setting $A_1 A_0$, put R/$\overline{\text{W}}$ to 1 and briefly pull $\overline{\text{ENABLE}}$ down to 0. While $\overline{\text{ENABLE}}$ is low, the bit that you want to retrieve will be put onto the data bus. Note that the last operation must always be the pulling low of $\overline{\text{ENABLE}}$; while D stays at 1, there is no danger that pulses will enter the clock terminals of the flip-flops.

RAM comes packaged in many different variations, and is quite cheap. Figure 9.10 illustrates the pinout of a 2112 RAM. It has eight address lines, so that means that $2^8 = 256$ different addresses can be fed in. There are four input/output lines, so the RAM stores four bit words. For these reasons, the 2112 is described as a 256 × 4 bit RAM.

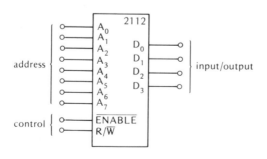

Figure 9.10 A 2112 256 × 4 bit RAM IC

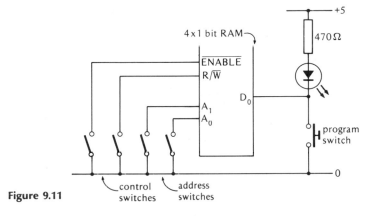

Figure 9.11

EXPERIMENT

Assembling a couple of RAM systems

Start off by building the 4×1 bit RAM shown in figure 9.7. Figure 9.11 shows how to set up the control, address and program switches. You had better check that each stage of the circuit works as you assemble it. So verify that the 7442 is doing the right things first. Then do an independent check on each of the four flip-flops; clock each one and check that it reads in 1's and 0's correctly. Then check that the tristate attached to each flip-flop is functioning. Finally knit the whole system together.

The system can be programmed to behave like any two input logic gate. You program it as follows. Set up an address with the two address switches. Set R/$\overline{\text{W}}$ to 0. Use the program switch to set up the bit to be stored; leave it for 1 and press it for 0. Then pull $\overline{\text{ENABLE}}$ low and return it high. You can check if the bit has been stored by releasing the program switch (if you had been pressing it) and setting R/$\overline{\text{W}}$ to 1. The LED will display the stored bit.

Program the system to behave like an EOR gate, so that when the two address switches are the same the LED comes on.

Now modify your circuit so that it becomes a 2×2 bit RAM. You will not need to add any components, apart from another LED to display the second bit. When you have got the system to function correctly, draw a circuit diagram for it and explain how it works.

PROBLEMS

1 Taking the 4×1 bit RAM of figure 9.11 (i.e. figure 9.7) as your basic unit, together with a NOT gate, show how you would construct an 8×2 bit RAM.

2 You are supplied with eight 2112 RAMs and a 7442 one-of-ten selector. Show how you would connect them to make a 1024×8 bit RAM with the same sort of control lines (i.e. $\overline{\text{ENABLE}}$ and R/$\overline{\text{W}}$) as a 2112 RAM.

3 Show how to connect two 7442 one-of-ten selectors and a NOT gate to make a one-of-sixteen selector.

4 The 2102 RAM is a 1024×1 bit RAM. It has separate input and output lines for the data, but has the same two control lines as the 2112 RAM.

a) How many address lines will the IC need?

b) Show how you would use a couple of active-low tristates and a NOT gate to attach a 2102 RAM to one line of a data bus.

INSIDE THE D FLIP-FLOP

By the time that you have worked through this section of the chapter you ought to have some idea of how a simple combinational logic unit like the NAND gate can be used to construct a D flip-flop. We will describe the operation of a master-slave D flip-flop; this is not quite the same as the circuit inside a 7474 IC, but it behaves in exactly the same way from the outside. It is also much easier to understand.

The clocked SR flip-flop

Look at figure 9.12. It shows a clocked SR flip-flop. You have, no doubt, recognised the $\overline{SR}$ flip-flop at the right hand side of the circuit. The two gates at the left hand side control the inputs of that flip-flop. The S and R terminals are used to set and reset the flip-flop; the C terminal allows you to choose the moment at which Q and $\overline{Q}$ respond to S and R.

Suppose that C is 0. Then both $\overline{S}.\overline{C}$ and $\overline{R}.\overline{C}$ are 1, whatever the states of S and R. You will recall that when the inputs of a $\overline{SR}$ flip-flop are both 1, its outputs do not change. Now further suppose that S = 1 and R = 0. When C is raised to 1, $\overline{S}.\overline{C}$ = 0 and $\overline{R}.\overline{C}$ = 1. This sets the $\overline{SR}$ flip-flop, so that Q = 1 and $\overline{Q}$ = 0. If C is now lowered back to 0, the outputs will remain frozen, whatever S and R do thereafter.

So the clocked SR flip-flop can only change its output state when its C terminal is held at 1. A logical 1 at S sets the flip-flop and a logical 1 at R will reset it. The C terminal cannot, on its own, alter the flip-flop outputs.

The D Latch

The addition of a single NOT gate converts a clocked SR flip-flop into a D latch; look at figure 9.13. In order to set Q to 1, $\overline{DATA . ENABLE}$ must = 0 for an instant. This requires both ENABLE and DATA to be 1, automatically ensuring that the other input of the SR flip-flop ($\overline{\overline{DATA} . ENABLE}$) is 1. Similarly, to reset the flip-flop (i.e. make Q = 0) we must have ENABLE = 1 and

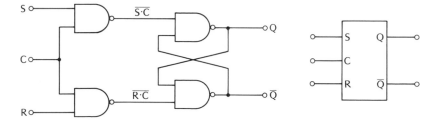

Figure 9.12 A clocked SR flip-flop

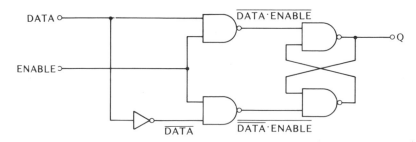

Figure 9.13 A D latch

DATA = 0. This makes $\overline{\text{DATA . ENABLE}} = 1$ and $\overline{\overline{\text{DATA}} . \text{ENABLE}} = 0$. When ENABLE is 0, both inputs of the $\overline{S}\overline{R}$ flip-flop are 1, so its outputs are frozen.

The D latch is a primitive memory cell. Indeed, it is sufficiently good to act as the basic memory cell of the RAM systems discussed in the last section. You present the bit to be stored at the DATA terminal, push ENABLE high for an instant and the bit is safely stored at Q. Provided ENABLE remains at 0, Q stays frozen regardless of the state of DATA.

The D latch does, however, suffer from one defect. It is transparent. When ENABLE is high, Q will follow DATA. (Convince yourself that this is true.) This fault is common to all clocked flip-flops; when the clock line is 1, the output of the flip-flop is set by the inputs and when it is 0 the outputs are frozen. On the other hand, edge-triggered flip-flops are not transparent. Their outputs can only change at the instant that their clock lines go from 0 to 1. As we shall see in the next section on binary counters, this property is crucial for a number of important sequential systems.

Transition gates

One method of converting a transparent flip-flop into an edge-triggered one is to insert a transition gate into the clock line. This device shoots out a brief pulse every time that a rising edge is fed into it, but ignores falling edges.

You could probably work out a suitable circuit based on an *RC* network; have a look at figure 2.29. A neater alternative circuit is shown in figure 9.14. Every time that a rising edge enters at A, a 15 ns pulse appears at Q. The system relies on the fact that the output of a TTL gate changes about 5 ns later than its input changes. So the chain of three NOT gates behaves like a single NOT gate with a built-in delay of about 15 ns. If you study the timing diagram in figure 9.14, you will see that this delay allows both inputs of the NAND gate to be logical 1 for a short instant after A goes from 0 to 1. So $\overline{Q}$ goes to 0 for 15 ns. A falling edge fed in at A produces no pulse at Q because one of the inputs of the NAND gate is always at logical 0 while the system is settling down into its new state.

Figure 9.15 illustrates a D flip-flop assembled from NAND gates which uses a transition gate inserted into the ENABLE line of a D latch. The flip-

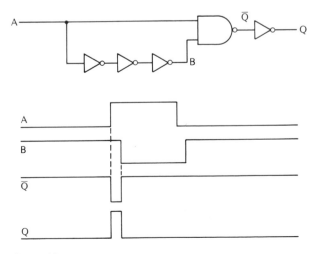

Figure 9.14 A transition gate

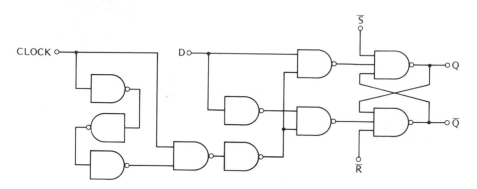

Figure 9.15 A D flip-flop

flop will only be transparent for 13 ns after each rising edge enters the
CLOCK input. The next section explains how a D flip-flop can be assembled
which is virtually never transparent.

The master-slave flip-flop

Figure 9.16 shows one way in which a D flip-flop can be assembled. It
consists of a pair of clocked SR flip-flops in series. They are clocked with
opposite signals, so that only one of the two is transparent at any one
instant.

Consider the master flip-flop first. If you look at the first three rows of
the timing diagram in figure 9.16, you will notice that when its clock line ($\bar{C}$)
is 1, its output (D′) follows its input (D). So when C is 0, the state of D′
follows that of D. On the other hand, when C is 1 the state of D′ is frozen.

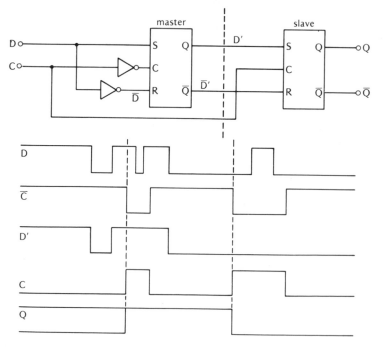

Figure 9.16 A master-slave D flip-flop

Now consider the slave flip-flop i.e. look at the last three rows of the timing diagram. Initially, C is 0 so Q is frozen at some unknown value. The input (D') changes (the master flip flop is transparent when C = 0) but the output of the slave does not. Then C rises to 1. The slave flip-flop is now transparent, so Q = D'. Because the master flip-flop is now opaque ($\overline{C}$ = 0), D' is frozen. So while C = 1 the state of Q remains frozen regardless of changes of D. Q takes the value of D at the instant that C rises from 0 to 1, and at no other time.

EXPERIMENT

Assemble the D latch shown in figure 9.13. Investigate its properties and summarise them in a timing diagram.

PROBLEMS

1 Draw the circuit of figure 9.15, replacing each NAND gate with a NOR gate, but leaving the NOT gates alone. Describe the properties of your circuit. Draw a timing diagram for it.

2 Figure 9.16 shows a circuit which behaves like a D flip-flop that is triggered by a rising edge entering C. Draw a similar circuit for a falling edge triggered D flip-flop.

3 The D flip-flops on a 7474 IC have $\overline{S}$ and $\overline{R}$ inputs. Show how you would adapt the D latch of figure 9.13 so that it had $\overline{S}$ and $\overline{R}$ inputs as well as DATA and ENABLE.

BINARY COUNTERS

A **binary counter** is a logic system which counts the number of digital pulses which are fed into it. The value of the count is fed out of the device in the form of a binary word. In order to build a binary counter you need to use an edge-triggered flip-flop. Two types of counter can be assembled from D flip-flops; we shall concentrate on the simpler ripple counter for the moment, leaving synchronous counters for Chapter Ten.

Figure 9.17 (overleaf) shows a one-bit binary counter. Since it has only one output (A) it can only count from 0 to 1. At first glance this is not very useful, but you can put one-bit counters in series to count up as far as you like.

How does the counter work? Look at the timing diagram of figure 9.17. A square wave is fed into $\overline{\text{CLOCK}}$. That signal is inverted before going into the clock terminal of the flip-flop. Since the flip-flop is triggered by rising edges, it will be triggered by the falling edges of $\overline{\text{CLOCK}}$. (You will find out later why binary counters always trigger on falling edges.) The output of the flip-flop is fed back to its input; $\overline{Q}$ is connected to D. So when it is triggered Q becomes what $\overline{Q}$ was the instant before. In other words, every time that $\overline{\text{CLOCK}}$ falls from 1 to 0 the state of A changes.

Put two one-bit counters in series and you have a two-bit counter. They have to be connected as shown in figure 9.18, with $\overline{Q}$ of the first flip-flop providing the clock pulses for the second one. The second flip-flop will

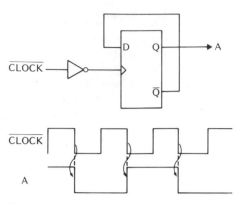

Figure 9.17 A one bit binary counter

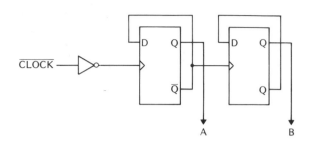

Figure 9.18 A two bit binary counter

change state every time that $\bar{Q}$ of the first flip-flop goes from 0 to 1. Since A = Q, B will change state every time that A falls from 1 to 0. In its turn, of course, A changes state every time $\overline{CLOCK}$ falls from 1 to 0.

The behaviour of the two-bit counter is summarised in the timing diagram of figure 9.19. Initially, BA = 00. The first falling edge of $\overline{CLOCK}$ swaps the state of A so that BA = 01. The next falling edge puts A back to 0; the falling edge out of A makes B swap state, so that BA = 10. The third falling edge puts A back up to 1 so that BA = 11. Finally, the fourth falling edge pushes A down to 0 which, in turn, pushes B down to 0 so that BA = 00. The four pulses fed into $\overline{CLOCK}$ have got us back to our starting point of BA = 00.

Number of clock pulses	B	A
0	0	0
1	0	1
2	1	0
3	1	1
4	0	0
5	0	1

If you study the truth table showing the values of A and B as a function of the number of clock pulses fed into the counter, it should be evident that BA holds the BCD value of the pulse count on a scale of 4. That is, BA tells you how many pulses have been fed into the counter since it was last reset to BA = 00. The counter can feed out 0, 1, 2 or 3 as the value of the count. When it tries to count 4 it resets to 0.

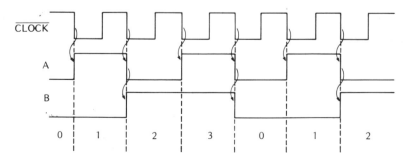

Figure 9.19 The timing diagram for a two bit binary counter

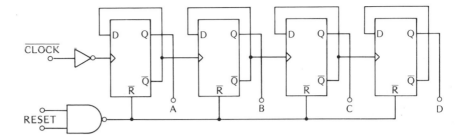

Figure 9.20 A four bit binary counter with reset

Obviously, a two-bit counter can count up to three pulses before resetting to zero. If you want to count more pulses, you just have to add more flip-flops to the counter chain. For example, figure 9.20 shows four flip-flops organised as a counter capable of counting up to 15 pulses before resetting to zero. The count is given by the number stored in the binary word DCBA ($1111_2 = 15_{10}$). Note the use of a NAND gate to reset all of the flip-flops before the counting is started.

Switch debouncing

So far we have concentrated on describing how the binary counter works without worrying too much about preparing pulses for it to count. Obviously it could count pulses fed out by an oscillator, but it would also be useful to be able to count the number of times that a switch was pressed. The business of generating suitable pulses for a binary counter is not trivial as they need to have fast clean falling edges.

Figure 9.21 shows a suitable pulse preparer which converts analogue waveforms into countable pulses. The analogue signal (a sine wave in our example) arrives at the left, its DC level is shifted up to $+2.5\,V$ and it is finally squared off with a CMOS Schmitt trigger. The size of the input waveform is not critical, provided that it is not so large that it blows the clamp diodes of the NOT gates.

Switches are not so easy to interface to a binary counter. When switches are closed they **bounce**. That is, the contacts bounce apart after their initial encounter with each other. A well designed switch may have a bounce time of less than 1 ms, but the contacts will bounce several times before they settle down. This means that it is virtually impossible to get a clean falling edge out of a push switch/resistor network. When the switch is closed it feeds out several falling edges instead of just one.

Figure 9.22 illustrates a possible solution to the problem. Suppose that the switch is initially open. Then $\overline{CLOCK}$ will be high as the 10 kΩ resistor

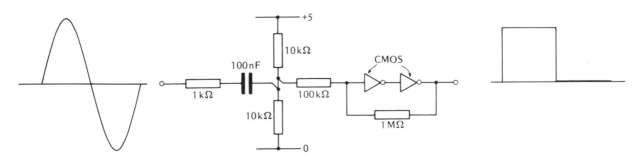

Figure 9.21 A sine-to-square wave converter

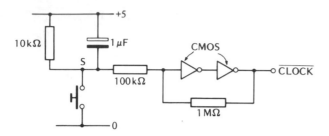

Figure 9.22 A simple switch debounce circuit

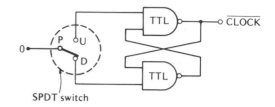
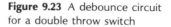

Figure 9.23 A debounce circuit for a double throw switch

pulls the input of the Schmitt trigger up to 1. When the switch is closed S goes immediately down to 0 (the capacitor charges up very quickly via the switch). So $\overline{CLOCK}$ goes cleanly down to 0. As the switch contacts bounce for the next millisecond or so, S will make several attempts to rise up to +5 V. The 10 kΩ resistor in parallel with the 1 µF capacitor ensure that S takes $0.7 \times 10 \times 1 = 7$ ms to get up to +2.5 V. So during the bounce time, S never gets anywhere near the switching threshold of the Schmitt trigger. $\overline{CLOCK}$ stays at 0. Of course, when the switch is released there will be a 7 ms delay before $\overline{CLOCK}$ rises up to 1; this is usually not important as only the falling edges of $\overline{CLOCK}$ are counted.

The debounce circuit we have just been discussing has one drawback. It puts pulses onto the supply lines when the switch is closed. Those pulses could upset other circuitry connected to the same supply. If this is a problem you can usually solve it by putting a large capacitor (at least 10 µF) across the supply rails close to the switch debouncer. It acts as a store for the charge needed to be dumped on the 1 µF capacitor when the switch is closed.

A neater solution to the problem of switch bounce is shown in figure 9.23. It does require the use of a more complex switch, but there are no problems with unwanted pulses on the supply rails. The switch goes by the grand title of **single pole double throw** (**SPDT**). The **pole** is the bit marked P; the two **throws** are labelled U and D. The pole is in contact with either one of the throws; when it bounces off a throw, the pole is not in contact with anything. It follows that the switch is able to pull one or other of the inputs of the $\overline{SR}$ flip-flop down to 0. Because the NAND gates are TTL, during bounces both inputs float up to 1.

The four diagrams in figure 9.24 show what happens as the switch contacts are changed over. As soon as the pole touches one of the throws $\overline{CLOCK}$ is set or reset. The flip-flop then remains frozen until the pole touches the other throw. It ignores any bounces of the pole against the throw.

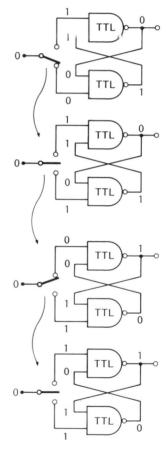

Figure 9.24 How the circuit of figure 9.23 works

EXPERIMENT
Investigating the behaviour of a two-bit binary counter

Assemble the circuit shown in figure 9.25. Check that it obeys the timing diagram of figure 9.19, in particular that the system counts the falling edges of $\overline{CLOCK}$. (Note that we have used a debouncing flip-flop to replace the usual NOT gate at the input of a binary counter.)

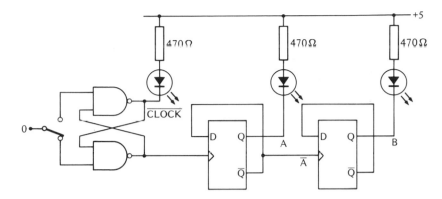

Figure 9.25

PROBLEMS

1 Draw the circuit diagram of a three bit binary counter. Draw a timing diagram to show what happens to the three outputs when eight pulses are fed into the counter; assume that all three outputs are 0 before the first pulse arrives.

2 How many flip-flops do you need to build a binary counter which can count up to 1000 pulses?

3 Look at figure 9.26. The circuit is a three bit down counter. Draw the two lines of the timing diagram which are shown; draw the three lines for the outputs underneath. Explain why the counter is called a down counter.

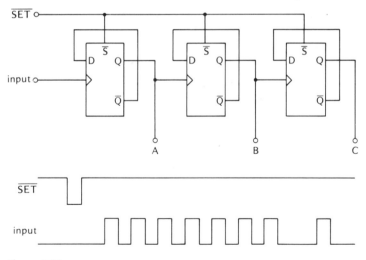

Figure 9.26

4 Draw a circuit diagram of an SPDT switch debounce circuit which employs a NOR gate bistable.

5 An electronic lap counter records the number of times that a bicycle goes around a circuit. A long rubber tube is stretched across the track; when it is squeezed the increased air pressure in the tube closes a switch. The lap counter indicates the count with a row of ten LEDs labelled 0 to 9. The lit LED tells you the value of the count. A master reset switch makes the counter show 0 when it is pressed. Draw a suitable circuit for the system; explain how it works. (Do not forget that a bicycle has two wheels!)

DECIMAL COUNTERS

Human beings can be taught to read binary numbers, but they are rarely at ease with them. (Which makes more immediate sense, 1101_2 or 13_{10} ?) Since the end result of most electronic counting operations has to be presented to a human being, it makes sense to use a decimal counting network rather than a binary one. More specifically, the value of the count has to be presented in decimal form; digital systems have to count in binary, but they can be designed to present the count in decimal!

The 7493 IC

Useful decimal counters need to be able to count up to several significant figures. Each decimal digit needs four flip-flops to store it, so a useful counter can easily contain over twenty flip-flops. A convenient way of condensing the number of units needed to build a decimal counter is to bunch the flip-flops together in groups of four. This is the grouping present in a 7493 IC, whose symbol is shown in figure 9.27. It is a four-bit binary counter, counting falling edges fed in at INPUT and feeding out the value of the count as the binary word DCBA. The two reset pins are combined by an internal AND gate to generate the RESET signal. A logical 1 on both reset pins makes DCBA = 0000.

The whole IC performs exactly like the four-bit counter shown in figure 9.20. The reasons for the apparently arbitrary reset arrangements will become clear later.

Cascading counters

Using the 7493 as your basic unit it is easy to build large binary counters. Figure 9.28 shows how two can be put in series to create an eight-bit

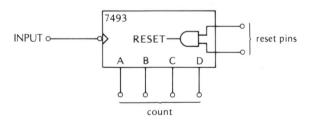

count

Figure 9.27 A 7493 TTL IC

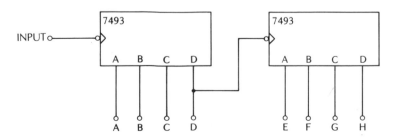

Figure 9.28 An eight bit binary counter

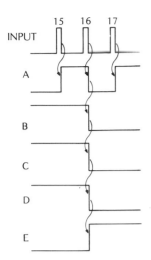

Figure 9.29

counter. Note how the **most significant bit** (**msb**) of the first counter (D) provides clock pulses for the second counter. Essentially, the second counter counts the number of times that the first counter is reset. Figure 9.29 makes this a bit clearer to see. It shows a portion of a timing diagram for the circuit of figure 9.28. Both counters were initially reset, so that HGFEDCBA = 00000000. Pulses were then fed into INPUT; figure 9.29 shows what happens when the fifteenth, sixteenth and seventeenth pulses go in. After the arrival of the fifteenth pulse HGFEDCBA = 00001111. The sixteenth pulse sets DCBA = 0000; as D falls from 1 to 0 E swaps from 0 to 1. So HGFEDCBA = 00010000. The seventeenth pulse sets HGFEDCBA = 00010001, etc. You can see that the value of HGFE is going to increase by 1 for every sixteen pulses fed into INPUT.

The need to be able to cascade counters in this fashion is the reason for making them count falling edges. So although individual flip-flops can be obtained in rising-edge or falling-edge configurations, binary counters are invariably triggered by falling edges.

The 0 to 5 counter

The 7493 comes into its own when you want to count fewer than fifteen pulses. By hooking the two reset pins to the counter outputs you can make the device reset on any value below 13 except 7 and 11. Extra logic gates allow you to set up the 7493 so that it resets to zero after any count you like.

As an example, consider the counter shown in figure 9.30. The reset pins

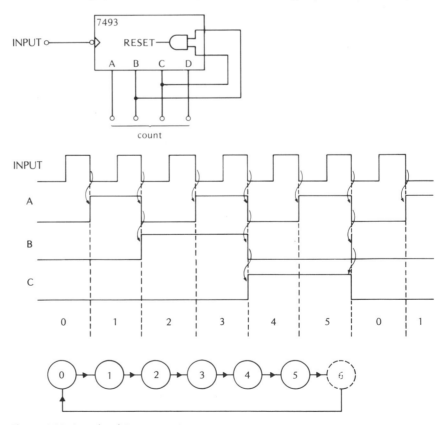

Figure 9.30 A scale of 5 counter

are connected to B and C, so the counter will reset DCBA to 0000 whenever B . C = 1.

Number of pulses	D	C	B	A	RESET
0	0	0	0	0	0
1	0	0	0	1	0
2	0	0	1	0	0
3	0	0	1	1	0
4	0	1	0	0	0
5	0	1	0	1	0
6	0	1	1	0	1

The truth table shows that if the counter is initially reset, then RESET = 1 immediately after the sixth pulse has been counted. That sixth pulse therefore resets the counter, i.e. makes DCBA = 0000 once more.

The operation of the counter may become clearer if you study the timing diagram of figure 9.30. The value of the count goes from 0 to 5, 6 exists for a small fraction of a microsecond and then the count goes from 0 to 5 again. This behaviour is summarised in the state diagram.

At first glance, you might think that the reset signal for a 0 to 5 counter ought to be given by:

$$\text{RESET} = \bar{D} . C . B . \bar{A}$$

(The BCD version of 6_{10} is 0110_2.) In fact, because the counter is counting up the binary scale, DCBA = 0110 is the first time that C . B = 1. So making RESET = C . B is sufficient to get the counter to reset on every sixth pulse.

BCD counters

If a counter is to display the value of the count in decimal form, then the binary word fed out by the chain of flip-flops must be juggled around by a set of logic gates to generate signals which can run a seven segment LED display. For example, a single BCD digit needs four bits to represent it, but a seven segment LED needs seven; a set of logic gates, known as a converter, is needed to combine the four bits of the BCD binary word and generate the seven signals needed to light up the appropriate LEDs. The 7447 IC is an example of such a converter. Feed in a BCD digit between 0 and 9 and it sinks current from the LEDs needed to display that digit (figure 9.35).

The counter system shown in figure 9.31 can count from 0 to 99. Initially, the RESET line needs to be briefly raised to logical 1. This resets both of the counters. Each counter feeds its value into a BCD to seven segment converter which, in turn, feeds signals into a seven segment LED display. So just after reset, both displays show 0.

The pulses to be counted are fed in at INPUT. As these are counted by the first binary counter, the LED display shows, in decimal, the value of the count. The tenth pulse resets the first counter to 0 ($1010_2 = 10_{10}$) and fires a falling edge into the second counter. So the tenth pulse leaves the LEDs on the right displaying 1, those on the left showing 0. It follows that the LEDs on the right display how many groups of ten pulses have entered INPUT. The LEDs on the left display the number of pulses left over after they have been grouped in tens. If you place the LED displays next to each other in the right order (i.e. units on the right, tens on the left) then the value of the count is displayed in standard decimal form.

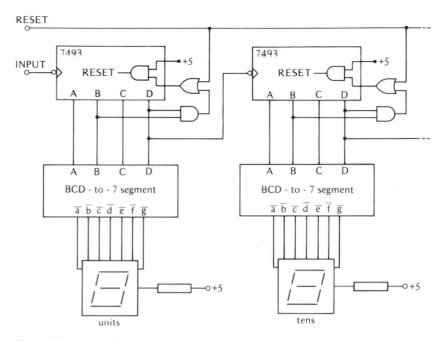

Figure 9.31 A 0 to 99 counter

Notice that the system which counts the tens is exactly the same as the one which counts the units. So a third identical BCD counter could be added to count hundreds of pulses, a fourth to deal with thousands etc.

EXPERIMENT

Using a 7493 binary counter and displaying the value of the count in three different ways

Assemble the circuit of figure 9.32. The astable is shown in more detail in figure 9.33; you can use a TTL version if you want to use larger capacitors, but you will not need a pull-up resistor for $\overline{E}$. Generate falling edges at Q by pressing the switch. Check that the 7493 counts those edges, i.e. that the LEDs change each second. If the counter does not work, check that at least one of the reset pins is at 0 and that you have included the external wire link between pins 1 and 12 of the 7493.

Press the switch until all four LEDs are lit, i.e. DCBA = 0000. Then briefly press the switch so that a single edge goes into the counter. Record the

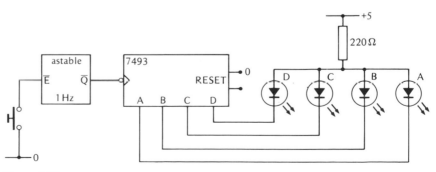

Figure 9.32

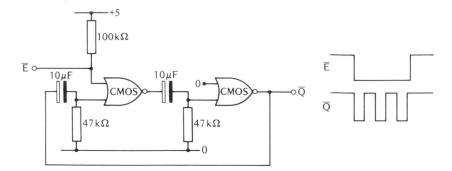

Figure 9.33

new value of DCBA. Carry on feeding in single edges, noting the value of DCBA each time, until DCBA = 0000 once more. (If you have bounce problems with your switch, slow down the astable and keep the switch pressed.) Press the switch and watch the LEDs count in binary. Then replace them with the BCD-to-decimal converter and ten LED array shown in figure 9.34. The value of the count should now be displayed in a more comprehensible form. Arrange the counter so that it only counts from 0 to 9; the ten LED array cannot display 10 to 15!

Finally, replace the 7442 and ten LED array with a 7447 BCD-to-seven segment converter and seven segment LED (figure 9.35).

Work out how to get the system to count from 0 to 5. Try out your design. Draw a circuit diagram of it, and explain how it works.

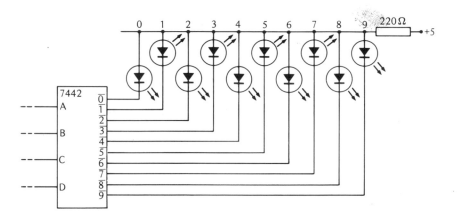

Figure 9.34

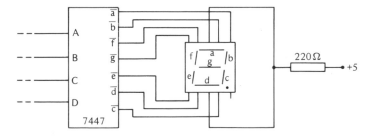

Figure 9.35

PROBLEMS

1 Draw diagrams to show how a 7493 can be converted into counters which count up to

a) 3   **b)** 12   **c)** 6   **d)** 14.

2 Have a look at figure 9.36. It shows the first two counting stages of a frequency meter. It displays the number of pulses which enter INPUT in each second. Study the circuit. Explain how it works and what is displayed on the seven segment LEDs as time progresses.

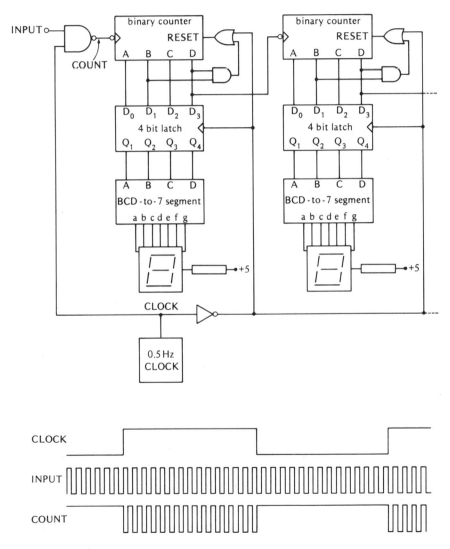

Figure 9.36 A frequency meter

3 An electronic metronome flashes an LED at a rate of exactly 1 Hz. It uses the mains 50 Hz waveform (suitably attenuated!) as its reference. Draw a circuit diagram of a system which will take in a 5 V peak sine wave at 50 Hz and flash the LED on and off at 1 Hz.

4 A digital clock runs off the mains supply and uses its frequency as its reference. It is American, so the mains frequency is 60 Hz. Design a counter system which will take in a 60 Hz square wave and will feed out one falling edge every minute. You should be able to do it with four 7493 ICs.

5 A digital clock displays hours and minutes on four seven segment LEDs. It is a 24 hour clock, i.e. it displays up to 23 hours 59 minutes before resetting to 00 hours 00 minutes. Design a counter system which will count falling edges which arrive at one minute intervals and display the time as described.

6 A period meter displays the period of a waveform that is fed into it in milliseconds. The period is shown on three seven segment LEDs, so the device can measure periods from 0 to 999 milliseconds. Given an oscillator which has a frequency of 1.000 kHz, design a suitable counting system for the period meter.

FREQUENCY DIVIDERS

When a binary counter is fed a square wave rather than a series of pulses, it performs the operation of frequency division. Look at figure 9.37. A signal with a frequency f is fed into the counter. If you study the timing diagram, you will see that A feeds out a signal with a frequency of $f/2$. A goes through one cycle for every two cycles of the input. Similarly, B has a frequency which is half that of A, i.e. $f/4$; C has a frequency which is half that of B etc.

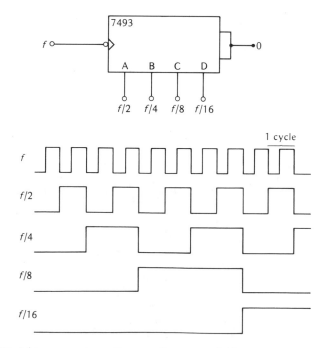

Figure 9.37 A binary counter acting as a frequency divider

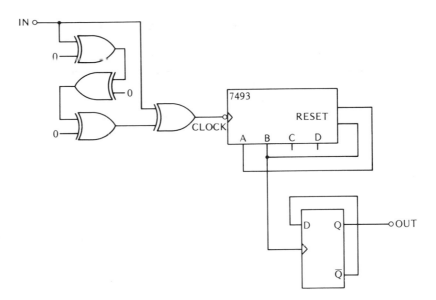

Figure 9.38 A frequency divider

What do you do if you want to divide the frequency by a number which is not a power of 2? For example, how do you get a waveform with a frequency of $f/3$ from a signal which has a frequency f? A possible solution is provided by the circuit of figure 9.38. The square wave at f is fed in at IN and a square wave at $f/3$ appears at OUT.

In order to appreciate how it works you need to study the timing diagram of figure 9.39. The signal at IN is first processed by a transition gate so that both its rising and falling edges feed pulses into the 7493 counter; the pulses are only 15 ns long, but their falling edges are crisp and clean. So each cycle of IN feeds two pulses into CLOCK. The counter resets when B . A = 1. So BA = 00 for every three pulses of CLOCK ($11_2 = 3_{10}$). Every reset of the counter is followed by a single rising edge from B; this makes

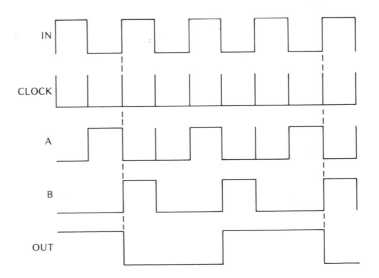

Figure 9.39 The timing diagram for a frequency divider

OUT change state. Therefore in order to make OUT change state twice (i.e. go through one cycle) the counter has to be reset twice. This requires six pulses of CLOCK or three cycles of IN.

Obviously, you can generate output waveforms with frequencies of f/n where n is any integer between 1 and 15 with this circuit; you simply make the 7493 counter reset when it has counted n pulses.

Many complex sequential systems, such as microprocessors, digital voltmeters and frequency counters require several timing signals. These have to synchronise the operations of different parts of the system so that binary words are shuffled around in the correct order. The timing signals have to keep exactly in step with each other; this is clearly impossible if each signal is generated by a different oscillator. It can only be done by deriving all of the timing signals from a single waveform. A single oscillator produces a high frequency square wave which is divided down to produce the various frequencies required. Then even though the frequency of the oscillator may drift up or down, the derived frequencies will have a constant relationship with each other.

As an example, consider an electronic organ. You could make it out of a large number of oscillators, each of which has to be separately tuned to the correct frequency. Each oscillator will drift, so the whole system will need to be tuned periodically. Alternatively, you could have a single high frequency oscillator, and derive all of the frequencies that you needed by using appropriate frequency dividers. A scheme which gives all of the white notes of the piano is shown in figure 9.40. A single oscillator, which can be tuned, feeds out a signal at 184 320 Hz. This is simultaneously fed into seven different frequency dividers to produce the top octave of the piano range. The next octave down is generated by seven divide-by-two units (i.e. one-bit binary counters). Successive octaves can be generated by dividing the frequencies of the previous octave by two. As all of the outputs are derived

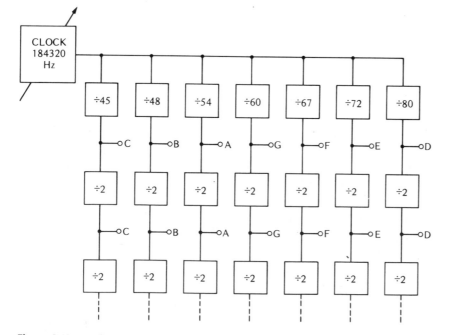

Figure 9.40 An electronic organ

from the same original signal, the organ will always be in tune with itself. In order to tune it to another instrument, you simply adjust the frequency of the oscillator.

EXPERIMENT

Assembling and testing a frequency divider

Start off by assembling a system which will convert a sine wave into a square wave whose edges can be reliably counted by a 7493. A suitable circuit is shown in figure 9.21; check that its output will clock a binary counter properly.

Then assemble the transition gate shown in figure 9.38. You will not be able to see the pulses that it feeds out when edges are pumped into it as they only exist for a fraction of a microsecond. But you can see if they clock a 7493 properly. It pays to keep all of the wires of the transition gate short and neat. You may need to put a TTL NOT gate in front of the transition gate to sharpen up the edges fed into it.

Finally assemble the rest of the frequency divider shown in figure 9.38. Feed in a sine wave (above 300 Hz) and check that the output has exactly one-third of its frequency. You might care to use a TTL NOT gate and loudspeaker to listen to the original and divided down square waves.

Now amend the system so that it feeds out a signal whose frequency is one tenth of the frequency of the sine wave fed into it. Find out the range of frequencies and range of amplitudes of the input that the system can cope with. Draw a circuit diagram of the whole system. Explain how it works, with the help of a timing diagram.

PROBLEM

Use a timing diagram to explain how the EOR transition gate of figure 9.38 feeds out a pulse every time its input changes state.

SEQUENCES

Combinational logic systems combine digital signals to create other digital signals. The ultimate combinational logic system is the ROM; the address pins take in digital signals and the data pins feed out other signals as a consequence. Digital counters, on the other hand, produce digital signals which change with time in an orderly fashion. By combining both sorts of system you can easily generate any time varying pattern of digital signals that you like. When you put a digital counter in front of a ROM, you generate a set of digital signals which follow a sequence.

CYCLIC SEQUENCES

Suppose that you want to generate a three bit sequence which has four steps that are endlessly repeated over and over again. The three bits are displayed by three LEDs called X, Y and Z. The four steps of the sequence are as follows. First X is lit, then Y, then Z, then Y again. The sequence is then repeated again, and again.

The first step in designing the system is to draw up a truth table. It will have four lines, one for each step; each line will be numbered with the binary word BA. We shall call the three outputs which run the LEDs X, Y and Z. Bearing in mind that you need a logical 0 to light an LED, the truth table looks like this:

B	A	X	Y	Z
0	0	0	1	1
0	1	1	0	1
1	0	1	1	0
1	1	1	0	1

Note that we have chosen BA so that it corresponds to the output of a binary counter. So we can generate the four values of BA in the order of the truth table without too much difficulty. Figure 9.41 gives you an idea of how to do it.

We can now write down expressions for X, Y and Z in terms of B and A.

$$\overline{X} = \overline{B} . \overline{A} \qquad \text{therefore} \quad X = \overline{\overline{B} . \overline{A}}$$
$$\overline{Y} = \overline{B} . A + B . A = A \qquad \text{therefore} \quad Y = \overline{A}$$
$$\overline{Z} = B . \overline{A} \qquad \text{therefore} \quad Z = \overline{B . \overline{A}}$$

Figure 9.41 shows how to implement these three expressions with NAND and NOT gates. That combinational logic system takes its inputs (B and A) from a binary counter which resets on every fourth clock pulse. Clock pulses are supplied to the counter at the rate of one per second, so the whole sequence lasts for four seconds.

Figure 9.41 shows clearly the two parts of the system. There is the digital counter which steadily counts up to 3 before resetting to 0. Then there is the combinational logic system which combines the counter outputs to generate X, Y and Z.

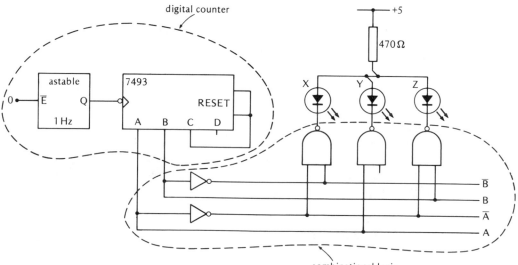

Figure 9.41 A sequence generator

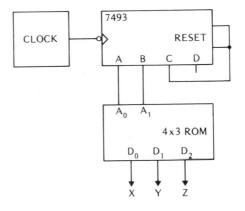

Figure 9.42 A general sequence generator

You can take a generalised view of the logic system and replace it with a 4 × 3 bit ROM as shown in figure 9.42. The two address pins of the ROM (A_1 and A_0) are connected to the counter outputs, and the three data pins provide the signals X, Y and Z. The ROM contains the truth table of the system; each of the four words stored within it represents one line of the truth table. Each word is stored at a location within the memory, and each location has a unique address (00, 01, 10 or 11). When the counter feeds an address into the ROM, the word stored at that location appears at D_2, D_1 and D_0.

A simple ROM which has been programmed with the truth table is shown in figure 9.43. A 7442 one-of-ten selector is used to decode the address; each address lowers one of the four outputs ($\bar{3}$, $\bar{2}$, $\bar{1}$ and $\bar{0}$) down to 0. Three quad NAND gates combine the signals from those four lines to generate D_2, D_1 and D_0. Consider the gate which generates D_0. D_0 will be 1 when any one of its inputs is 0 (the inputs only go to 0 one at a time anyway). So when the address is 01, 10 or 11, $D_0 = 1$. The address 00 leaves all four inputs at 1 so D_0 is 0. Similarly, $D_1 = 1$ when the address is 00 or 10, and $D_1 = 0$ when the address is 01 or 11. (Convince yourself.)

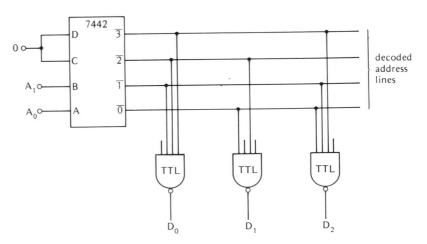

Figure 9.43 A ROM for figure 9.41

You may be asking yourself the question: why bother with ROM when the original set of logic gates was so much less complicated? Well, using ROM requires more hardware but it does make the problem easier to solve. Once you have written down the truth table you simply write it straight into the memory. Each of the decoded address lines corresponds to one line of the truth table; hooking a NAND gate input to one of the lines stores a 1, and the absence of a connection stores a 0. No Boolean algebra at all!

The advantages of the ROM approach to generating a cyclic sequence will become clearer when we look at the next example.

Address			Stored word		
C	B	A	X	Y	Z
0	0	0	0	1	1
0	0	1	0	0	1
0	1	0	0	0	0
0	1	1	1	0	0
1	0	0	1	1	0
1	0	1	1	0	0
1	1	0	0	0	0
1	1	1	0	0	1

The truth table above shows the sequence that we want to produce. X, Y and Z drive LEDs as before, so a 0 in the table corresponds to a lit LED. The

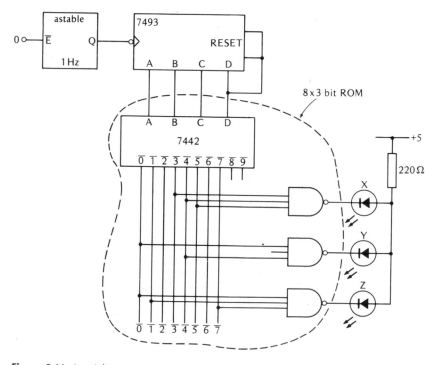

Figure 9.44 An eight step sequence generator

whole sequence has eight steps, so we need a three bit address CBA to store the whole table in ROM. Each of the eight locations within the ROM must, of course, store the three bit word XYZ appropriate to its address.

Figure 9.44 shows the circuit needed to make the sequence. The binary counter feeds each of the addresses in turn into the 7442 one-of-ten selector. The links between the NAND gates and the eight decoded address lines are made according to the truth table. For example, X is 1 when the address is 011, 100 or 101. Its NAND gate is therefore hooked up to $\bar{3}$, $\bar{4}$ and $\bar{5}$.

In this example, the ROM approach gives a much simpler circuit than the custom-built combinational logic system does. Take a look at figure 9.45. You can check for yourself that it converts the binary counter outputs into the correct signals to run the LEDs. It should be clear that designing the system requires a certain amount of experience with handling Boolean algebra, whereas putting the truth table into the ROM is very straightforward. Furthermore, the ROM circuit can be easily adapted to generate an enormous number of different sequences; the same is not true for the network of gates in figure 9.45.

EXPERIMENT
Investigating a sequence generator

Assemble the circuit of figure 9.44. Check that it generates the correct sequence. Then adapt it so that it becomes the system of figure 9.43, using triple input NAND gates. Check that it generates the correct sequence.

Call the three LEDs R, A and G. Adapt the system so that the LEDs go through a traffic light sequence, i.e. R, R and A, G, A, R, R and A, G etc. When you have got it to work, draw out the circuit diagram and the truth table.

Finally, arrange to have two LEDs go through this sequence; L, −, L and R, −, R, −, L, −, L and R, etc. Draw out the circuit diagram and truth table.

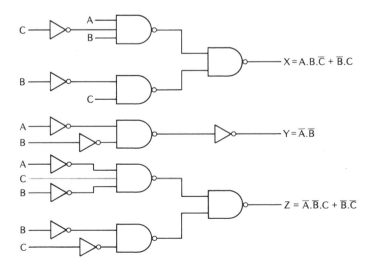

Figure 9.45 A collection of gates to replace the ROM of figure 9.44

PROBLEMS

For each of the systems described below you have to design a suitable circuit. Make sure that you show all the steps in your thinking (truth tables etc.) and explain how the system works as well as drawing the circuit diagram. (The circuit of figure 9.46 might be useful.)

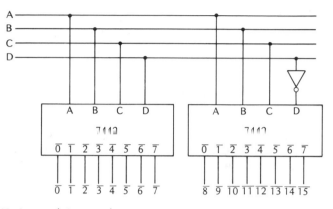

Figure 9.46 A one-of-sixteen selector

1 A counter has a row of nine LEDs to display the value of its count. The number of lit LEDs is equal to the number of pulses counted. Design a suitable 0 to 9 counter.

2 Seven LEDs are arranged so that they can display the numbers 1 to 6 in the format used in dice. The numbers are displayed in order, from 1 to 6, for one second each. The sequence repeats itself continuously.

3 A four bit latch and a push switch are added to the system of question 2 to convert it into an electronic die. When the switch is pressed and released a number from 1 to 6 is displayed by the LEDs; the actual number which appears is random.

4 An electronic circuit controls a piece of machinery. The machine can do four operations on material fed into it. They are called reeling, writhing, uglifying and distracting; each operation is controlled by an active-low input. The circuit makes the machine go through a one minute cycle of operations which is repeated. It reels for the first thirty seconds and the last five seconds. Ten seconds after the start it writhes for twenty seconds. Uglifying takes up the last fifteen seconds. The distracting is done in the middle twenty seconds.

5 An egg-packing machine has a seven segment LED which displays how many eggs have to be added to a box of six to fill it up. When an egg arrives on the conveyor belt, it triggers a switch as it falls in the box. When the box is full, the belt stops until it has been removed. The electronic system has to control the belt, count the eggs, run the display and sense the presence of the box.

6 An automatic SOS bleeper contains a sequence generator which controls a loud oscillator/speaker system. The sequence generator feeds the oscillator with dots and dashes, so that SOS is continually fed out by the speaker.

TRIGGERED SEQUENCES

Sequences which repeat themselves time after time are useful, but there is also a need for systems which only go through a sequence once. For example, the traffic lights at a set of crossroads go through a cyclic sequence, endlessly repeating the same pattern. On the other hand, the lights at a pedestrian crossing only need to go through one cycle of their sequence when someone wants to cross the road.

For a sequence generator to switch itself off at the end of the sequence, the binary counter needs to be able to switch off its clock pulses. Look at figure 9.47a; it shows an example of a circuit which can be triggered into feeding out only one cycle of a sequence. Let us assume that initially all of the flip-flops are reset, i.e. $\bar{Q} = 1$, DCBA = 0000. The slow astable will be off, so DCBA will be static at 0000. The fast astable will also be off because $\bar{0}$ will be low, forcing $\bar{N}$ high. The whole system is waiting to be triggered.

You trigger it by pressing the switch and releasing it. The D pin of the flip-flop is held at $+5\,$V, so as soon as the switch is released, $\bar{Q} = 0$. (Note that $\bar{9}$ is high at this point, so that $\bar{R} = 1$; had it been 0, the pressing of the switch would have had no effect on $\bar{Q}$.) The slow astable is enabled, feeding a train of falling edges into the 7493.

The counter will increase the value of DCBA by 1 each second until DCBA = 1001. This address puts $\bar{9}$ low. As $\bar{R}$ is pulled low by $\bar{9}$, the flip-flop is immediately reset and the slow astable stops oscillating. A short while later the 7493 is reset, so that DCBA = 0000 once more. Finally $\bar{9}$ goes back up to 1 and $\bar{0}$ goes down to 0. The system is back in its initial state. In the meantime, each of the 7442 outputs ($\bar{1}$ to $\bar{8}$) have spent one second at logical 0, and $\bar{N}$ has been pulled low in three one-second bursts and a final two-second one. ($\bar{N}$ goes low whenever all four inputs of the NAND gate go high.)

The behaviour of the whole system is summarised in the timing diagram of figure 9.47b (overleaf). Note that the flip-flop has to be reset before the binary counter. If the binary counter resets first, the sequence starts all over again as $\bar{9}$ will return to 1. Since it takes about 5 ns for the output of a TTL gate to respond to its input, there is a delay of about 15 ns between $\bar{9}$ going low and going back high again; this is ample time for the flip-flop to reset.

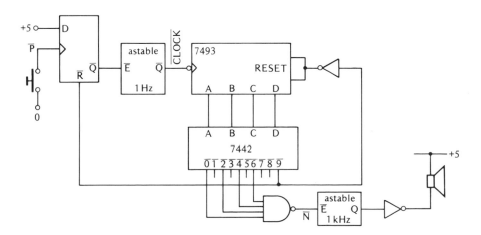

Figure 9.47a A triggered sequence generator

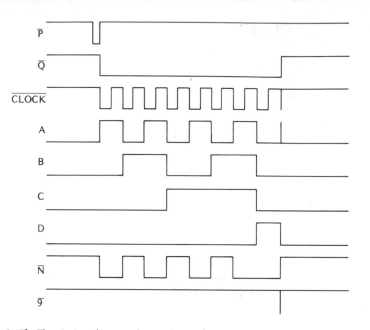

Figure 9.47b The timing diagram for a triggered sequence generator

EXPERIMENT
Investigating a triggered sequence generator

Assemble the circuit shown in figure 9.47a. Check that it behaves as it ought to.

Adapt the system so that it produces two one second buzzes when the switch is pressed. There has to be a three second delay between the buzzes. Draw the circuit diagram of the successful circuit and explain how it works.

EXPERIMENT
Investigating a programmable monostable

The circuit of figure 9.48 is a programmable monostable. You program the length of the pulse at OUT by opening or closing the four switches A, B, C and D. Assemble the circuit, triggering it with the waveform from a 100 Hz astable; look at OUT with a CRO. Find out how the pulse length depends on the state of the four switches (have at least one closed at any one time).

Explain how the circuit works and state how to work out the pulse length from the state of D, C, B and A.

Now design a monostable which feeds out a number of 0.1 ms pulses when triggered. The number of pulses generated is set by four switches; up to 15 pulses can be given out. Try out your circuit. When you have got it to work, draw the circuit diagram. Explain how the system works with the help of a timing diagram.

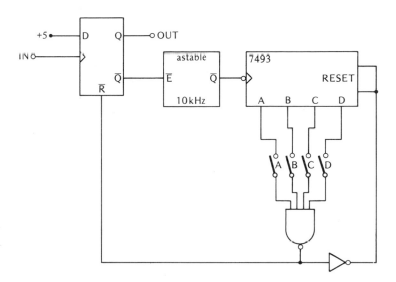

Figure 9.48 A programmable monostable

ANALOGUE-TO-DIGITAL CONVERSION

Analogue signals cannot be handled by digital systems directly. If you want to feed an analogue signal into a digital system you are going to have to convert it into a binary word first. This section deals with systems which convert analogue signals into digital ones. As you will see, they are usually based on a triggered sequential system.

Flash conversion

The object of an **analogue-to-digital converter** (**ADC**) is to feed out a binary word whose value is proportional to the size of the voltage fed into it. The fastest type of ADC (known as a flash converter) relies on combining the outputs of a chain of comparators to generate the binary word.

A simple two bit flash converter is shown in figure 9.49 (overleaf). A chain of resistors is used to establish reference signals of $+1\,V$, $+2\,V$ and $+3\,V$. Three comparators compare the input signal with those three reference ones. Each comparator behaves like a 741 at the input end and like a logic gate at the output end. (Figure 9.50 shows how an op-amp output can be made TTL compatible.) The three comparator outputs (X, Y and Z) depend on the size of V_{IN}, as shown in the table below:

V_{IN} (V)	X	Y	Z	B	A
0–1	0	0	0	0	0
1–2	1	0	0	0	1
2–3	1	1	0	1	0
3–5	1	1	1	1	1

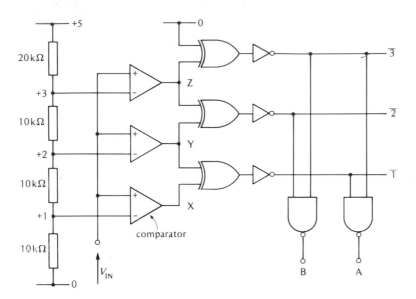

Figure 9.49 A flash conversion ADC

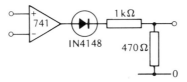

Figure 9.50 A TTL compatible comparator

The binary word fed out by the converter has been chosen so that its value tells you how big V_{IN} is. Thus if BA = 10, you know that V_{IN} is between 2 V and 3 V.

We could, of course, use straight Boolean algebra to combine X, Y and Z to generate B and A. A more roundabout method gives a circuit which can easily be stretched to generate larger binary words. You use a series of EOR gates to compare the outputs of adjacent comparators, as shown in figure 9.49. You will recall that the output of an EOR gate is only 1 when its inputs are different. So when V_{IN} lies between $+2\,V$ and $+3\,V$, the line labelled $\bar{2}$ will be 0 and the other two ($\bar{1}$ and $\bar{3}$) will be 1. B and A are generated by NAND gates hooked up to the three lines $\bar{1}$, $\bar{2}$ and $\bar{3}$; when only $\bar{2}$ is 0, B = 1 and A = 0. Convince yourself that the circuit shown does indeed generate the correct signals for B and A for all values of V_{IN} between 0 V and $+5$ V.

Flash converters are very good if you want to transform an analogue signal into a digital one quickly. The comparators tend to be the slowest components in the system, but you can easily get BA to follow V_{IN} with a delay of less than 50 ns. Their great disadvantage is the lack of resolution; in our example, the resolution is one part in four, or 25%. You improve the resolution by putting in more comparators. Thus another four comparators allow you to convert V_{IN} into a three bit word; the resolution is now one part in eight, or 13%. If you want a resolution that is less than 1% of the

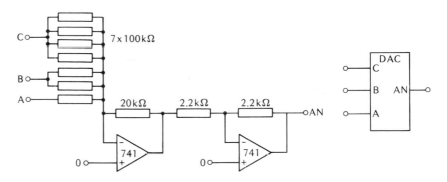

Figure 9.51 A three bit DAC

possible range of V_{IN}, then you are talking about a seven bit word and 127 comparators! Although extending the circuit of figure 9.49 so that it can process the outputs of 127 comparators is quite straightforward, the resulting circuit is unwieldy and expensive.

DACs

The secret of economical A to D conversion is to use a D to A converter. This device reads in a binary word and feeds out a voltage which is proportional to it. It is essentially (as its name implies) an A to D converter run backwards.

A simple three bit DAC based on a summing op-amp is shown in figure 9.51. The word to be converted is placed on the terminals marked C, B and A; the analogue output appears at AN. It should be obvious that the most significant bit of the binary word (C) feeds four times more current into the 20 kΩ resistor than the least significant bit. All seven input resistors for the summing amplifier have to be identical; it is therefore a good idea to have them all on the same IC.

It can easily be shown that the analogue output is given by:

$$AN = 4C + 2B + A$$

(Assuming that logical 1 is $+5$ V.) Thus when CBA $= 011$, AN $= +3$ V.

EXPERIMENT

Assembling a three bit DAC and evaluating its performance

Build the circuit of figure 9.52. The DAC is the circuit shown in figure 9.51; if you do not have a group of seven 100 kΩ resistors, you can try using one

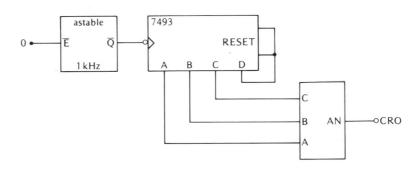

Figure 9.52

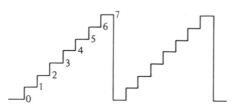

Figure 9.53

each of 100 kΩ, 47 kΩ and 22 kΩ. Use a 5 kΩ potentiometer as a variable feedback resistor for the inverting amplifier. Be careful about mixing 15 V supplies and 5 V supplies on the same board; two separate breadboards might be safer!

Look at the output of AN with a CRO; adjust the value of the feedback resistor of the inverting amplifier so that the waveform goes up in steps of 1 V (figure 9.53).

Can you work out how to get AN to ramp down instead of up?

ADCs via DACs

A three bit ADC which is built around the DAC discussed in the last section is shown in figure 9.54. The signal to be converted into binary is V_{IN}; the binary word is displayed on a seven segment LED as a number between 0 and 7.

The converter has to be triggered into action by a pulse fed into $\overline{START}$. When $\overline{START}$ goes low then high again it clocks the flip-flop. So the astable starts to fire pulses into the counter, but these are not recognised by it until $\overline{START}$ goes back up to 1. When this happens, CBA is increased by 1 for every pulse which comes out of the astable. So AN goes up in steps of 1 V, starting from 0 V.

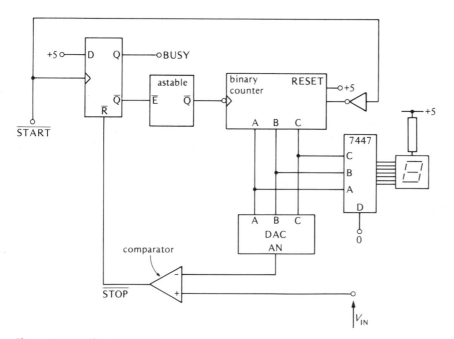

Figure 9.54 A three bit ADC

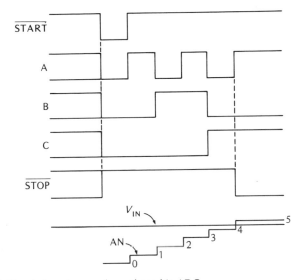

Figure 9.55 The timing diagram for a three bit ADC

Now AN is compared with V_{IN} by a comparator. If V_{IN} is greater than $0\,V$, then the output of the comparator ($\overline{STOP}$) is initially 1. As soon as AN is greater than V_{IN}, $\overline{STOP}$ falls to 0. This resets the flip-flop, and $\overline{Q}$ becomes 1, switching off the astable. The binary counter is now frozen, with the binary word equivalent to V_{IN} sitting on C, B and A. The word stays there until the next time that $\overline{START}$ goes low and resets the counter.

The timing diagram of figure 9.55 shows what happens when V_{IN} is $+4.5\,V$. A short while after $\overline{START}$ has returned to 1, the count freezes at 5.

The converter has a resolution of about 14% of the range of analogue signals that it can measure. To improve that resolution, you have to increase the length of the binary counter. DACs which deal with more than three bits are not very satisfactory when they are built from discrete resistors, so it is general practice to use a DAC IC. (It is possible to obtain very close matching of resistors when they are built on the same slice of silicon. Furthermore, they will all drift together as they heat up and cool down.) Unfortunately, this increase in resolution is bought at the expense of a longer conversion time. For example, the three bit converter of figure 9.54 may need up to seven clock pulses before the digital output is stable. If it had a resolution of less than 1%, it would need up to 127 clock pulses. It follows that each bit added to the binary word generated by the converter doubles the conversion time!

The 507 IC

The 507 IC is a cheap and simple unit which contains all the awkward bits of an ADC. We are going to describe its contents, show how to incorporate it into a five bit ADC and suggest that you assemble and evaluate the system for yourself.

Look at figure 9.56 (overleaf). It shows the pinout of the IC and a circuit which mimics its behaviour. As you can see, the IC has three inputs (CLOCK, INPUT and RESET) and one output (OUTPUT). (Note that two of the terminals need external pull-up resistors.) The function of the CLOCK and RESET inputs is obvious. The output of the seven bit counter controls a DAC. This is an op-amp organised as a summing amplifier, with seven

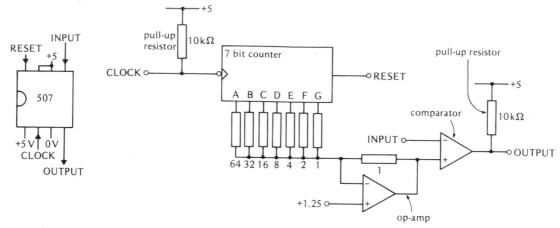

Figure 9.56 The interior of a 507 IC

closely matched resistors to give the correct binary weighting. Each counter output is 0 V at logical 0 and +1.25 V at logical 1. So when the counter has been reset by pushing RESET high for an instant, the op-amp output is $1.25 + 1.25 \times (1 + 1/2 + 1/4 \ldots + 1/64) = +3.75$ V. As clock pulses are fed into the counter, the op-amp output goes down by about 20 mV for each pulse. When GFEDCBA = 1111111, the op-amp output is +1.25 V. The comparator compares the op-amp voltage with that of an analogue signal fed into INPUT. You can see that OUTPUT will be logical 1 when INPUT is less than the output of the summing amplifier.

Figure 9.57 shows one way in which the 507 can be used to make a five bit ADC. The system is triggered by pressing and releasing the START

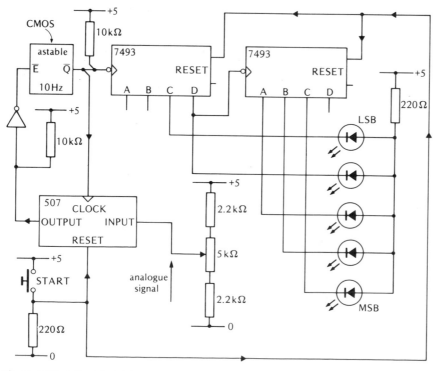

Figure 9.57 A five bit ADC using a 507 IC

button. This resets the external counter chain (2 × 7493) and the 507. As you can see, the astable is controlled by the output of the 507; it will fire out pulses until the analogue input from the potentiometer matches the signal generated inside the 507. The subsequent change of OUTPUT switches off the astable, leaving the number of pulses frozen on the five LED display (in binary, of course!).

EXPERIMENT
Investigating a five bit ADC using a 507 IC

You are going to assemble and evaluate the circuit shown in figure 9.57. It is fairly complex, so you would be wise to assemble and test it in stages.

Get the astable and binary counter chain to work first. The astable must be a CMOS one because the 507 is very fussy about its clock pulses. Check that the system counts properly, and that the display works. Put in the START switch; it should reset the display to 00000 when it is pressed. Then assemble the potentiometer network, and check that its output can be varied from 3.8 V to 1.2 V. Finally insert the 507.

When you press and release the START switch, the count displayed on the LEDs should steadily go up and then freeze. If it does not stop counting (and the conversion time is about 13 seconds) your analogue input may be too low.

First investigate how the value of the binary word fed out by your counter depends on the value of the analogue input. Assuming that the five bit word represents a decimal number from 0 to 31, plot a graph of analogue input voltage against binary output.

Then speed up the conversion time and see how fast you can make it.

Finally, modify the system so that it automatically converts the analogue input at one second intervals, with the conversion taking place in much less than a second.

PROBLEMS

1 Draw the circuit diagram of a three bit flash converter with the following properties, shown in the table.

Input voltage (V)	Output		
	C	B	A
0.0–0.5	0	0	0
0.5–1.0	0	0	1
1.0–1.5	0	1	1
1.5–2.0	0	1	0
2.0–2.5	1	1	0
2.5–3.0	1	1	1
3.0–3.5	1	0	1
3.5–4.0	1	0	0

2 Draw the circuit diagram of a simple digital voltmeter capable of reading from 0 to 1.27 V. The voltage must be displayed on seven segment LEDs. Assume that you have access to a 507 IC; remember that it will "accept" 0 clock pulses for an input of +3.75 V and 127 pulses for an input of +1.25 V. The meter should update its measurement at least once a second. Explain how your circuit works.

REVISION QUESTIONS

1 Draw the circuit diagram of a D flip-flop. Explain the function of each terminal and explain what it does.

2 Show how to make a four bit latch out of D flip-flops. Discuss the properties of the circuit.

3 An IC is described as being a 1024 × 8 bit RAM. Discuss its properties.

4 Draw the circuit diagram of a D latch made out of NAND gates. Use a timing diagram to illustrate its behaviour.

5 Draw a three bit binary counter based on D flip-flops. Draw a timing diagram to illustrate its properties.

6 Describe the various ways in which a switch can be debounced.

7 Show how a four bit counter can be made to count from 0 to 9 and how the value of the count can be displayed on a seven segment LED.

8 Draw a circuit diagram of a decimal counter which can count from 0 to 99. The display is done with a pair of seven segment LEDs and there is a reset pin.

9 Draw the circuit diagram of a system which will feed out a signal which has a frequency that is 1/7 of the frequency fed into it.

10 Draw the circuit diagram of a two bit down counter using D flip-flops. State how you would modify the system so that it is reset to 3 every time that it counts down to 0.

11 Explain what an ADC is supposed to do, and discuss the relationship between conversion time and word length for one.

12 Draw the circuit diagram of a four bit DAC based on a pair of op-amps. Explain how it works.

13 An up-down counter looks like an ordinary four bit binary counter except that it has an extra pin labelled $U/\bar{D}$. When $U/\bar{D}$ is 1 the system counts up in binary; when $U/\bar{D}$ is 0 it counts down. Show how the addition of four EOR gates converts a four bit up counter based on D flip-flops into an up-down counter.

14 Draw the circuit diagram of a system which could control the lights at a pedestrian crossing. Explain how it works.

15 With the help of a circuit diagram, explain how a DAC can be used to make an ADC.

10

Shift registers

A **shift register** is an array of D flip-flops which is a fundamental building block of microprocessor systems. Shift registers allow binary words to be stored, processed and moved from one part of a system to the other.

After introducing shift registers, we shall explain how to make them generate cyclic sequences and random numbers. Then we shall discuss their use in the serial and parallel processing of binary words, including the operations of addition and subtraction. You will then find out how to use shift registers to convert binary words into a stream of bits which can be sent down a single line, and how that stream of bits can be reassembled into a binary word at the other end of the line. Finally we shall deal with synchronous counters.

REGISTERS

A four bit shift register is shown in figure 10.1. Notice the way in which the output of each flip-flop provides data for the next flip-flop along in the chain. All of the clock terminals go to a common line, called CLOCK. So when a rising edge enters CLOCK, the flip-flops will simultaneously transfer the signals at their inputs to their outputs. The result is a simultaneous rightwards shuffle of the binary word ABCD held in the register.

The operation of the shift register will become clearer if you study the timing diagram of figure 10.2 (overleaf). Initially all the flip-flops are reset, so that ABCD = 0000. The SERIAL INPUT line is held high during one rising edge of CLOCK and then let down to logical 0. That logical 1 is then shifted down the register, moving one place to the right for each rising edge on the CLOCK line. So ABCD goes through the sequence 1000, 0100, 0010, 0001 and 0000. Subsequent clocking pulses leaves ABCD = 0000 as 0 is being continually fed into SERIAL INPUT.

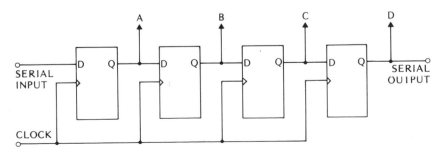

Figure 10.1 A four bit shift register

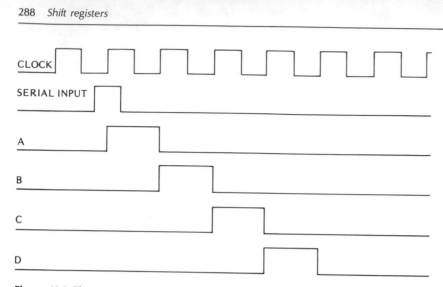

Figure 10.? The timing diagram for a four bit shift register

THE 7496 IC

You could assemble shift registers from individual D flip-flops, but since a useful register has to have several bits you are faced with using more than one 7474 IC. The 7496 IC is a useful building block, containing five D flip-flops wired together to make a five bit shift register. Figure 10.3 shows its inputs and outputs. The SERIAL INPUT, CLOCK, A, B, C, D and E terminals correspond to the terminals of a five bit register assembled on the lines of the four bit one shown in figure 10.1. The other terminals permit the register to act like a 1 × 5 bit RAM, so that a five bit binary word can be loaded into it. You feed the word into the five parallel inputs (a, b, c, d and e), reset all the flip-flops by pulsing $\bar{R}$ low and then push the word onto A, B, C, D and E by pulsing EN high.

Figure 10.4 shows the circuit diagram of a system which behaves just like the 7496. The five flip-flops have a common reset line ($\bar{R}$), so they can be simultaneously reset by briefly pulling it low. That is, if you pull $\bar{R}$ down to 0, ABCDE = 0000. Obviously, the flip-flops will only respond to rising edges on the CLOCK line if $\bar{R}$ = 1! (The IC is TTL so you can let $\bar{R}$ float up to 1 on its own.) Similarly, the EN line has to be held at 0 in normal operation otherwise the $\bar{S}$ inputs of the flip-flops will be 0 and their outputs will be forced to 1. It should be fairly obvious that once the flip-flops have been reset, if EN is taken high then ABCDE = abcde.

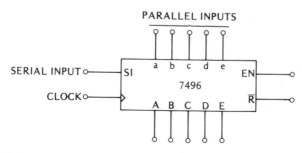

Figure 10.3 A 7496 TTL IC

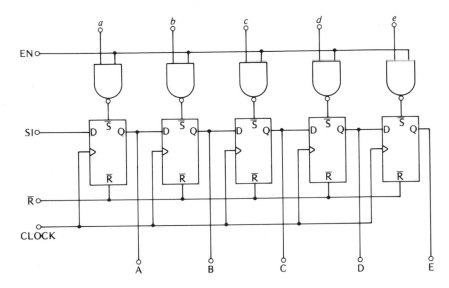

Figure 10.4 The circuitry inside a 7496 IC

The parallel-load facility of a shift register is extremely useful. It permits binary words to be converted into a sequence of bits, and allows words to be simultaneously transferred from one register to another.

EXPERIMENT
Exploring the properties of a five bit shift register
and using it to generate some cyclic sequences

Assemble the circuit shown in figure 10.5. Leave room for a couple of ICs, and put the LEDs in a row as shown; a 10 LED array might be useful.

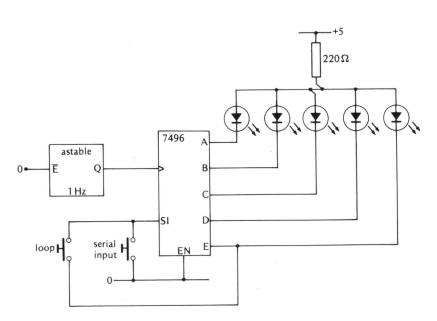

Figure 10.5

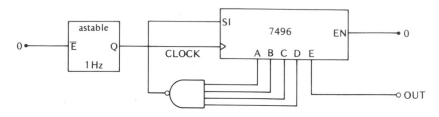

Figure 10.6 A sequence generator

Press the serial input switch and watch the lit LEDs chase along the line. If you press and release the serial input switch in sequence you should be able to generate interesting patterns that chase along the line of LEDs. You can make a pattern cycle continuously by pressing the loop switch; this feeds E into SI so that the five flip-flops are connected in one continuous loop.

Now alter the circuit so that it becomes the system shown in figure 10.6. You should find that the LED that monitors E comes on for one second every five seconds. Explain why.

Work out how to modify the circuit so that E is lit once for two seconds in every five seconds. Try it out; draw the circuit diagram and explain how the system works.

EXPERIMENT
Generating noise with a digital circuit

Assemble the circuit shown in figure 10.7. Note the use of EOR gates to act as inverters (the one feeding into EN) and buffers (the one feeding the loudspeaker). The third EOR gate feeds a signal into the serial input (SI) of the shift register. As a consequence, the five bit word ABCDE goes through a sequence of 31 different words in a pseudo-random fashion. This means that E spends almost as much time at 1 as it does at 0, with an almost equal probability of it ending up at 0 or 1 after every clock pulse.

Look at E with a CRO and listen to it with a loudspeaker. If your system gets stuck in the state ABCDE = 00000, you will have to break the supply connections and start it up again. (The *RC* network and inverter hold EN at logical 1 for about 10 ms after the circuit is powered up, so that ABCDE starts off at 11111. Thereafter, EN is held at 0, so the shift register can operate normally.)

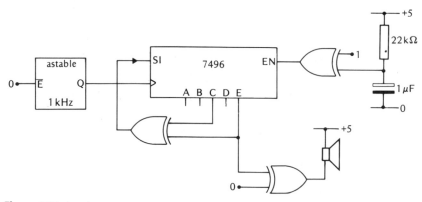

Figure 10.7 A noise generator

Work out how to modify the system so that it displays a sequence of almost random numbers from 0 to 7 on a seven segment LED. Make the numbers appear at a rate of one per second. Try it out. Draw the circuit diagram and describe its behaviour. Check that the sequence of numbers is 31 long.

PROBLEMS

1 Suppose that the system of figure 10.6 had a quad NOR gate instead of a quad NAND gate. Describe and explain its behaviour.

2 A five bit register must start off with ABCDE = 10011 when it is powered up. Draw a circuit diagram of a suitable circuit using a 7496 IC and an *RC* network similar to the one shown in figure 10.7. Remember that the register must be reset before you load a word from the parallel inputs. Explain how your system works.

3 A four bit register, like the one shown in figure 10.1, is arranged so that its serial output is connected to the serial input. The four flip-flops are therefore connected in a loop; clock pulses fed into the CLOCK line rotate the word ABCD from left to right along the register. Draw the circuit diagram of a four bit register which can rotate its word to the right or to the left depending on the state of a terminal called RIGHT/$\overline{\text{LEFT}}$. Use tristates.

4 A three bit shift register has its serial input = B $\oplus$ C. If ABC is initially 111, draw up a truth table to show the value of ABC for the next seven clock pulses. If ABC = 000, the system gets stuck in that state; explain what this means. Can you draw the circuit diagram of the system with a modification so that if ABC = 000 at any time, it is fed back into the pseudo-random sequence at the next clock pulse?

PROCESSING WORDS

Shift registers are used extensively to process binary words. For example, they are used to store numbers in calculators and to feed them into systems which combine those numbers e.g. multiply them together.

The standard way of drawing a shift register when it is being used to hold a word is shown in figure 10.8. The system is that of figure 10.1 i.e. a four bit register, but the labelling of the outputs is different. The word stored in the register is DCBA, with the **most significant bit** (**msb**) on the left; this is so that you can read the word straight out of the diagram. The clock line is not shown at all; the usual convention is that all of the registers in a diagram have the same clock line unless otherwise indicated. Obviously, each clock pulse moves the contents of the register one place to the right; the **least significant bit** (**lsb**) is lost and the signal at SERIAL IN becomes the msb. The lsb (A) is fed out of the serial output terminal all the time.

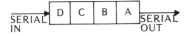

SERIAL IN | D | C | B | A | SERIAL OUT

Figure 10.8 A representation of a register

SERIAL PROCESSING

An example of the use of shift registers to process a binary word is shown in figure 10.9. The system inverts the word fed into the register on the left.

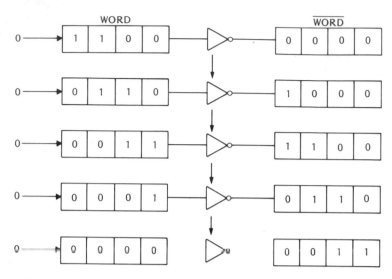

Figure 10.9

Initially the word 1100 is loaded into that register. Four clock pulses later the right hand register contains 0011, the complement of 1100.

The system is called a **serial processor** because each bit is processed in turn. When the first clock pulse arrives the lsb is processed by the NOT gate and the result of that operation stored as the msb of the $\overline{\text{WORD}}$ register. Each subsequent clock pulse moves successive bits through the NOT gate into the $\overline{\text{WORD}}$ register, until the whole word has been processed.

It has probably occured to you that the same result could be effected by using four NOT gates and processing all four bits simultaneously. Nevertheless, serial processing comes into its own when you want to perform complex operations on long binary words. It turns out that they represent a large saving in hardware at the expense of a longer processing time.

COMBINING SYSTEMS

Consider the system shown in figure 10.10. It shows how shift registers can be used to feed two words into a combining system and store the word resulting from that operation. For example, the two registers A and B could be loaded with two nibbles. The combining system could be a collection of logic gates which compares the lsb of one register with that of the other, pushing the result of that comparison into the msb of register C.

Such a system is illustrated in figure 10.11. A and B are initially loaded with 1010 and 1100 respectively. These two nibbles are combined by an

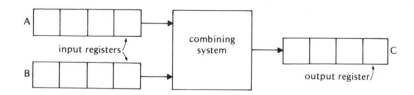

Figure 10.10 A serial processor

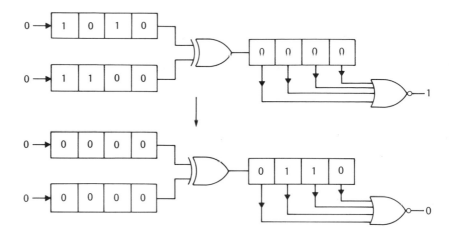

Figure 10.11 A serial nibble comparator

EOR gate and the result pushed into C; after four clock pulses C contains A $\oplus$ B, i.e. 0110. The output of the quad NOR gate is 0 unless all of the bits of register C are 0. So if that terminal is 1 after the four clock pulses, then the two nibbles were identical.

This sort of system could be used as part of an electronic security lock. Suppose that the output of the quad NOR gate controlled a lock mechanism, with a logical 1 opening the lock. Then the lock will only open if both A and B were originally loaded with the same word. The word fed into A could be set up with four switches by the person wishing to open the lock and parallel-loaded into the register. B could similarly be parallel-loaded with a word stored on a bank of four switches inside the circuit. A push switch would be needed to trigger the system into generating four clock pulses; if the output of the NOR gate was 1 after the fourth clock pulse, the lock would open.

PARALLEL PROCESSING

The system shown in figure 10.12 has the same function as the system of figure 10.11. It combines the contents of two registers (A and B) with four EOR gates, storing the resulting word in a third register (C). But instead of

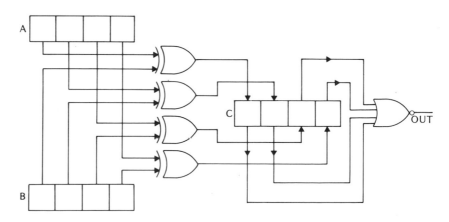

Figure 10.12 A parallel nibble comparator

processing the two words one bit at a time, all eight bits are processed simultaneously. A single pulse parallel-loads the outputs of the four EOR gates into the C register.

It should be obvious that this sort of parallel processor requires more hardware than its serial equivalent. On the other hand, it only needs one clock pulse rather than four. So it is faster. You gain speed at the expense of size. When speed is not important, you can keep things simple by using a serial processor. If you need results fast then you must be prepared to put up with the extra complexity of a parallel processor.

PROBLEMS

1 Look at figure 10.13. What is the state of each of the shift registers E and F after

a) 4

b) 8

c) 12 clock pulses?

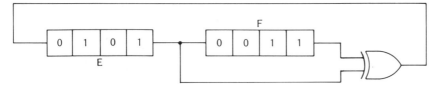

Figure 10.13

2 The system shown in figure 10.14 is used to combine two numbers. If A and B are loaded with 9 and 5 (in BCD) respectively, what is the number stored in C four clock pulses later?

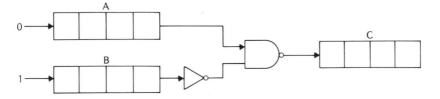

Figure 10.14

3 This question concerns the system shown in figure 10.15. If J and K are

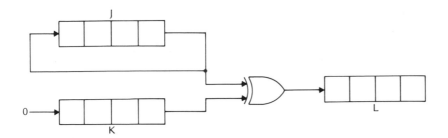

Figure 10.15

initially loaded with 1111 and 1010 respectively, what are the contents of J, K and L after

a) 4 clock pulses

b) 8 clock pulses?

4 Draw the circuit diagram for the parallel processor which has the same function as the system shown in figure 10.14.

PROCESSING NUMBERS

We are now going to explain how systems based on shift registers can be used to perform arithmetic operations on binary numbers. But before we introduce these systems we need to spend a little time discussing binary addition and how negative numbers are represented in binary.

BINARY ADDITION

Suppose that you want to add 0110_2 to 0011_2; how do you do it?

$$
\begin{array}{rll}
 & 0\ 1\ 1\ 0 & A \\
+ & 0\ 0\ 1\ 1 & B \\
\hline
= & 1\ 0\ 0\ 1 & \text{sum} \\
\hline
 & 0\ 1\ 1\ 0\ - & \text{carry}
\end{array}
$$

Let us analyse the sum as it is written out above. Consider the addition of the two least significant bits of A and B. $0 + 1$ gives a sum of 1 and a carry of 0. Now we deal with the next column on the right; $1 + 1$ gives a sum of 0 and a carry of 1. That carry has to be taken into account when we deal with the next column. So that column gives us $1 + 0 + 1$, a sum of 0 and a carry of 1. The final column is $0 + 0 + 1$, giving a sum of 1 and a carry of 0.

So the result of adding 0110 to 0011 is 1001. You can check for yourself that the decimal value of the sum is what you would expect it to be.

A device that generates sum and carry signals when two bits are fed into it is called a **half adder**. It has the truth table shown below:

B	A	SUM	CARRY
0	0	0	0
0	1	1	0
1	0	1	0
1	1	0	1

So $\text{SUM} = \bar{B} \cdot A + B \cdot \bar{A} = B \oplus A$ and $\text{CARRY} = A \cdot B$. The circuit diagram of a half adder assembled from NOT and NAND gates is shown in figure 10.16 (overleaf).

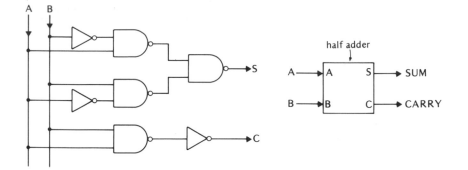

Figure 10.16 A half adder

NEGATIVE NUMBERS

Up till now we have assumed that the value of the word 1001 is 9. It certainly is 9 if we assume that the word represents a number coded according to the rules of BCD. But it can mean different numbers if we use a different code. For example, if we assume that the word represents a number using the two's complement code, then 1001 is −6!

The **two's complement** code is not the only way of representing negative numbers, but it is probably the most natural. The table below shows the numbers represented by a three bit word using the two's complement code.

Binary word	Two's complement
011	+3
010	+2
001	+1
000	0
111	−1
110	−2
101	−3
100	−4

We shall explain how to convert a decimal number into its two's complement in a moment. First, there are some important things to note about two's complement words. To start with, the msb of the word tells you what the sign of the number is. A 1 means a negative number, a 0 means a positive number. Then all of the positive numbers are identical to their BCD equivalents. If you add 1 to any of the numbers in the table (except +3) you get the number above it. So if you start a three bit binary counter at CBA = 100 (i.e. −4) it will steadily count up to 011 (i.e. +3) as clock pulses are fed into it. Standard binary counters will therefore count up or down correctly with the two's complement coding.

How do you find the two's complement representation of a negative number? Consider −2 for example. You start off by writing down the BCD version of +2, i.e. 010. Then you find its complement. To do this, you

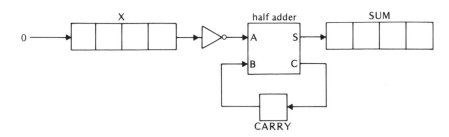

Figure 10.17 A times-minus-one serial processor

invert each bit of the word so that you end up with 101. Finally, you add 1 to the result to get the two's complement, i.e. $101 + 001 = 110$.

If you perform exactly the same sequence of operations (invert then add one) on a two's complement number which is negative you get its positive equivalent. Furthermore, if you try the operation on zero, you end up with zero again. So the two's complement coding behaves properly as far as the negation operation is concerned.

Figure 10.17 illustrates how a half adder can be used to change the sign of a number. The number is fed into the X register in its two's complement coding. The CARRY register is set to 1, and four clock pulses later the SUM register contains the two's complement code for the number times minus one.

The operation of the system is shown in detail in figure 10.18. Initially X

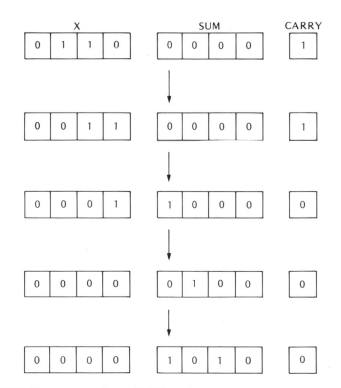

Figure 10.18 The system of figure 10.17 in action

holds $+6$, SUM is reset to 0 and CARRY is set to 1. The half 'adder adds the lsb of $\bar{X}$ to CARRY. The sum output (S) is pushed into the SUM register at the first clock pulse. At the same instant, the carry output (C) is pushed into the CARRY register. So at each clock pulse one bit of $\bar{X}$ is added to the carry from the addition of the previous two bits.

Work through the effects of all four clock pulses, and convince yourself that you end up with the two's complement code for -6.

ADDING NUMBERS

Figure 10.19 shows a system which can be used to add two numbers together. Since there is only a single processing unit—the **full adder**—it must be a serial processor. The two numbers that are to be added together are loaded into registers A and B. After the CARRY register has been reset to 0, the two numbers are added together one bit at a time. C_{IN} is the carry from the addition of the previous two bits. The sum of the two numbers is fed into the SUM register.

Since the addition of two four bit numbers can result in a five bit number, the final answer of the addition is stored in both CARRY and SUM, with CARRY holding the msb. If you want to add two's complement numbers accurately then you have to ensure that the addition of two positive numbers will not leave the msb of SUM equal to 1. So if you have four bit registers, you can only load numbers between $+3$ and -4 into A and B; the answer of the addition will be stored in SUM and you can ignore the contents of CARRY.

A useful adding system is going to have to be able to deal with large numbers, with several decimal digits. This means that the shift registers are going to have to be very long. For example, a ten bit register can hold numbers between $+511$ and -512. You will need an eighteen bit register to hold numbers between $+100\,000$ and $-100\,000$. With such long registers the processing time for the addition is going to be long; you need a clock pulse to add each pair of bits together.

You can speed up the process of addition by using a parallel processor. A four bit parallel adding system is shown in figure 10.20. It contains four

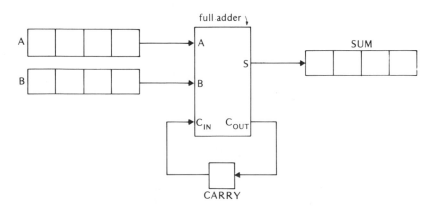

Figure 10.19 A serial adder

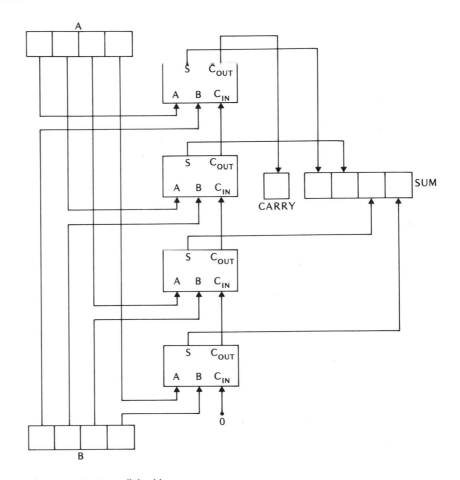

Figure 10.20 A parallel adder

full adders, one for each pair of bits in registers A and B. A single clock pulse loads the outputs of the full adders into the SUM register. Note the way in which the carry from one full adder is fed into the one above it.

Of course, parallel processing requires much more hardware. This gets expensive! The usual way around this is to compromise, by mixing serial and parallel processing together. For example, suppose that you want to add the contents of two sixteen bit registers together. You can save time by adding four bits at a time using a parallel adder like the one shown in figure 10.20. (Of course, you have to take account of the carry from the previous addition (stored in CARRY).) The net result is that the sum is done in four clock pulses instead of sixteen, at the expense of a fourfold increase of processing hardware.

This approach is the one used in microprocessor systems. The standard size of registers in a microprocessor is eight bits. Their contents can be combined together in parallel. If you need to process words that are more than one byte long, you have to deal with them one byte at a time.

PROBLEMS

1 Here is the truth table for a full adder.

A	B	C_{IN}	SUM	C_{OUT}
0	0	0	0	0
0	0	1	1	0
0	1	0	1	0
0	1	1	0	1
1	0	0	1	0
1	0	1	0	1
1	1	0	0	1
1	1	1	1	1

Write down the Boolean expressions for SUM and C_{OUT} as a function of A, B and C_{IN}. Draw circuit diagrams to show how they can be implemented in NAND gates.

2 Draw the circuit diagram of a parallel processing system which has the same function as the system shown in figure 10.17.

3 This question concerns the system shown in figure 10.19. If A and B are loaded with +2 and +3, and CARRY is reset, draw the contents of SUM and CARRY after 1, 2, 3 and 4 clock pulses.

4 Repeat question 3, for A and B loaded with +3 and −4.

5 Write down the five bit two's complement representations of the numbers between +9 and −9.

SERIAL TRANSMISSION

We are now going to show how shift registers can be used to convert binary words into streams of bits which can be sent down a transmission line. The actual mechanism of sending digital data down a transmission line will be dealt with in Chapter Fourteen; at the moment we are only concerned with the business of parallel-to-serial conversion.

A glance at figures 10.21 and 10.22 should make things clearer. If you want to transmit a four bit word from one system (the word generator) to another (the word processor) you can do it two ways. You can send the word down four lines in parallel, one bit per line. This obviously gets expensive on cable if the word has to be transmitted an appreciable distance. Alternatively you can send the four bits of the word down a single line one after the other; this **serial transmission** obviously takes longer than **parallel transmission**, but does save on cable. It also requires two extra bits of

Figure 10.21 Parallel transmission of a nibble

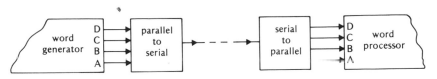

Figure 10.22 Serial transmission of a nibble

hardware, namely, a parallel-to-serial converter and a serial-to-parallel converter. This section of the chapter is devoted to explaining how these may be built out of shift registers.

PARALLEL-TO-SERIAL CONVERSION

We are going to describe the operation of a four bit parallel-to-serial converter. It transmits each bit of a word for a fixed length of time, with a special "start" bit preceding the word. For example, look at the bottom line of the timing diagram shown in figure 10.23. The SERIAL OUTPUT line starts off at logical 1. It then goes to 0 for a certain time and then feeds out D, C, B and A (the four bits of the word) one after the other. After all four bits have been transmitted, the line goes back to 1 and stays there until the next start bit.

A circuit which reads in the nibble and feeds it out one bit at a time is shown in figure 10.24 (overleaf). On the right is the shift register which pulls in the nibble and shunts it out in serial form. The rest of the circuit provides the clock pulses, the parallel-load pulses and the reset signals for the register.

Have a look at the timing diagram of figure 10.23. To start the transmission, the $\overline{\text{SEND}}$ line has to be pulled low for an instant. This does two things. It loads the nibble into the shift register, with a 0 as the msb to act as the start bit. (Note that the 5 ns delay through the NOT gate means that EN goes to 0 *after* $\bar{\text{R}}$ goes to 1.) The $\overline{\text{SEND}}$ pulse also clocks the flip-flop. So the astable starts to fire falling edges into the 7493 counter. The A output rises immediately, and thereafter pushes rising edges into the clock line of the 7496 shift register at regular intervals. The first rising edge is ignored by the register as it is being reset, but it responds to all the subsequent ones. So 0, D, C, B and A appear at the SERIAL OUTPUT terminal in turn. When the counter has counted 12 falling edges the flip-flop is reset, so that the astable stops oscillating. A few nanoseconds later the counter itself is reset. Since five rising edges have been sent into CLOCK, the shift register has been completely filled with 1, the signal fed in at the serial input (SI). So the system feeds a 1 out of SERIAL OUTPUT until the next time that $\overline{\text{SEND}}$ is pulled low.

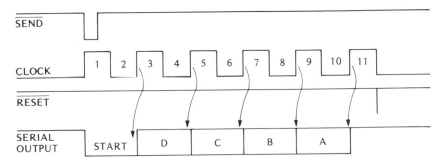

Figure 10.23 The timing diagram for a parallel-to-serial converter

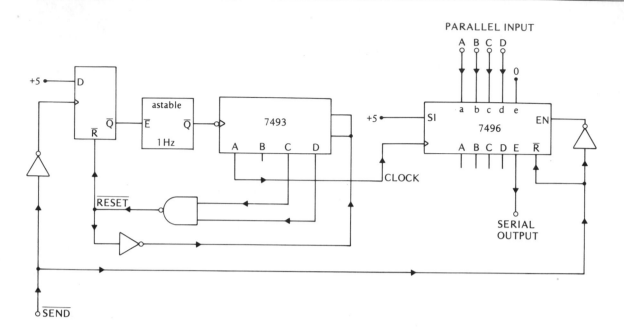

SERIAL-TO-PARALLEL CONVERSION

Now for the problem of reading and storing the four bits as they appear on the line. The difficulty is ensuring that the receiver is correctly synchronised with the transmitter. The receiver has to know when the four bits are present on the transmission line so that it can shunt them into a shift register and recreate the original nibble. More precisely, the receiver has to be able to recognise the start bit, and use that for the synchronisation.

A receiver which is compatible with the transmitter we have just described is shown in figure 10.25; its behaviour is illustrated in the timing diagram of figure 10.26. The SERIAL INPUT signal is shunted into the SI input of the shift register, so that the nibble DCBA eventually appears at the parallel outputs. The falling edge of the start bit is cleaned up by a Schmitt

Figure 10.24 A parallel-to-serial converter

Figure 10.25 A serial-to-parallel converter

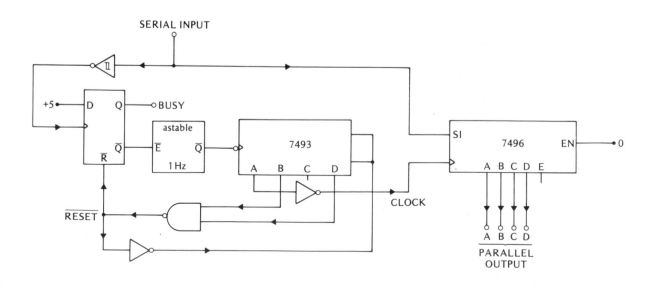

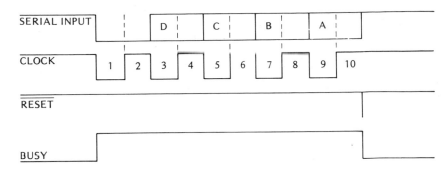

Figure 10.26 The timing diagram for a serial-to-parallel converter

NOT gate (edges tend to be degraded when they are sent down long lengths of cable) and used to clock the flip flop. The CLOCK line subsequently goes high five times, so that 0, D, C, B and A are read into the SI input of the register. Finally, when the counter has detected ten falling edges the whole system is reset.

The BUSY line feeds out a falling edge after the whole nibble has been received. This can provide a signal to a word processor that a nibble has just arrived and is sitting stably on the PARALLEL OUTPUT terminals. It also sits at logical 1 when a nibble is being received, warning the word processor that the data on the PARALLEL OUTPUT terminals is not stable.

The transmitter and receiver are only matched if their astables run at the same frequency. A little thought should convince you that provided the frequencies are within about 10% of each other, the systems are matched. The serial transmission of longer words places greater constraints on the matching of the clock frequencies. Thus the transmission of a byte needs the astables to be matched to within 5%. If you want to transmit a word that is longer than one byte, it is usual practice to transmit it as several bytes in order to avoid the use of super-precise oscillators.

PROBLEMS

1 Draw the circuit diagram of a serial transmitter which will transmit a three bit binary word. The word is set up with three switches, and transmitted when a switch is pressed. The start bit must be 1 ms long. Explain how your system works.

2 Draw the circuit diagram of a serial receiver which is compatible with the transmitter of question 1. The received word must be displayed as a number from 0 to 7 on a seven segment LED, with a 3 bit latch between the shift register and the display circuitry. Explain how your system works.

SYNCHRONOUS COUNTERS

Think about a three bit ripple counter, made out of three one bit counters in series. It is called a ripple counter because the effect of the clock pulse fed into the first counter ripples from left to right through the chain. So any

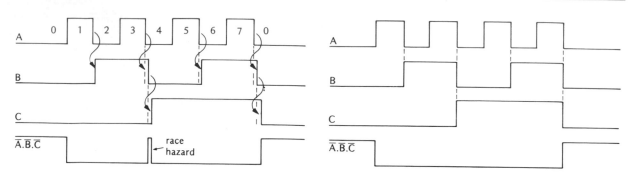

Figure 10.27 An example of a race hazard arising from the use of a ripple counter

Figure 10.28 The output of a synchronous counter

change in the output of a counter in the chain will take place at least 5 ns after the change in the output of the previous counter.

The consequences of this ripple are shown in the timing diagram of figure 10.27. The clock frequency is high so that the propagation delay of each flip-flop is easily detected. At the bottom we have drawn the graph for the function $\overline{A} . \overline{B} . \overline{C}$, i.e. the output of a NOR gate fed from the three outputs of the counter. It should only be high when A, B and C are all 0, i.e. every eighth clock pulse. But because of the ripple of information down the counter chain, $\overline{A} . \overline{B} . \overline{C} = 1$ for about 5 ns just before C rises from 0 to 1. This glitch is known as a **race hazard**. It does not exist for long, but can cause chaos if $\overline{A} . \overline{B} . \overline{C}$ is used to clock a flip-flop!

Race hazards are avoided if you use a **synchronous counter**. This is a collection of one bit registers and logic gates; since all of the flip-flops have a common clock line their outputs change simultaneously. So a three bit synchronous counter will have a timing diagram like the one shown in figure 10.28, and there is no possibility of a race hazard.

BINARY COUNTERS

Apart from their synchronous nature, synchronous counters have the advantage that they can be designed to display the count in any code you like. This flexibility will become clear after we have shown you how to design a two bit binary counter.

You start off with a pair of flip-flops arranged as shown in figure 10.29. They have a common clock line, with a NOT gate to invert IN. This is so that the system responds to falling edges in the same way that ripple up

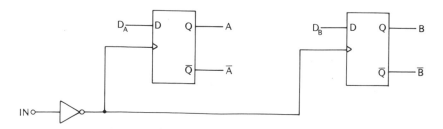

Figure 10.29 The skeleton of a two bit synchronous counter

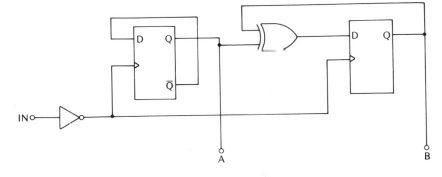

Figure 10.30 A two bit binary up counter

counters do. Then you write down a truth table to show the sequence of outputs that you want out of the counter.

Pulse	B	A	D_B	D_A
0	0	0	0	1
1	0	1	1	0
2	1	0	1	1
3	1	1	0	0

You then enter D_B and D_A into the truth table, so that B and A change as specified when the next pulse is fed into IN. For example, consider the first line of the truth table. $BA = 00$. After the next clock pulse we want $BA = 01$. So we set $D_B = 0$ and $D_A = 1$. Similarly, if you consider the last line of the table you will see that the next clock pulse will reset $BA = 00$.

Finally, you can use the truth table to work out how to generate D_B and D_A from the current values of B and A.

$$D_B = \bar{B} \cdot A + B \cdot \bar{A} = B \oplus A$$
$$D_A = \bar{B} \cdot \bar{A} + B \cdot \bar{A} = \bar{A}$$

Figure 10.30 shows the final system assembled with the combining logic. Its behaviour is summarised with a state diagram in figure 10.31.

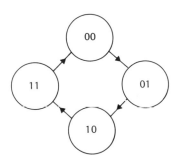

Figure 10.31 The state diagram for a two bit binary up counter

A NON-BINARY COUNTER

As we stated previously, a synchronous counter can count in any code you like. Suppose that you had used the logic system of figure 10.32 to generate

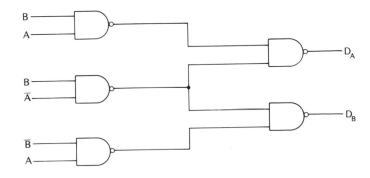

Figure 10.32

the inputs of the flip-flops in figure 10.29. Then we get the following expressions for D_A and D_B:

$$D_A = B \cdot A + B \cdot \bar{A}$$
$$D_B = B \cdot \bar{A} + \bar{B} \cdot A$$

We can use these expressions to write out the sequence of outputs from the counter. Start off with the counter in an arbitrary state, e.g. BA = 11. Then $D_B = 0$ and $D_A = 1$, so the next state will be BA = 01 etc.

Pulse	B	A	D_B	D_A
0	1	1	0	1
1	0	1	1	0
2	1	0	1	1

There are only three states in the sequence of outputs from the counter; BA = 00 does not appear. In fact, if the system gets into the state BA = 00, it gets stuck there. Both D_B and D_A will be 0 so every subsequent pulse feeds the system back into the same state. This is shown clearly in the state diagram of figure 10.33.

The existence of a "stuck" state obviously creates difficulties. It means that you have to be careful to avoid BA = 00 when you power up the circuit. Furthermore, should a stray pulse in the power supply accidentally jolt the system into BA = 00 it will stay there until the power is switched off. So when you design a synchronous counter you must ensure that all of the states feed into your desired sequence. This is shown in figure 10.34. State 00 feeds into 11, so that if the system gets into 00 it returns to the sequence at the next clock pulse. (The count will be wrong, but at least the system is still counting!) Figure 10.35 shows how the logic system has to be modified in order to unstick the system. It does not use the minimum number of NAND gates, which is two; work it out for yourself!

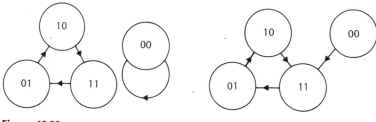

Figure 10.33   **Figure 10.34**

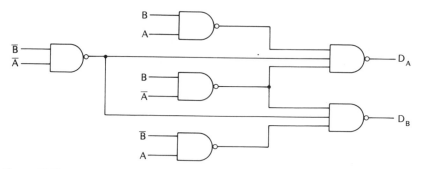

Figure 10.35

EXPERIMENT
Designing and evaluating synchronous counters

You are going to design a number of synchronous counters. You will then assemble them and evaluate their performance. Draw the circuit diagram for each one, with any associated algebra. Use a 1 Hz astable to provide the clock pulses and a pair of LEDs to monitor the outputs of the counters.

1) A binary down counter

Pulse	B	A
0	1	1
1	1	0
2	0	1
3	0	0

2) A three state counter

Pulse	B	A
0	0	1
1	0	0
2	1	0

3) A Gray code counter

Pulse	B	A
0	0	0
1	0	1
2	1	1
3	1	0

PROBLEMS

1 Write out the sequence of outputs for the counter shown in figure 10.36.

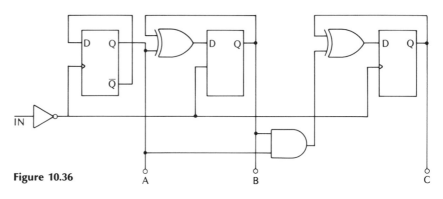

Figure 10.36

2 Design a synchronous counter which goes through the following sequence before resetting.

Pulse	C	B	A
0	1	1	1
1	1	1	0
2	1	0	0
3	0	0	0
4	0	0	1
5	0	1	1

Make sure that it cannot get stuck!

REVISION QUESTIONS

1 Show how a four bit shift register can be built out of D flip-flops. Explain what it does.

2 Explain how a shift register can be used to transmit a byte in serial form, and reconvert it into parallel form at the other end of a transmission line.

3 Draw a circuit diagram to show how a couple of 7496 shift register ICs and some logic gates can be used to make a single lit LED move continuously from left to right along a ten LED array. Explain how the system works.

4 Explain what a full adder is, and how it is used in the serial addition of two binary numbers.

5 How do you work out the two's complement code for a decimal number between $+9$ and -9?

6 Discuss the relative advantages of parallel and serial processing of long binary words.

7 State how the output of a synchronous counter differs from that of a ripple counter.

8 Draw the circuit diagram of a four bit synchronous binary up counter.

11

The microprocessor

The microprocessor is one of the wonders of the world. It is the ultimate systems component of electronics, capable of being moulded into an enormous number of different devices with only a little design effort. As you will see, a microprocessor can be used to make all of the digital systems that you have met so far in this book. Furthermore, you will be using it to design systems which would be very complex and unwieldy if built out of discrete shift registers, tristates and flip-flops.

So what is it that makes the microprocessor so useful? It is **programmable**. That is, a microprocessor is built so that it has no particular function. The function has to be programmed into it in the form of a sequence of instructions. That sequence (called a program) is held in the memory of the microprocessor and makes it behave in a specific way. Once you have had a bit of practice, programming a microprocessor becomes quite straightforward. The device is user-friendly, being fiendishly complex inside in order to attain an apparently simple exterior!

A MICROPROCESSOR SYSTEM

In this chapter we are going to try and show you how to use a microprocessor as a chunk of electronics. We are not going to discuss how a microprocessor can be made into the thinking heart of a computer, despite the importance of this application. Instead we shall look at the vast field of electronics in which a microprocessor can perform a function better and cheaper than conventional circuits.

Figure 11.1 (overleaf) illustrates the important parts of a small microprocessor system. Notice that it has six component blocks, interconnected by various buses. The principal block is the central processing unit, or CPU. It is the brain of the system, feeding out signals which control some of the other blocks. For instance, it can read bytes held in the memory and write bytes into it. It can also feed bytes out into the outside world via the output block, and can read in bytes via the input block. These bytes are shuffled towards and away from the CPU along the data bus, a set of eight parallel wires.

The memory performs two distinct functions. As we have stated above, it can be used to store bytes written into it by the CPU. It also contains the **program**. This sequence of bytes resides in the memory, and is used to tell the CPU what it should do. (We shall look at programs in more detail later.)

The final two units are the clock and the reset network. The CPU is an assembly of shift registers, and clock pulses are needed to shift bytes from

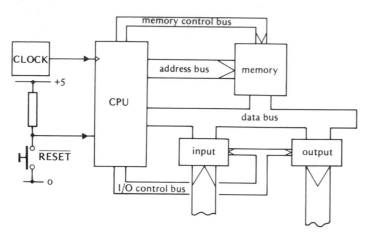

Figure 11.1 The architecture of a simple microprocessor system

one register to another. The reset network simply resets every flip-flop inside the CPU.

The current state of electronic technology means that a microprocessor system like the one shown in figure 11.1 can be built out of a few ICs. The CPU comes complete in a 40 pin IC package; we shall be looking at an example of one a little later on. You can get quite a lot of memory in a single IC (it increases every year); a widely used ROM (the 2716) holds 2048 bytes and is comparatively inexpensive. All you need is an astable, a bank of tristates, an eight bit latch and you have a useful system. Furthermore, it will be a blank system. Its precise function will be decided by the contents of the memory. So the CPU can be produced in vast quantities and therefore be very cheap despite its complexity.

PROGRAMMABLE SYSTEMS

All of the systems that you have met in the other chapters of this book have had their function decided by the connections between their components. For example, four D flip-flops can be made into a binary counter, a shift register or a latch, depending on how you wire them up. We say that the function of such circuits is **hard-wired** into them. You can only change the function of such a system by changing connections and components. A **programmable system**, on the other hand, can have its function changed without altering a single wire.

For example, look at the system shown in figure 11.2. It is just three shift registers and a large chunk of ROM. The ROM has 11 address lines (A_0 to A_{10}), so it contains $2^{11} = 2048$ locations. Each location holds a four bit word, or nibble. At any instant, only one of those nibbles is fed out of the data lines (D_0 to D_3); the location of the nibble being fed out is dictated by the contents of the A, B and I registers.

Think of the whole system as a parallel processor rather like the system of figure 10.20. The nibbles held in B and A are combined by the ROM. Each clock pulse loads the result of that combination process into A. The exact way in which A and B are combined depends on the contents of the I register. Obviously I can contain eight different three bit words, so the ROM can combine A and B in eight different ways.

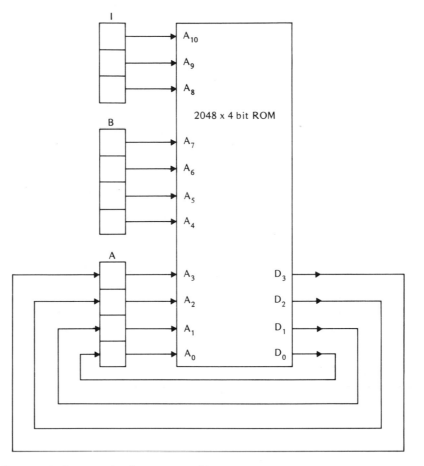

Figure 11.2 An example of a programmable system

Table 11.1

Nibble	Hex
0000	0
0001	1
0010	2
0011	3
0100	4
0101	5
0110	6
0111	7
1000	8
1001	9
1010	A
1011	B
1100	C
1101	D
1110	E
1111	F

In order to understand this idea a little better, you need to consider how the contents of the ROM may be organised. Each location in the ROM has a unique address, an eleven bit word. Eleven bit words are cumbersome and awkward; you can easily get lost in them! In order to make the addresses easier to handle we code them according to the **hexadecimal number system**. (Usually called **hex coding**.) This cuts down the number of digits from eleven to three. Table 11.1 shows the hex coding for each of the sixteen possible nibbles. So if the address of a location is 110 1110 0111, we write it as $6E7_H$. Similarly, the address $0D4_H$ is really 00011010100_2.

Figure 11.3 (overleaf) shows how the ROM is organised. It is split into eight **pages**, each page containing 256 locations. Since 256 locations need an eight bit binary word to specify their addresses, a two bit hex word can provide that address. Page 0 contains locations whose addresses are between 000_H and $0FF_H$; the msb gives the address of the page and the two lsb the address within the page.

Now if you return to figure 11.2 and consider it for a moment you should see that I provides the page address. That register (the instruction register) controls the three msb of the ROM address. The address within the page is provided by B and A. At each location within the page there resides a one bit hex word; that word has been written into the memory.

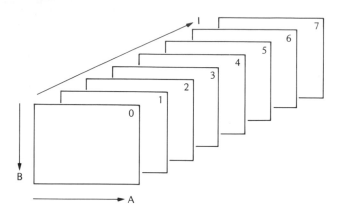

Figure 11.3 The eight pages of a 2 K memory

Suppose that you want the ROM to add the contents of A to B when I is 0_H. Then you fill page 0 with the truth table for A plus B. So location 023_H contains 5_H, location 093_H holds C_H, location $0F4_H$ holds 3_H etc. (Think about it!) Whatever the contents of A and B are, the result of the operation A plus B will appear at the data pins of the ROM ready to be loaded into A at the next clock pulse.

You can see that each page in the ROM can hold the truth table for a different operation. This is illustrated in the memory map of figure 11.4. The memory is presented as a column, with the locations piled one on top of the other. The highest address location is at the top, the lowest address at the bottom. The function stored in each page is shown at the relevant place in the map. For example, locations between 400_H and $4FF_H$ hold the truth table for A EOR B.

So the system of figure 11.2 is really eight systems in one. The eight possibilities are written into the ROM during its manufacture, and you select them by loading the appropriate three bit word into the instruction register. You can therefore transform the system from being a nibble adder into a nibble inverter with a single clock pulse. You could even store a stack of three bit words in a memory and sequentially load them into I so that the ROM acts like a series of different devices as time progresses; you could make it add, then subtract, then invert, then increment etc.

	800_H
A ← A + 1	
	700_H
A ← A - 1	
	600_H
A ← NOT A	
	500_H
A ← A EOR B	
	400_H
A ← A OR B	
	300_H
A ← A AND B	
	200_H
A ← A - B	
	100_H
A ← A + B	
	000_H

Figure 11.4 A possible memory map for the system of figure 11.2

THE Z80 CPU

It is unfortunate that each type of CPU has its own distinct pathology. Whereas all op-amps behave in basically the same fashion, all CPUs are unique. Despite common features, such as the way that they operate, each CPU has a unique set of registers and a unique set of instructions. So we have decided to stick to one CPU (the Z80) for all of our examples. Once you really understand how to use one particular CPU you should have no problems learning how to use any other.

The popularity of the Z80 as a CPU can be measured by its low price. It is designed to be the central processor of a fully fledged computer and has a very wide repertoire of functions. We are only going to tell you about some of those functions, so that you are not swamped by them. You will learn enough to be able to get the Z80 to perform any useful electronic

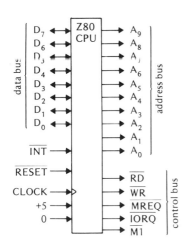

Figure 11.5 A partial pinout diagram of a Z80 CPU

function; the missing instructions have to do with arithmetic or duplicate what you can already do. Furthermore, we are only going to reveal some of the wonders which reside inside the Z80. This is because we are using the Z80 to teach you about the behaviour of CPUs in general; there is little point in telling you about the untypical parts of the Z80's makeup.

Figure 11.5 shows the input and output pins of the sawn-off version of the Z80 which we are going to be using from now on. We have only shown 28 of the 40 pins it actually has. Those 28 pins are what we need to incorporate the Z80 into a simple microprocessor system of the type shown in figure 11.1. That system needs to be described in a bit more detail before we deal with the task of programming the Z80.

The µP system

All of the examples of microprocessors in action in this chapter are going to assume a particular microprocessor system, what we shall call the µP system. It is essentially the system of 11.1, fleshed out with a sawn-off Z80 for the CPU. This section is going to describe how the Z80 controls the memory, the input and the output.

Start off with the memory. The µP system contains four pages of memory, each page containing 256 bytes or $\frac{1}{4}$K. Each page is contained within a separate IC, three in RAM and one in ROM; page 0 is the one in ROM. Figure 11.6 shows how the first two pages are hooked up to the data and address buses. Each page is enabled by bringing its $\overline{CE}$ pin low.

The CPU sends out three signals to control the memory. $\overline{MREQ}$ goes low when it wants to read or write bytes from or to memory. If it wants to read it also pulls $\overline{RD}$ low. Similarly, if it wants to write it pulls $\overline{WR}$ low instead.

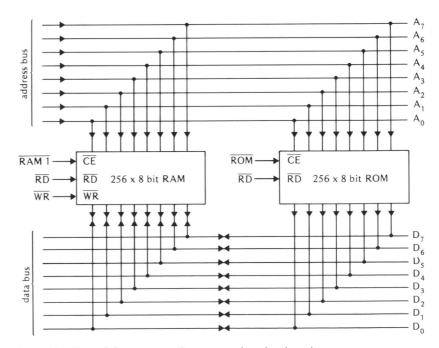

Figure 11.6 Two of the memory ICs connected to the three buses

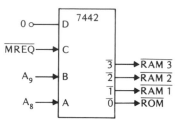

Figure 11.7 Generating the enable signals for the memory ICs

Figure 11.7 shows how a 7442 one-of-ten selector can be used to combine $\overline{\text{MREQ}}$ with A_8 and A_9 to generate the page address. So the CPU can enable one of the four pages of memory by sending the appropriate address along A_8 and A_9 and pulling $\overline{\text{MREQ}}$ low. It can then use the other address lines (A_0 to A_7) on the address bus to select a location within that page.

Figure 11.8 shows how the input and output ports are hooked up to the data bus. The output port is a set of eight D flip-flops organised as an eight bit (or octal) latch, and is available on a single IC (the 74LS374). The input port is just eight parallel tristates (74LS244). The signals which activate these two ports ($\overline{\text{IN}}$ and $\overline{\text{OUT}}$) can be obtained from the Z80 via a one-of-ten selector (figure 11.9). When the CPU wishes to read from or write to a port it pulls $\overline{\text{IORQ}}$ low. It also pulls $\overline{\text{RD}}$ low if it wants to read in a byte, and pulls $\overline{\text{WR}}$ low when it wants to feed out a byte.

It follows that the CPU has to make sure that it does not try to send bytes down the data bus in opposite directions at the same time. That is, it must only activate the pages of memory and the ports one at a time!

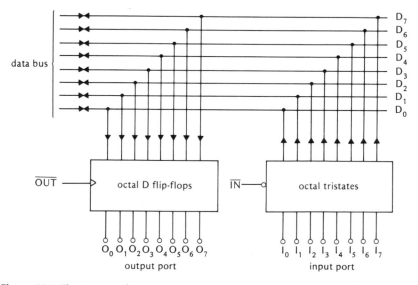

Figure 11.8 The input and output ports

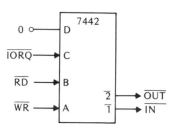

Figure 11.9 Generating the port enable signals

CPU OPERATION

We are now in a position to explain how the CPU goes about its business within the μP system. As an example, let us suppose that the memory is loaded with this sequence of bytes, starting at address 00 of page 0:

$$2F \quad ED \quad 79 \quad C3 \quad 00 \quad 00$$

(These are all in hex of course. The H subscript will only be used from now on where really necessary.) That sequence of bytes is a program which makes the μP system behave like a 167 Hz square wave generator. This is how it does it.

You first of all have to reset the CPU; this is done by pulling the $\overline{RESET}$ pin low. This fills every shift register inside the CPU with 0_{10}. At the moment we are only concerned with two of the registers. The first is called the program counter (or PC) and is sixteen bits long. The other is the accumulator (or A) and is eight bits long. So when $\overline{RESET}$ goes high again, we have the following situation within the CPU:

$$PC = 0000 \qquad A = 00$$

Now the CPU goes to work. It puts the contents of PC onto the address bus. Then it pulls $\overline{MREQ}$ and $\overline{RD}$ low so that the byte held at location 0000 in memory is put onto the data bus. The CPU then loads that byte into another of its registers, the instruction register (or I). So I is loaded with 2F. This is usually written as $I \leftarrow 2F$.

The I register performs the same function as the I register of the system in figure 11.2. Its contents decides what the CPU will do next. We can find out what the byte 2F will make the CPU do by looking it up in the instruction set (table 11.2). Opposite the op code of 2F we find the statement:

$$A \leftarrow \overline{A}$$

Before the CPU proceeds to obey this instruction it adds 1 to the contents of PC. That is, $PC \leftarrow PC + 1$. Then the instruction is obeyed. In this case, each bit of the accumulator is inverted. So at the end of this first **machine cycle** we have the following situation:

$$PC = 0001 \qquad A = FF$$

The machine cycle is summarised in the flowchart of figure 11.10. The CPU endlessly repeats machine cycles until it is stopped; the time taken for a machine cycle depends on the speed of the clock in the μP system. For the sake of simplicity we are going to assume that our system takes 0.5 ms to do each machine cycle. In practice the length of a machine cycle depends on the instruction which the CPU has to obey. Furthermore, the Z80 can be run at a hundred times this speed.

In the next machine cycle the address bus will have 0001 placed on it. So the byte stored at this location (ED) will be placed in the I register. A glance at table 11.2 (p. 316) will show you that there are four op codes which start off with ED. They are sixteen bit instructions, so they need to be loaded into I in two bytes. The CPU recognises ED as the first byte of a two byte instruction and proceeds to the next machine cycle. Meanwhile, the program counter has been incremented, so the next byte to be loaded into I is the one at location 0002, i.e. 79.

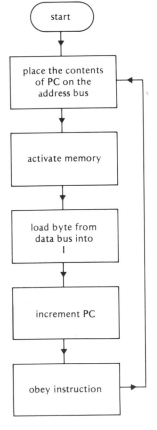

Figure 11.10 One machine cycle

start

place the contents of PC on the address bus

activate memory

load byte from data bus into I

increment PC

obey instruction

Table 11.2 The instruction set in alphabetical order

Op code	Function
00	no operation
04	B ← B + 1
05	B ← B − 1
06 n	B ← n
07	rotate left
0C	C ← C + 1
0D	C ← C − 1
0E n	C ← n
0F	rotate right
10 e	B ← B − 1, if B ≠ 0 then PC ← PC + e
14	D ← D + 1
15	D ← D − 1
16 n	D ← n
18 e	PC ← PC + e
1C	E ← E + 1
1D	E ← E − 1
1E n	E ← n
20 e	if not zero or equal then PC ← PC + e
21 n m	HL ← mn
23	HL ← HL + 1
28 e	if zero or equal then PC ← PC + e
2B	HL ← HL − 1
2F	A ← Ā
31 n m	SP ← mn
32 n m	(mn) ← A
3A n m	A ← (mn)
3C	A ← A + 1
3D	A ← A − 1
3E n	A ← n
47	B ← A
4F	C ← A
57	D ← A
5F	E ← A
76	halt
77	(HL) ← A
78	A ← B
79	A ← C
7A	A ← D
7B	A ← E
7E	A ← (HL)

Op code	Function
B8	A = B?
B9	A = C?
BA	A = D?
BB	A = E?
C1	BC ← (SP)
C2 n m	if not zero or equal then PC ← mn
C3 n m	PC ← mn
C5	(SP) ← BC
C6 n	A ← A + n
C9	PC ← (SP)
CA n m	if zero or equal then PC ← mn
CD n m	(SP) ← PC, PC ← mn
D1	DE ← (SP)
D5	(SP) ← DE
D6 n	A ← A − n
E1	HL ← (SP)
E5	(SP) ← HL
E6 n	A ← A AND n
ED 4D	return from interrupt
ED 56	establish interrupt mode
ED 78	A ← IN
ED 79	OUT ← A
EE n	A ← A EOR n
F1	AF ← (SP)
F5	(SP) ← AF
F6 n	A ← A OR n
FB	enable interrupts
FE n	A = n?

The instruction set tells you that ED 79 makes the CPU perform the following operation:

$$OUT \leftarrow A$$

That is, the CPU takes the contents of the accumulator, puts them on the data bus and latches them onto the output port. So the byte that is in A appears on the eight output pins of the μP system and stays there until further notice. Throughout all this, the contents of the accumulator remain unchanged.

The next machine cycle (the fourth) fetches the op code C3 from memory location 0003. The CPU recognises this as the first byte of a direct jump instruction; the next two bytes that it will read in from memory will eventually be placed in the program counter. The instruction set states that C3 nm has the following effect:

$$PC \leftarrow mn$$

m and n are each one byte long. m becomes the most significant byte of PC (the high order address) and n becomes the least significant byte (the low order address). So the fifth and sixth machine cycles fetch n and m from locations 0004 and 0005 in memory, stitch them together to make the sixteen bit word mn and finally place mn in PC.

The seventh machine cycle is therefore going to find that PC = 0000. It will therefore be identical to the first machine cycle; at the end of it we will find that:

$$PC = 0001 \qquad A = 00$$

A little thought should convince you that the CPU will continue to run round in circles indefinitely, inverting the byte in the accumulator every time that it goes round. The overall behaviour of the system is shown in the flowchart of figure 11.11. Since the CPU has to go through twelve machine cycles for the output pins to go through one cycle (i.e. 1 to 0 to 1 again), the whole system behaves like an oscillator with a frequency of $1/(12 \times 0.5) = 0.167\,kHz$. So if the circuit of figure 11.12 were attached to any one of the output pins (O_7 for example), a continuous buzz would come out of the speaker.

Table 11.3 contains a listing of the program which we have just been discussing. It shows the op code stored at each relevant memory location and indicates what each op code makes the CPU do. Note that some of

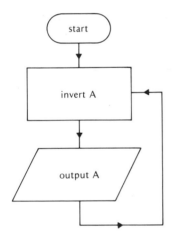

Figure 11.11 A flowchart for the program of table 11.3

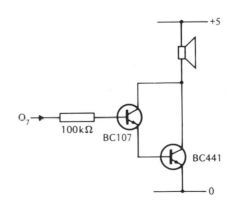

Figure 11.12

Table 11.3

Address	Op code	Function
0000	2F	A ← Ā ←
01	ED	
02	79	OUT ← A
03	C3	
04	00	
05	00	jump to ___

the op codes are one byte instructions (e.g. 2F), some are two byte instructions (e.g. ED 79) and that some are addresses (e.g. 00 00). Since the CPU fetches the op codes from the memory in sequence, you can use a program listing to work out what the μP system does as a function of time. The CPU executes the instructions one after the other starting from the one at the top of the listing, until it jumps back to the start again.

USING RAM AND ROM

By now you will have grasped how the overall behaviour of a microprocessor system is fixed by the sequence of op codes, addresses and numbers stored in its memory. You have looked in detail at a program which made the system behave like a square wave generator; that program is the first of many which you will either study or write yourself. But how did the program get into the memory in the first place?

Take a look at the memory map of the μP system, shown in figure 11.13. You will recall that the system has four pages of memory, with page 0 being ROM and the other three being RAM. Page 0 is always ROM because that is where the CPU looks for its first instruction after a reset. The contents of ROM are not lost when its supply is switched off. So the CPU is always sure of finding the same instruction at 0000 after it has been reset. If the program were stored in RAM, then it would certainly be lost whenever the supply was disconnected as the contents of RAM are **volatile**. You would have to fill the RAM with the program before allowing the CPU to fetch its first instruction!

Most applications of microprocessors require that the CPU be able to get on with the business of running through the program immediately its power supply is connected. So it is general practice to have at least page 0 held in ROM, with an *RC* network attached to the reset pin to provide an automatic reset. So the program has to be written into ROM once, perhaps during its manufacture, so that it sits there year after year regardless of whether the supply is attached or not.

We have included three pages of RAM so that the μP system can read in bytes and store them. The CPU always needs a certain amount of RAM so that it can perform certain functions (more about this later).

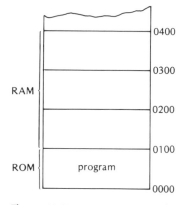

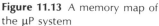

Figure 11.13 A memory map of the μP system

CPU REGISTERS AND OP CODES

The rest of this chapter will explain, in detail, how to use the various op codes for the Z80 CPU which are listed in table 11.4. This will be done by leading you by the hand through several program listings. These programs

Table 11.4 The instruction set in function order

Accumulator operations

3E n	A ← n	2F	A ← Ā
3C	A ← A + 1	E6 n	A ← A AND n
3D	A ← A − 1	F6 n	A ← A OR n
C6 n	A ← A + n	EE n	A ← A EOR n
D6 n	A ← A − n	07	rotate left
FE n	A = n?	0F	rotate right

8 bit register operations

	B	C	D	E
R ← n	06 n	0E n	16 n	1E n
R ← A	47	4F	57	5F
A ← R	78	79	7A	7B
R ← R − 1	05	0D	15	1D
A = R?	B8	B9	BA	BB
R ← R + 1	04	0C	14	1C

16 bit register operations

31 n m	SP ← mn	23	HL ← HL + 1
21 n m	HL ← mn	2B	HL ← HL − 1
CD n m	(SP) ← PC, PC ← mn	C9	PC ← (SP)

Accumulator transfers

32 n m	(mn) ← A	3A n m	A ← (mn)
77	(HL) ← A	7E	A ← (HL)
ED 79	OUT ← A	ED 78	A ← IN

Jumps

	PC ← mn	PC ← PC + e
jump	C3 n m	18 e
jump if zero or equal	CA n m	28 e
jump if not zero or equal	C2 n m	20 e
B ← B − 1, jump if B ≠ 0		10 e

Push and pop operations

	BC	DE	HL	AF
(SP) ← R	C5	D5	E5	F5
R ← (SP)	C1	D1	E1	F1

Miscellaneous

00	no operation
76	halt
ED 56	establish interrupt mode
FB	enable interrupts
ED 4D	return from interrupt

will not only show you how to program a microprocessor system, but also give you some examples of possible applications. You will also be encouraged to try to write programs of your own. However, first of all you will need to have some basic information about the registers within the CPU and its set of op codes (the instruction set). So now we are going to run quickly through what registers are within a sawn-off Z80 and how to decode some of its instruction set.

REGISTERS

The CPU contains a number of registers which can be used in a program. There are three 16 bit registers, the program counter (PC), the stack pointer (SP) and the memory pointer (HL). All of them are designed to hold addresses. There are four 8 bit general purpose registers (B, C, D and E) which can hold bytes transferred from the accumulator (A). Finally, there is the flag register (F).

Figure 11.14 summarises, in pictorial form, the registers described above. A full Z80 contains many more registers, but those shown in figure 11.14 will be ample for all of our examples.

THE INSTRUCTION SET

The set of op codes which we shall be using is shown in table 11.4 (p. 319). It is the instruction set of a sawn-off Z80, with the op codes grouped together according to function.

Accumulator operations

Each of these twelve operations changes the contents of the accumulator in some way. n is a byte stored in memory.

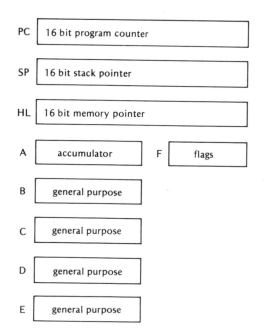

Figure 11.14 Some of the registers inside a Z80 CPU

The first five operations are arithmetic in nature. For example, C6 n adds the byte n to the contents of the accumulator using the standard rules of binary arithmetic and places the result in the accumulator.

FE n compares the contents of A with n, but leaves A unchanged. If n exactly matches the byte stored in A, then the Z flag is set. The Z flag is one bit of the F register.

The rest of the operations are logical ones. Thus E6 n takes each bit of A, ANDs it with the corresponding bit of n and places the resulting bit back into A. Suppose that A contains 0A. Then the instruction E6 06 puts 08 into A:

$$(00001010) \quad AND \quad (00001100) = 00001000$$

All of the accumulator operations set the Z flag if the result of that operation places zero in the accumulator.

8 bit register operations

These instructions allow the four general purpose registers to be loaded directly with a byte (n), loaded from the accumulator, pushed back into the accumulator, decremented, incremented and compared with the accumulator.

The last three operations will set the Z flag if the register is left containing zero or, in the case of the compare instructions, if the contents of the register matches that of the accumulator.

Accumulator transfers

This set of instructions allows bytes to be transferred between A and either memory or the input and output ports. There are two types of memory transfer operation. 32 n m uses **direct addressing**; n and m are the two bytes of a 16 bit address mn, where m is the most significant byte, or page address. So the instruction 32 0F 10 places the contents of A into the memory location whose address is 100F provided that it is RAM of course! The op code 77 uses **indirect addressing**. The 16 bit register HL holds the address of the location to which the byte in A is to be transferred.

Note the use of brackets to denote a byte held in memory. For example, A ← (HL) means the following; take the address being held in HL, find the byte stored at that address in memory and dump it into A.

WASTING TIME

There are two ways in which you can slow down a program. The easiest is illustrated in the listing of table 11.5; a more useful, but complex technique is used in table 11.6.

Have a look at table 11.5 first. The first three op codes invert A and feed it out of the output port. The next two op codes are 00. If you look this up in table 11.2 (the instruction set in "alphabetical order") you will see that it makes the CPU do nothing at all. So the op codes at locations 03 and 04 (forget the page address for convenience) make the μP go through two machine cycles without changing anything. As a machine cycle is $\frac{1}{2}$ ms long, this gives a time delay of 1 ms.

Table 11.5

Address	Op code	Function
0000	. 2F	A ← Ā ←
01	ED	
02	79	OUT ← A
03	00	no operation
04	00	no operation
05	18	
06	F9	jump to ─────

Table 11.6

Address	Op code	Function
0000	3E	
01	FF	A ← FF
02	ED	
03	79	OUT ← A
04	06	
05	FF	B ← FF
06	10	
07	FE	B ← B − 1, if B ≠ 0 jump to ─────
08	3D	A ← A − 1
09	18	
0A	F7	jump to ─────

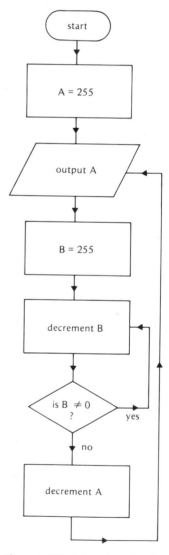

Figure 11.15 A flowchart for the program of table 11.6

The last two op codes of the listing make up a **relative jump instruction**. 18 is the instruction itself, and F9 is treated by the CPU as a two's complement number which is to be added to the contents of PC. Since the value of PC is increased by 1 immediately after each byte has been fetched from memory, PC = 0007 when both op codes (18 F9) have been read in. The decimal equivalent of F9 is −7 (work it out), so the net result of the whole operation is to leave PC = 0000. The program jumps back to the beginning, inverts A, pumps it out of the output port, wastes a millisecond etc.

Relative jumps are useful, but limited. They only need two bytes (an **instruction** and a **displacement**) as opposed to the three bytes of a direct jump (instruction, low order address and high order address). So they take up less space in ROM and are faster to execute. Their principal limitation is that they do not allow you to jump very far, whereas a direct jump can take you anywhere you like.

It follows that if you want to waste more than a few milliseconds in a program you are going to have to fill a lot of memory space with 00 op codes. The alternative strategy is to use a **delay loop**. Essentially, you give the CPU a number and ask it to subtract 1 from it until it gets to 0; only then can it fetch the next instruction in the program.

The flowchart of figure 11.15 shows in more detail how to set up a delay loop in a program. The whole program makes the µP behave like an 8 bit counter which is being fed a 4 Hz square wave. If a bank of LEDs is hooked onto the output port (as shown in figure 11.16) and a lit LED represents a logical 1, then the LEDs will count up in binary with the lsb

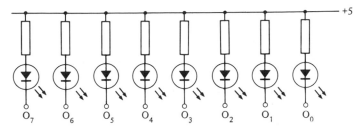

Figure 11.16

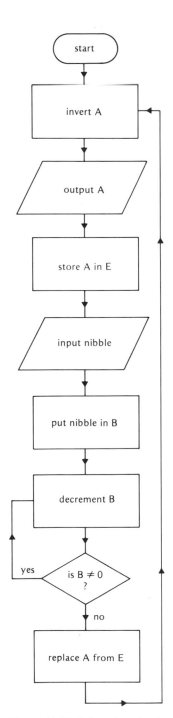

Figure 11.17 A flowchart for the program of table 11.7

changing four times a second. It is helpful to look at the flowchart in detail before studying the program listing.

Initially, A is loaded with 255_{10} i.e. FF_H. Then A is fed out of the output port to the LEDs; none of them light up of course. We now need to waste some time. So B (a general purpose register) is loaded with the largest number it can take; $B \leftarrow FF$. B is then decremented until it reaches zero; this requires the CPU to chase around the inner loop in the flowchart 255 times, giving a delay of about a quarter of a second. Only then can the next instruction (decrement A) be obeyed. The new value of A is fed out through the output port (A = FE or 11111110) and one LED lights up. Obviously, the CPU never stops chasing round the outer loop, and for one turn around that loop it does 255 turns around the inner loop.

Now study the listing in table 11.6. The delay loop is made up of the op codes stored in locations 04, 05, 06 and 07 on page 0. 06 FF puts FF into B; note that this does not affect the contents of A in any way. 10 FE are the codes for the jump instruction. This starts off by decrementing B. It then gets the CPU to look at the Z flag to see if B is zero yet. If it finds that Z has not been set (i.e. $B \neq 0$) then FE is added to the current value of PC (i.e. 0008). As FE is the two's complement coding for -2, this makes PC = 0006 and the program loops back to the 10 FE instruction. Each of these loops takes 1 ms (two machine cycles), so the CPU has to waste about 250 ms before it gets to the instruction at location 0008.

Saving the accumulator

When you want to push bytes out of the μP system or read them in, you have to do it via the accumulator. For example, the instruction ED 79 takes the contents of A and places them on the terminals of the output port. Similarly, ED 78 pushes the current state of the terminals of the input port into A. But you also need to use the accumulator if you want to perform operations on a byte, such as inversion or addition. In fact, the accumulator is a unique register in terms of what you can do with its contents. It often happens that a program will have to perform operations on several bytes and only one, obviously, can be in the accumulator at any one instant. The other bytes have to be stored somewhere, awaiting their turn.

Bytes can be stored in two distinct ways. They can be dumped into RAM or into a general purpose register. We will be showing you how to use RAM in a later section, but if you look at the instruction set you will find that the op codes 32 n m and 3A n m enable you to place and retrieve A in RAM. If you store A in a general purpose register you only need one machine cycle as opposed to the three needed for storage in RAM.

The program listed in table 11.7 illustrates the technique of saving bytes in registers. Figure 11.17 is a flowchart which may help you to unscramble

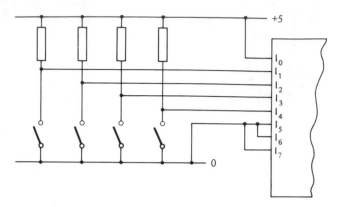

Figure 11.18

Table 11.7 Programmable square wave generator

Address	op code	Function
0000	2F	$A \leftarrow \bar{A}$ ←
01	ED	
02	79	OUT ← A
03	5F	E ← A (store byte in E)
04	ED	
05	78	A ← IN
06	47	B ← A
07	10	←
08	FE	$B \leftarrow B - 1$, if $B \neq 0$ jump to ─┘
09	7B	A ← E (return byte from E)
0A	18	
0B	F4	jump to ─────

how the program works. The µP system behaves like a programmable square wave generator, with the external programming done by the four switches shown in figure 11.18. The nibble which is fed into I_1, I_2, I_3 and I_4 is used to tell the CPU how many times it has to go around the delay loop. When the nibble is 0000 the output terminals of the µP oscillate at 83 Hz; as the value of the nibble is increased the rate of oscillation slows down.

Study the listing of table 11.7 and convince yourself that it makes the µP behave as described above!

PROBLEMS

1 The following set of op codes is placed at the bottom of page 0 of memory.

3E   FF   ED   79   5F   06   FF   10   FE   D6   10   18   F5

Write out a full listing for the program, and explain what it makes the µP do to a bank of LEDs hooked up to its output port.

2 The op code 07 moves every bit of the accumulator one place to the left, with the msb being moved to where the lsb used to be. For example, if A = 10001100, this 07 transforms it to 00011001. The program below makes use of this facility:

3E   FE   ED   79   00   07   18   FA

a) Write out a listing of the program.

b) If the usual bank of LEDs is used to monitor the output port, state and explain what will happen to them as the program runs.

c) Modify the program so that it can be slowed down by the bank of switches shown in figure 11.18. The program should run at full tilt when all switches are closed and at its slowest when they are all open. Write out your program listing in full, and describe how it works.

3 Write a program which will make the μP behave as follows. A bank of eight LEDs displays, in binary, how many times a push switch has been pressed and released. The push switch pulls all eight terminals of the input port down to 0 when it is closed; the terminals share a common pull-up resistor. Draw up a flowchart first. Then write out the listing. Carefully explain how it works.

THE ALERTNESS TESTER

The program that you are going to analyse in this section makes the μP behave like an alertness tester, or dead man's handle. That is, we shall assume that the μP is able to control a device (such as a train or piece of

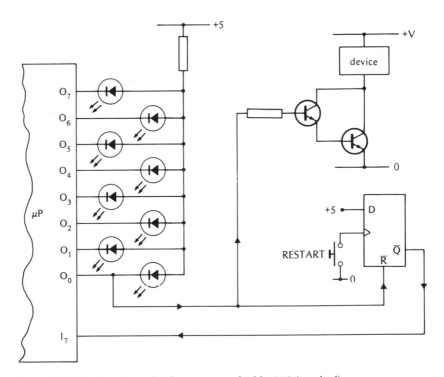

Figure 11.19 The hardware for the program of table 11.8 (overleaf)

Table 11.8 The alertness tester program

Address	Op code	Function
0000	3E	
01	FE	A ← FE
02	ED	
03	79	OUT ← FE (1111 1110)
04	3E	
05	7F	A ← 7F (0111 1111)
06	16	
07	08	D ← 08 (set counter)
08	ED	
09	79	OUT ← A
0A	0E	
0B	04	C ← 04 (set delay)
0C	06	(start ¼ second delay loop)
0D	FF	B ← FF
0E	10	
0F	FE	B ← B − 1, if B ≠ 0 jump to
10	0D	C ← C − 1
11	20	
12	F9	if C ≠ 0 jump to
13	5F	E ← A (store byte in E)
14	ED	
15	78	A ← IN
16	E6	
17	80	A ← A AND 80 (mask off 7 lsb)
18	28	
19	E6	if A = 0 jump to
1A	7B	A ← E (return byte from E)
1B	0F	rotate right
1C	15	D ← D − 1
1D	20	
1E	E9	if D ≠ 0 jump to
1F	76	halt

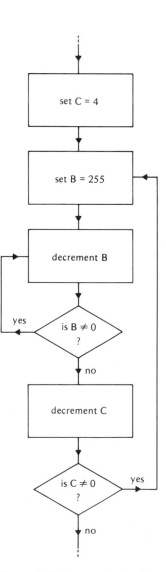

Figure 11.20 A nested pair of delay loops

machinery) via its output port. It will keep that piece of machinery going provided that it receives periodic signals from a human being through its input port. If those signals fail to arrive, the μP promptly switches off the device and goes to sleep.

The hardware attached to the ports is illustrated in figure 11.19 (p. 325). O_0 controls the device via a pair of transistors put together as a Darlington pair. Current can only flow through the device when O_0 is 1; if the transistors cannot sink enough current to run the device directly, then they can run it with a relay instead. The operator communicates with the system via a flip-flop. When he releases the RESTART switch a logical 0 is fed into I_7. The μP responds by momentarily pulling O_0 low and resetting the flip-flop (it is assumed that the device will not notice the momentary drop of current). Then each of the LEDs comes on in turn, starting with the one at the top. If, eight seconds later, the bottom LED is allowed to light up, it stays lit up for good and the device is switched off. The operator must therefore press the RESTART button at least once every eight seconds if the device is to continue functioning. Note that by using a flip-flop, which

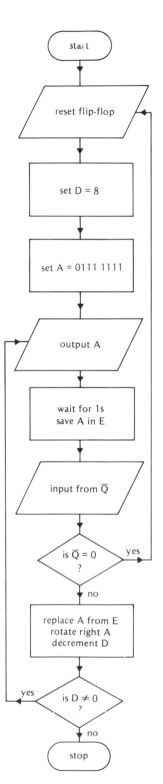

Figure 11.21 A flowchart for the program of table 11.8

responds to edges, the operator cannot get away with just pushing the button all of the time (or getting a brick to do it!).

The listing of the program is shown in table 11.8. Before you have a go at working through it in detail, we would like to draw your attention to a couple of techniques used in it.

Wasting more time

The instructions stored in locations 0A to 12 inclusive waste the CPUs time for about 1 s. This comparatively long delay is accomplished by nesting two delay loops one within the other. If you look at the flowchart of figure 11.20 you will be able to see how this works. The inner loop makes the CPU subtract 1 from 255 until it gets to 0. The outer loop makes the CPU enter the inner loop four times. The net result is that the CPU is forced to waste time for about 2000 machine cycles, or 1 s. You can see that this double delay loop technique can be used to hold up the CPU for about a minute if needs be.

Masking

Only I_7 is attached to $\bar{Q}$ of the flip-flop; the other seven bits of the input port are left to their own devices. So when the CPU reads in a byte from the input port, it must only look at the msb of that byte. So just after the byte has been placed in the accumulator, the uninteresting bits are set to zero. This is done by using the AND operation:

$$(\bar{Q}XXXXXXX) \quad AND \quad (10000000) = \bar{Q}0000000$$

So the instruction E6 80 **masks off** the seven lsb of the byte. This leaves the value of A completely dependent on the value of $\bar{Q}$; A will only be zero if $\bar{Q}$ is zero.

The behaviour of the whole program is summarised in the flowchart of figure 11.21. Use it to help you work through the listing in table 11.8; in particular, make sure that you understand how the D register is used. The op code 76 makes the CPU waste time indefinitely; it goes through machine cycles repeatedly without altering the value of PC, so it never moves onto the next instruction! Not dead, but sleeping?

AN ELECTRONIC DIE

A program which makes the µP behave like a die is shown in table 11.9. The output is displayed by seven LEDs, arranged in the usual dice fashion, as shown in figure 11.22 (p. 329). Each time that the THROW button is pressed and released a random number from 1 to 6 is displayed on those LEDs. The program illustrates three techniques, namely handshaking, the use of a look-up table and how to toggle bits of a byte.

Handshaking

The flip-flop in figure 11.22 acts as an intermediary between the dice thrower and the CPU. It responds immediately when the THROW button is pressed by pulling $\bar{Q}$ low. Some time later the CPU takes a look at $\bar{Q}$; if it finds that $\bar{Q}$ is 0 it responds by resetting the flip-flop, and then gets on with the rest of the program.

Table 11.9 Dice throwing program

Address	Op code	Function
0000	3E	
01	FF	A ← FF
02	ED	
03	79	OUT ← A ($\bar{R}$ = 1, all LEDs off)
04	21	
05	23	
06	00	HL ← 0023 (start address)
07	06	
08	06	B ← 06 (set counter)
09	ED	
0A	78	A ← IN (sample $\bar{Q}$)
0B	E6	
0C	01	A ← A AND 01 (mask)
0D	28	
0E	05	If A = 0 jump to ─────
0F	2B	HL ← HL − 1 (decrement address)
10	10	
11	F7	B ← B − 1, if B ≠ 0 jump to ─────
12	18	
13	F0	jump to ─────────────→
14	7E	A ← (HL) (load byte from table) ←
15	ED	
16	79	OUT ← A
17	EE	
18	80	A ← A EOR 80 (toggle msb)
19	ED	
1A	79	OUT ← A (flip-flop reset)
1B	18	
1C	E7	jump to ─────
1D	—	
1E	08	0000 1000 (6)
1F	14	0001 0100 (5)
20	1C	0001 1100 (4)
21	36	0011 0110 (3)
22	3E	0011 1110 (2)
23	77	0111 0111 (1)

If we did not use the flip-flop there would be problems in synchronising the CPU with the dice thrower. Because of the nature of the program, the CPU has to spend nearly all of its time counting from 1 to 6; it can only look at the state of I_0 from time to time. So the dice thrower would have to keep pressing the THROW button until the CPU noticed that he had done so! Furthermore, the CPU would have to study patiently the state of I_0 until it was satisfied that the THROW button had been released again. Both the CPU and human being would be hampered by each others slowness.

By using a hand-shaking flip-flop, information is safely transferred from human being to CPU with the minimum of inconvenience for both.

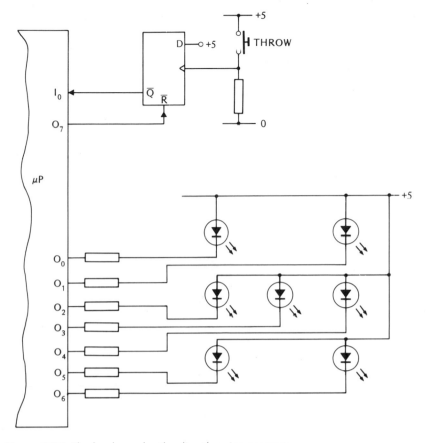

Figure 11.22 The hardware for the dice throwing program

Look-up tables

When the CPU displays a number it has to use the standard dice coding. Since a binary-to-sevenspot decoder has yet to become widely available, this decoding has to be done by the program itself. The code for each number is stored in a **look-up table** in ROM; whenever the CPU has to feed out a number it fetches the appropriate code from the table.

The look-up table is between locations 1E and 23 of page 0 (see table 11.9). Each of the six bytes held in the table generates one of the dice displays; convince yourself that they are correct. In order to pull a byte out of the table so that it can be pushed out of the output port, the program makes use of the HL register, the **memory pointer**. Initially, HL $\leftarrow$ 0023, the highest address at the top of the look-up table. The CPU then decrements HL so that the address steadily moves down the table. The B register records how many times HL is decremented so that when the bottom of the table is reached HL can be reset so that it points to the top of the table once more. If, at any time, the CPU detects that the THROW button has been pressed, it executes the instruction A $\leftarrow$ (HL) and feeds that byte to the output port.

Toggling bits

A very useful technique is employed between locations 15 and 1A of page 0. After the CPU has fed out a number to be displayed by the LEDs it needs to reset the handshaking flip-flop. It does this by taking O_7 low for an instant, without changing the state of any of the other output terminals. The byte that is fetched from the look-up table has 0 as its msb. So when this is shoved out of the output port, $\bar{R}$ is pulled low. The CPU then performs the operation $A \leftarrow A$ EOR 80. This has the effect of inverting only the msb of A, leaving all of the other bits intact. For example:

$$(00111110) \quad EOR \quad (10000000) = 10111110$$

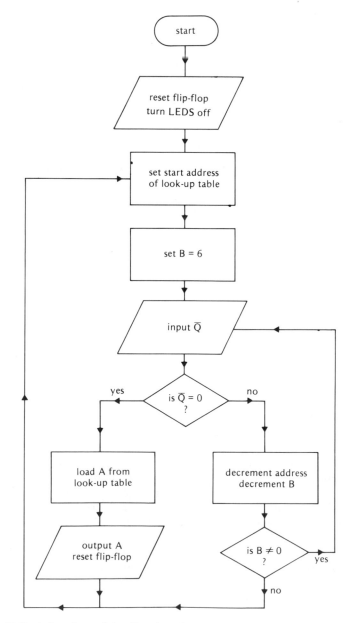

Figure 11.23 A flowchart of the dice throwing program

The new byte in A is then pushed out of the output port so that $\bar{R}$ is pushed up to 1; the flip-flop is now ready to respond to pressure on the THROW button.

The operation of inverting a bit is called **toggling**. You can use the logical operations EOR, AND and OR to manipulate each bit of A independently of all the other bits. For example, you have already seen how the AND operation can be used to selectively reset bits to 0. In order to selectively set bits to 1 you have to use the OR function. Finally, you can change the state of a bit by using the EOR function.

Figure 11.23 shows a flowchart for the whole of the program. Work through the listing, making sure that you understand why each step is written the way it is. Then try your hand at writing some of the programs below.

PROBLEMS

1 Write a program that will make the µP behave as follows. A seven segment LED is hooked up to the 7 lsb of the output port. It displays the numbers 0 to 9, changing the display once a second, before starting the sequence all over again.

2 Write a program, which will make the µP light up a number of LEDs, which depends on the decimal value of the three bit word fed into the three lsb of the input port. So if $I_2I_1I_0 = 011$, the first three of a row of seven LEDs are lit up. Use a handshaking flip-flop to tell the CPU when the three bit word can be read in; use I_7 to let the CPU look at $\bar{Q}$, and O_7 to reset the flip-flop. You may find this sequence of instructions useful!

$$\cdots \quad ED \quad 78 \quad E6 \quad 03 \quad 47 \quad 21 \quad 30 \quad 00 \quad 04 \quad 2B \quad 10 \quad FD \quad \cdots$$

THE DUCK SHOOTING GAME

Take a look at figure 11.24 (overleaf). It shows the hardware connected to the µP to make it into a duck shooting game. The duck is represented by a lit LED in a row of seven; the lit LED continuously runs along the row from bottom to top. If you press the FIRE button at the instant that the middle LED is lit then the score displayed on the seven segment display is incremented. The score is initially 0 when the CPU emerges from a reset. After you have pressed the FIRE button nine times your score is frozen on the display and the CPU goes to sleep.

You will have noticed that the CPU uses the $\overline{INT}$ terminal to sample the state of the handshaking flip-flop. As you are about to find out, this is, in some respects, a better way of handshaking than via one terminal of the input port. When the $\overline{INT}$ pin on the CPU is brought low, it is forced to fetch its next instruction from location 0038. So instead of waiting for the program to get round to looking at the state of $\bar{Q}$, it responds immediately. The price that has to be paid for this gain in speed is a longer and more complicated program. So before you make a start at analysing the listing of the program, we shall have to explain some of the finer points of using this **interrupt facility**.

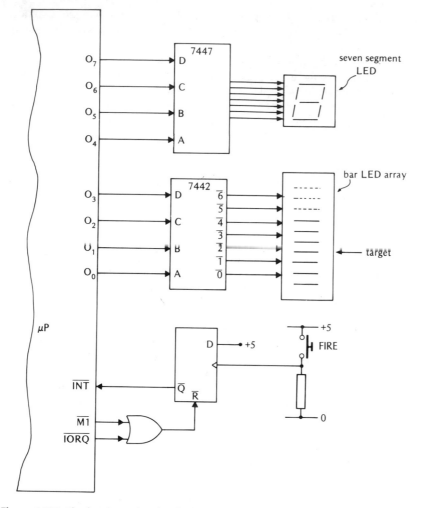

Figure 11.24 The hardware for the duck shoot program

INTERRUPTS

Before the $\overline{\text{INT}}$ pin will affect the operation of the CPU you have to prepare some of the registers. The reasons for this will become obvious later. So near the start of the program you need to have the following sequence of op codes:

$$\cdots \quad 31 \quad FF \quad 03 \quad ED \quad 56 \quad FB \quad \cdots$$

The first three bytes put an address (03FF) into the **stack pointer** (SP). The particular address is not very important, provided that it corresponds to somewhere in a page of RAM. We have chosen to place the stack pointer right at the top of RAM; there has to be plenty of blank RAM underneath it. The pair of op codes ED 56 instructs the CPU to be prepared to respond to interrupt signals by placing 0038 in PC. The final byte (FB) tells the CPU to look at the state of the $\overline{\text{INT}}$ pin at the end of every machine cycle.

An interrupt signal is anything which pulls the $\overline{\text{INT}}$ pin low. It is usual practice to use a handshaking flip-flop to mediate between $\overline{\text{INT}}$ and the rest of the world. Figure 11.24 shows how it is done. When the CPU detects an interrupt signal it simultaneously pulls $\overline{M1}$ and $\overline{IORQ}$ low for an instant so that the flip-flop is reset immediately; this signals to the rest of the world that the CPU is responding to the interrupt.

The flowchart of figure 11.25 describes what the CPU does as soon as it detects an interrupt. The FB instruction is cancelled so that subsequent interrupts will not mess things up. Then the flip-flop is reset. The CPU now has to store the current value of PC on the **stack**; this is so that it can go back to where it left off in the program, when it has finished dealing with

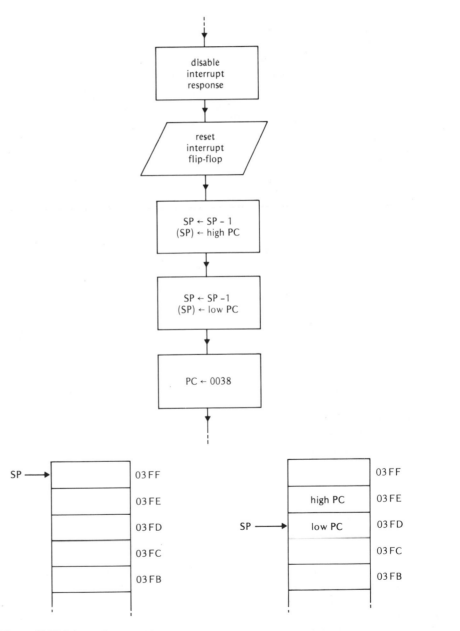

Figure 11.25 Interrupt response

the interrupt. Saving PC is done in two stages. SP is first decremented, and the most significant byte of PC placed at (SP). SP is then decremented again so that the least significant byte of PC can be stored just below the other byte. Finally, 0038 is placed in PC, and the CPU returns to the start of a normal machine cycle.

The next instruction to be obeyed by the CPU will obviously be the one at location 0038. It will be the start of the **interrupt service routine**. When that routine has been performed these three op-codes will return the CPU to the place where it was when it was interrupted:

$$\cdots \quad FB \quad ED \quad 4D \quad \cdots$$

The first op-code instructs the CPU to pay attention to the state of $\overline{INT}$ from now on. The next two make the CPU take two bytes off the bottom of the stack and place them in PC. The current value of SP is 03FD (look at figure 11.25). So the byte at that location is transferred to the lower half of PC, and SP is incremented. Then the byte at 03FE is transferred to the upper half of PC and SP incremented again. It should be fairly obvious that the next instruction obeyed by the CPU is the one that it was about to tackle when it was interrupted.

Pushes and pops

One of the complications of using an interrupt is that the CPU can be almost anywhere in its program when the interrupt arrives. By saving the current value of PC in the stack before going off to the interrupt service routine, the CPU can ensure that it picks up the program exactly where it left off. However, it is not always sufficient to ensure that the value of PC is preserved; you often need to save the current contents of some or all of the registers as well. This is because the interrupt service routine is sure to alter the contents of some of these, and those changes could cause chaos when the CPU goes back to the main program.

So the first thing that the interrupt service routine must do is store the contents of sensitive registers in the stack. This is known as pushing registers onto the stack. Thus the op code F5 pushes the registers A and F onto the stack, decrementing SP twice in the process so that it ends up pointing to the bottom of the stack again. Other op codes make the CPU push the memory pointer and the four general purpose registers onto the stack.

At the end of the interrupt service routine, the CPU has to pop the contents of the registers off the stack in the correct order. As bytes are popped off the stack, SP is incremented automatically so that it points to the next byte up. So if the service routine started off with this sequence of op codes:

$$\cdots \quad F5 \quad E5 \quad D5 \quad C5 \quad \cdots$$

it has to end with this sequence:

$$\cdots \quad C1 \quad D1 \quad E1 \quad F1 \quad FB \quad ED \quad 4D \quad \cdots$$

Since the register pair BC was the last to be pushed onto the stack it must be the first popped off the stack.

The duck shoot program is listed in tables 11.10 and 11.11. The second listing is that of the interrupt service routine. Use the flowcharts of figure 11.26 (p. 336) to help you work through the listings.

Table 11.10 Main program

Address	Op code	Function
0000	31	
01	FF	
02	03	SP ← 03FF
03	0E	
04	09	C ← 09
05	ED	
06	56	establish interrupt mode
07	FB	enable interrupts
08	16	←
09	07	D ← 07
0A	7B	A ← E
0B	E6	
0C	F0	A ← A AND F0
0D	5F	E ← A
0E	7B	A ← E ←
0F	ED	
10	79	OUT ← A
11	06	
12	3F	B ← 3F
13	10	←
14	FE	B ← B − 1, if B ≠ 0 jump to ——
15	1C	E ← E + 1
16	15	D ← D − 1
17	20	
18	F5	if D ≠ 0, jump to ——
19	18	
1A	ED	jump to ——

Table 11.11 Interrupt service routine

Address	Op code	Function
0038	F5	(SP) ← AF (Push A, F on stack)
39	7B	A ← E
3A	E6	
3B	0F	A ← A AND 0F (mask high nibble)
3C	FE	
3D	03	A = 03?
3E	20	
3F	04	if A ≠ 03, jump to ——
40	7B	A ← E
41	C6	
42	10	A ← A + 10
43	5F	E ← A
44	0D	C ← C − 1 ←
45	28	
46	05	if C = 0, jump to ——
47	F1	AF ← (SP) (Pop A, F off stack)
48	FB	enable interrupts
49	ED	
4A	4D	return from interrupt
4B	—	
4C	76	halt ←

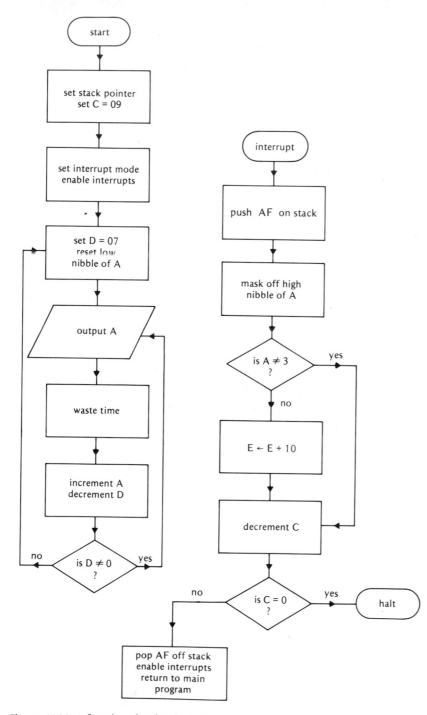

Figure 11.26 A flowchart for the duck shoot program

PROBLEMS

1 Write a program that will make the µP behave like an event counter. The state of the count is displayed, in decimal, by a pair of seven segment LEDs. Each event clocks a handshaking flip-flop which feeds $\overline{Q}$ into $\overline{INT}$. (Note that the CPU will still respond to an interrupt when it is obeying the halt instruction, 76.)

2 You are going to write a program which will make the µP into a programmable buzzer. The CPU makes one terminal of the output port oscillate with a period determined by the value of a three bit word fed into the input port. That word is read in whenever $\overline{INT}$ is brought low. Draw a circuit diagram of any hardware which needs to be attached to the µP system.

THE TELEPHONE DIAL

Our last example is a fairly long program, but it demonstrates a powerful technique that makes long and complex programs easy to write. By using

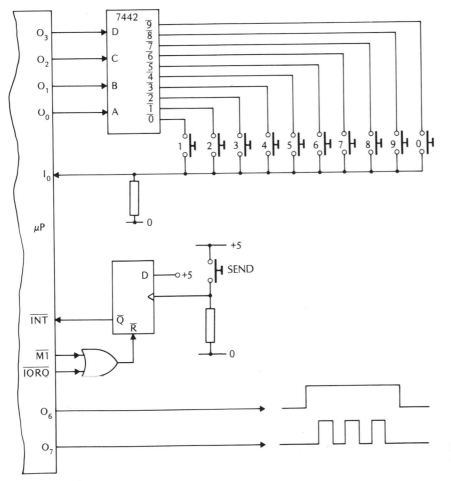

Figure 11.27 The hardware for the telephone dial program

subroutines the program can be split into small self-contained chunks. Each chunk can be tried separately before they are sewn together to make the whole program.

The hardware which needs to be attached to the µP system is shown in figure 11.27. At the top of the diagram is a keyboard, with the keys numbered from 1 to 0 in the style of a telephone dial. The CPU uses a 7442 one-of-ten selector to scan that keyboard, and work out which of the ten switches is being pressed. When it has found the switch which is closed, a number between 1 and 10 is pushed onto a special stack in RAM (the pile). Every time that a switch is pressed, another number is pushed onto the pile. So if a telephone number is punched out onto the keyboard it will be stored in the pile.

In the centre of the diagram is the familiar handshaking flip-flop. When the SEND switch is pressed the CPU transmits the telephone number stored on the pile. Each digit of the number is sent out as a series of pulses via the O_7 line. While each digit is being transmitted the O_6 line goes high; figure 11.27 shows the timing diagram for O_6 and O_7 while the number 3 is being transmitted.

The software needed is outlined in the memory map of figure 11.28. You will have noticed that there are gaps in memory, places where no program is stored. Those gaps separate the various subroutines which go to make up the whole program. The main program is stored between 00 and 23, with the interrupt service routine between 38 and 51. There are also three subroutines. The time delay subroutine is held between 30 and 34, the keyboard subroutine between 60 and 72, and the pulse output subroutine between 80 and 90.

SUBROUTINES

Table 11.12 is the listing of a very short subroutine. It is a chunk of program with a purpose; it makes the CPU waste time for about 125 ms. In order to use the subroutine, the following instructions have to be used:

$$\cdots \quad \text{CD} \quad 30 \quad 00 \quad \cdots$$

When the op code CD is encountered, the CPU stores the current value of PC on the stack. It then places the next two bytes (30 00) in PC to give the address of the next instruction (i.e. 0030). So the CPU jumps to the subroutine and obeys the instructions contained in it. At the end of the subroutine the CPU comes across the op code C9. This makes it take two bytes off the bottom of the stack and place them in PC, so that it goes back to where it was before.

The beauty of a subroutine is that the same set of instructions can be used in many places in a program. Every time that the CPU gets to the end of a subroutine it returns to where it left the main program. So if a set of

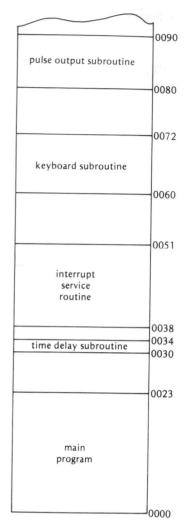

Figure 11.28 The memory map of the telephone dial program

Table 11.12 Time delay subroutine

Address	Op code	Function
0030	06	
31	7F	B ← 7F
32	10	←
33	FE	B ← B − 1, if B ≠ 0 jump to ─┘
34	C9	return to main program

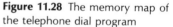

instructions is used many times in a program, it pays to write that set as a subroutine. A lot of memory space can be saved in this way.

But subroutines can do more than just save memory space. They can make writing a complex program a simple task. The whole program is split into a number of smaller sub-programs. Each sub-program is written as a subroutine. That subroutine can then be tried out and tested, to see that it does what it is supposed to do. When the subroutine is perfect, it can be placed in a convenient location in memory. It is important that a subroutine uses only relative jump instructions, so that the exact location of the subroutine in memory is not important. The use of **relocatable code** means that a tried and tested subroutine can be used in a number of programs without any alterations.

Table 11.13 shows the listing of another subroutine used in the telepone dial program. It is the pulse output subroutine, generating a number of 125 ms pulses and feeding them out of O_7. It uses the contents of register D to decide how many pulses have to be fed out. Note that, although it is a subroutine itself, it calls upon the services of another subroutine.

The main program is listed in table 11.16. Its function is to set up the initial addresses of the stack pointer and the memory pointer, set up the interrupt response and look at the keyboard. Every time that the CPU detects a switch being pressed, it calls upon the keyboard subroutine (table 11.15) to work out which one. The subroutine returns to the main program with the switch value stored in register E. This is then pushed onto the pile (i.e. (HL) ← A) and the memory pointer decremented. Once the CPU is satisfied that the switch has been released it proceeds to wait for another one to be pressed.

The interrupt service routine is listed in table 11.14. It runs through the numbers which have been stored in the pile, using the pulse output subroutine (table 11.13) to feed each one out in serial form. (Note the use of the 07 op code to multiply the contents of A by two.) Register C is used to tell the CPU when it has got to the bottom of the pile.

Table 11.13 Pulse output subroutine

Address	Op code	Function
0080	ED	
81	79	OUT ← A
82	CD	←
83	30	
84	00	use time delay subroutine
85	EE	
86	80	toggle bit 7
87	ED	
88	79	OUT ← A
89	15	D ← D − 1
8A	20	
8B	F6	if D ≠ 0, jump to
8C	EE	
8D	40	toggle bit 6
8E	ED	
8F	79	OUT ← A
90	C9	return to main program

Table 11.14 Interrupt service routine

Address	Op code	Function
0038	F5	push A, F onto stack
39	21	
3A	FE	
3B	02	HL ← 02FE
3C	7E	A ← (HL) ←
3D	07	A ← A × 2
3E	57	D ← A
3F	3E	
40	4F	A ← 4F
41	CD	
42	80	
43	00	use pulse output subroutine
44	CD	
45	30	
46	00	use time delay subroutine
47	2B	HL ← HL − 1
48	0D	C ← C − 1
49	20	
4A	F1	if C ≠ 0, jump to
4B	21	
4C	FF	
4D	02	HL ← 02FF
4E	F1	pop A, F off stack
4F	FB	enable interrupts
50	ED	
51	4D	return from interrupt

Table 11.15 Keyboard subroutine

Address	Op code	Function
0060	1E	
61	FF	E ← FF
62	7B	A ← E ←
63	3C	A ← A + 1
64	5F	E ← A
65	ED	
66	79	OUT ← A
67	ED	
68	78	A ← IN
69	E6	
6A	01	look at bit 0
6B	20	
6C	F5	if A ≠ 0, jump to
6D	1C	E ← E + 1
6E	3E	
6F	0F	A ← 0F
70	ED	
71	79	OUT ← A
72	C9	return to main program

Table 11.16 Main program

Address	Op code	Function
0000	31	
01	FF	
02	03	SP ← 03FF
03	21	
04	FF	
05	02	HL ← 02FF
06	0E	
07	00	C ← 00
08	3E	
09	0F	A ← 0F
0A	ED	
0B	79	OUT ← A
0C	ED	
0D	56	establish interrupt mode
0E	FB	enable interrupts
0F	ED	
10	78	A ← IN
11	E6	
12	01	look at bit 0
13	28	
14	FA	if A = 0, jump to
15	CD	
16	60	
17	00	use keyboard subroutine
18	ED	
19	78	A ← IN
1A	E6	
1B	01	look at bit 0
1C	20	
1D	FA	if A ≠ 0, jump to
1E	2B	HL ← HL − 1
1F	7B	A ← E
20	77	(HL) ← A
21	0C	C ← C + 1
22	18	
23	EB	

PROBLEMS

1 Write subroutines with the following function.

a) The current value of A is pushed onto the stack. Once I_7 is 0, A is popped off the stack. Return to the main program.

b) O_7 is initially 1. It must briefly be pulled low to reset a flip-flop. Assume that A equals the state of the output port.

c) The nibble $I_5 I_4 I_3 I_2$ is read in from the input port and is stored in register E.

2 Figure 11.29 shows an arrangement which allows the µP to scan ten switches and work out which one has been pressed. Initially all of the output lines are reset to 0. Then if either I_0 or I_1 go to 0 a switch must be pressed. By holding the output lines at 0 one at a time and looking at I_1 and I_0 the CPU can establish which switch is pressed. The appropriate number can be stored in register E before resetting the output port and waiting for the switch to be released.

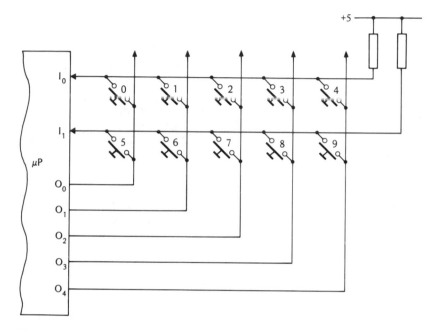

Figure 11.29 The CPU scanning a decimal keyboard

Write a subroutine that will let the CPU scan the keyboard as described. Then write another subroutine which will emit a number of bleeps via a loudspeaker; the number of bleeps is governed by the contents of register E. Finally write a short main program which uses both subroutines; the whole program should make the µP emit a number of bleeps when a switch is pressed. (0 gives no bleeps, 9 gives nine bleeps etc.)

3 Microprocessors can be used to replace a lot of hardware with software. For example, figure 11.30 illustrates how a µP system can use a single 7447 BCD-to-seven segment decoder to run four seven segment LEDs. The CPU can enable the displays one at a time via O_7 to O_4, and present a BCD number to the decoder via O_3 to O_0. By rapidly displaying the four digits one after the other, the CPU can make all four digits appear simultaneously.

Write a program to make the CPU display four BCD digits stored in a pile in RAM; an interrupt service routine will be added later to put the four digits into RAM.

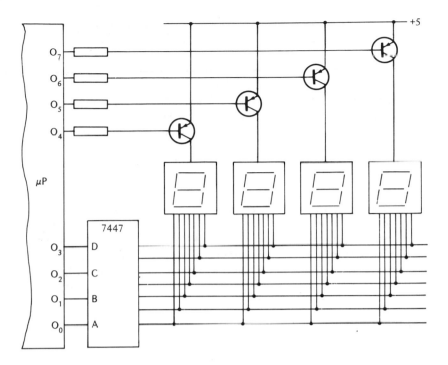

Figure 11.30 A multiplexed output display

4 Figure 11.31 illustrates a 507 ADC (see p 283) connected to a μP system. Write a subroutine that will make the CPU read in the value of INPUT and place the seven bit word in register E.

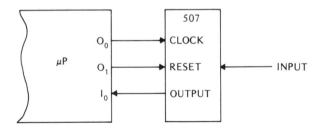

Figure 11.31

5 Write a subroutine which will transmit a seven bit word in serial form via O_7. The word is initially held in the seven lsb of register E; the msb of E is 0.

6 Combine your answers to 4) and 5) to make a program which will read in the value of an analogue signal and transmit it in serial form once every second or thereabouts.

HARDWARE VERSUS SOFTWARE

Within certain limits, you can use a microprocessor to build any digital system that you care to think of. The only constraint is that of speed; the CPU can only perform operations at a limited rate. For example, the Z80 has a maximum clock frequency of 4 MHz. So if you want a fast digital system, you will have to assemble it from discrete counters, gates etc. Fast systems consist solely of hardware.

On the other hand, if you can get the CPU to run through its program fast enough, the microprocessor solution to a problem has many advantages. By having enough input and output ports, by having enough RAM and ROM, you can more or less make a microprocessor system do anything you want. Furthermore, the behaviour of the system is dominated by its software i.e. by the program sitting in ROM. The same hardware, i.e. the same circuit, can be made to perform a completely different function by pulling out the ROM and replacing it with a different one. Thus our µP system is a general purpose circuit, capable of replacing any system which requires up to eight inputs and outputs.

More ports

The Z80 can have up to 512 ports, half of them for reading in bytes and the other half for writing them out. The C register is used to guide a particular byte towards a port; its contents are placed on the lower half of the address bus during an input/output instruction cycle. Some hardware is needed to decode that address and activate the appropriate port, but it is easily done if you do not want too many ports. For example, a 7442 one-of-ten selector hooked up to the lower four address lines will allow you to have up to ten input ports and ten output ports (80 bits in and 80 bits out!).

An alternative approach to the problem of port expansion is to use **memory-mapped input/output**. (Indeed, for some CPUs it is the only strategy!) You make the ports look like memory, so that the CPU thinks that it is reading and writing from RAM rather than a port. This allows you to have an enormous number of ports if you want, as sixteen address lines allow the CPU to address 64 K locations (1 K = 1024). The address decoding can be complicated if you need to use a lot of RAM as well.

Compromise

A real application of a microprocessor will involve a compromise between software and hardware. In order to gain speed, dedicated chunks of hardware such as ADCs will be hooked up to the ports. These will be controlled by the CPU, which will periodically instruct these chunks to perform their function. The CPU will act as the central control, organising the hardware attached to it, storing the results of their operations in RAM, making sure that the different chunks act in the right order etc. Obviously, some flexibility is lost by anchoring extra hardware to the ports. A µP system that has ADCs attached to it can easily be reprogrammed to read in analogue signals, but it cannot easily be converted into a frequency meter!

REVISION QUESTIONS

1 Draw a diagram of a simple microprocessor system to show how its component parts are connected to each other. Explain the function of each component part.

2 A microprocessor system has 8 K of memory, consisting of 2 K of ROM and 6 K of RAM. Explain what this means. Draw a memory map for the system. If the memory is built out of ICs which contain 1024 bytes, show how they can be connected up to the address bus of the microprocessor system.

3 Discuss the importance of the program counter by describing what happens during a machine cycle. Explain how the stack is used to save the value of the program counter when the CPU is interrupted.

4 Explain what the stack pointer, memory pointer, accumulator and general purpose registers are and what they can be used for.

5 With the aid of two examples explain the difference between direct and indirect addressing.

6 Explain the difference between a direct jump and a relative jump. Why does the use of the latter give relocatable code, and why is this a desirable feature of a subroutine?

7 A CPU which has sixteen address lines is hooked up to 1 K of memory with the lowest ten lines. It has a memory-mapped input port at address 0300 and an output port at address 0400. Show how these two ports could be assembled from flip-flops, tristates and logic gates. You may assume that the CPU has the usual control lines of a sawn-off Z80 namely $\overline{MREQ}$, $\overline{RD}$ and $\overline{WR}$. Make sure that the CPU does not get confused between memory and port!

8 Explain the meaning of the following terms; instruction set, data bus, CPU, input port, address bus, subroutine, interrupt, stack, memory map, reset, program, jump instruction, op code and clock.

9 With the aid of the listing of a simple program, explain what is meant by hand shaking. Why is it necessary?

10 Describe what happens when the CPU is interrupted. Explain how the CPU returns to the main program when it has completed the interrupt service routine.

11 Write a program which will make the μP system continually shift a pattern across a set of eight LEDs from left to right with a 1 s delay between each shift. When the system is interrupted a new pattern is read in from a bank of eight switches.

12 What is a look-up table? Illustrate its use by writing a program which will make the μP system into a set of traffic lights.

12

Negative feedback

You will be right in thinking that the title of this chapter is familiar. You have indeed met the idea of negative feedback before. It played a large part in the chapters on audio systems and analogue systems. There are, however, many other areas of electronics where negative feedback can be used to great advantage; this chapter will cover some of them.

CLOSED LOOP GAIN

We are going to run through the derivation of a formula which can be used to calculate the closed loop gain of any system which employs negative feedback. The result will look familiar; you have already seen it derived for a special case in Chapter Eight.

The essence of a system which employs negative feedback is illustrated in figure 12.1. The circle stands for a **comparator**, a device which looks at two signals and feeds out a signal which is equal to their difference. It is easily fabricated with a 741, as shown in figure 12.2; provided that all four resistors are identical, $V_E = Y - X$. The triangle is an amplifier. It can be purely electronic (like the 741 amplifier shown in figure 12.3) or it can amplify the ERROR signal in other ways. For example, if you look at figure 12.4 you will see that part of the amplifier might include a motor which is mechanically linked to the wiper of a potentiometer. In that case, the output signal of the whole amplifier is the voltage of the wiper. In general, the amplifier (whatever its mode of action) will feed out a signal which is equal to $A \times$ ERROR. The quantity A is called the **open loop gain** of the system.

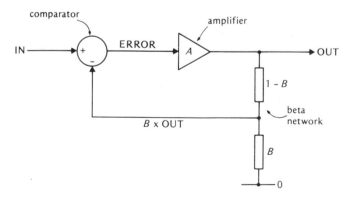

Figure 12.1 A block diagram of a system which uses negative feedback

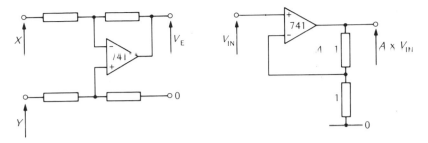

Figure 12.2 A comparator   **Figure 12.3** An amplifier

Returning to figure 12.1, you will see that the signal which emerges from the amplifier (OUT) is processed by a voltage divider known as the **beta network**. Its function is to feed back a fraction of the output signal, so that it can be compared with the input signal. The consequences of this feedback arrangement will become clearer after we have juggled with the algebra which describes the system. For the comparator:

$$ERROR = IN - B \times OUT$$

For the amplifier:

$$OUT = A \times ERROR$$

Combining these two equations to eliminate ERROR, we get:

$$ERROR = IN - B \times OUT = \frac{OUT}{A}$$

therefore
$$IN = \frac{OUT}{A} + B \times OUT$$

therefore
$$IN = OUT \frac{(1 + AB)}{A}$$

therefore
$$\frac{OUT}{IN} = \frac{A}{1 + AB}$$

The quantity OUT/IN is the **closed loop gain** of the system, G. So the system of figure 12.1 is described by this equation:

$$G = \frac{A}{1 + AB}$$

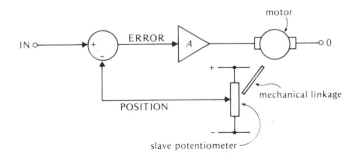

Figure 12.4 A position control system

The secret of successfully using negative feedback is to arrange matters so that the loop gain (AB) is always much much larger than 1. Then $G = 1/B$:

$$G = \frac{A}{1 + AB} = \frac{A}{AB} = \frac{1}{B} \quad \text{if } AB \gg 1$$

But the value of B is fixed by the relative sizes of the two resistors in the beta network. So the closed loop gain will not depend on the properties of the amplifier (i.e. the value of A) when AB is large enough. The comparator is continually comparing the signal fed back from the beta network with the signal coming in; any difference is amplified in such as way as always to reduce ERROR to zero. Use the equations above to convince yourself that ERROR = 0 when the loop gain is very large.

The application of negative feedback to a system gives it stability. Without negative feedback, OUT = A × IN. So any drifts in the value of A will result in a change in the value of OUT; this is especially serious when the amplifier has a non-electronic component like a motor or a heater. Changes in the physical state of the amplifier will cause changes in the value of OUT. When negative feedback has been inserted, OUT = IN/B ($AB \gg 1$) almost independently of the properties of the amplifier.

POSITION CONTROL

The conclusions of the last section are best appreciated when illustrated with a real application. The first of many such applications is position control. The system is outlined in figure 12.4. The motor is attached to the wiper of the slave potentiometer, so that as it rotates the POSITION signal changes. That signal is compared with another signal (IN) by the comparator. The ERROR signal is amplified and fed into the motor. Provided that everything has been connected so that the feedback is negative the motor will rotate the potentiometer until ERROR is zero. In fact, the motor will always rotate the potentiometer in such a way as to make POSITION = IN.

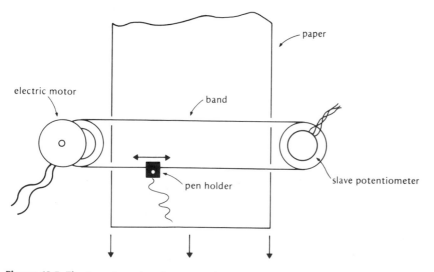

Figure 12.5 The top view of a chart recorder

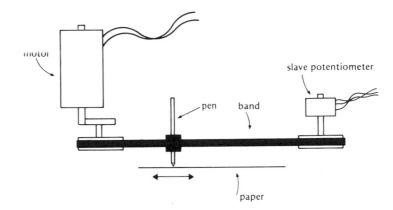

Figure 12.6 A side view of the chart recorder

So if IN varies with time, the potentiometer will be rotated so that the value of POSITION tracks that of IN.

At first glance, the system which has been described appears to have little to do with position control. All it does is rotate a shaft to a position determined by the size of a signal fed in by some external system. It is really a voltage-to-angle system. It can, however, be easily adapted to make a true voltage-to-position system. Look at the device shown in figures 12.5 and 12.6. It is a simple chart recorder. The pen is moved across the paper in response to an external signal; as the paper is moved past the pen at a constant rate, the resulting trace left by the pen is a record of how the external signal changes with time. Note how the slave potentiometer senses how far the pen has been moved across the paper.

Overshoot

Now let us apply what we know about negative feedback to our position control system. There is no beta network, so $B = 1$. The closed loop gain will therefore be $A/(1 + A)$. Provided that A is big enough, G will be virtually 1 and POSITION = IN at all times. The bigger A is, the closer is the match between POSITION and IN—or is it?

Suppose that we have a very large open-loop gain. This is illustrated in figure 12.7. The 741 produces a signal which is equal to

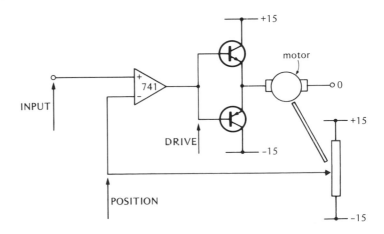

Figure 12.7 An on-off feedback system

100 000 × (INPUT − POSITION); that signal is buffered with a complementary follower before being fed into a DC motor. The op-amp acts as both comparator and amplifier; the open loop gain of the system will be enormous. This means that the slightest difference between INPUT and POSITION forces DRIVE to saturate at either +13 V or −13 V. So the motor will be either driven hard forwards or hard backwards; it will never sit still.

The behaviour of this type of system is shown in the graphs of figure 12.8. INPUT remains constant throughout, and POSITION starts off at some value other than INPUT. Immediately DRIVE shoots up to +13 V, so that POSITION approaches INPUT. As soon as POSITION is greater than INPUT, DRIVE dives down to −13 V. The motor, however, cannot respond to this change of voltage instantaneously. In common with all physical systems it possesses inertia. So when it is spinning fast in one direction there will be a time delay before it can be set spinning in the other direction. Thus POSITION carries on increasing for a short while even though DRIVE is trying to change the direction of the motor. By the time the motor has stopped turning, POSITION will have definitely overshot its target value of INPUT. Meanwhile, DRIVE is still −13 V, so the motor starts to push POSITION back towards INPUT. This time POSITION will undershoot INPUT. It follows that POSITION is going to continue to oscillate about its target value INPUT.

This phenomenon of oscillation in a feedback system is called **hunting**. It is obviously not a good thing. The amount of overshoot depends on several factors. If the motor has to drive a heavy load there will be a lot of inertia. Very light systems will respond faster to changes of DRIVE than heavy

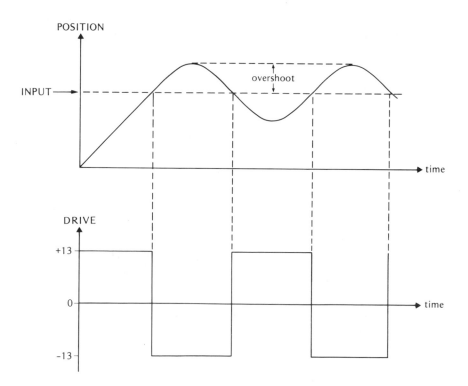

Figure 12.8 The response of an on-off feedback system

systems, so they will have smaller amounts of overshoot. A strong motor is going to be able to reverse the direction of the system faster than a weak one. On the other hand, a strong motor will make the system go faster than a weak one, tending to increase the overshoot.

The oscillation is eventually damped out by friction, but it can take a long time. Friction slows the system down so that the amplitude of each oscillation is smaller than the previous one; eventually the overshoot becomes negligible. On the other hand, friction makes the system sticky; a lot of friction means that it is difficult to get the system to stop in precisely the right place.

Proportional feedback

The defects of the on-off feedback system we have just been describing can be eliminated if we can make the open-loop gain smaller. By reducing the value of A we can make DRIVE get smaller as POSITION approaches INPUT, so that the motor automatically slows down before it has a chance to overshoot.

A proportional feedback system is shown in figure 12.9. The op-amp on the left acts as the comparator. Its output (ERROR) is amplified by the other op-amp; note how the inclusion of a complementary follower within the feedback loop allows the amplified ERROR signal to be fed directly into the motor.

The behaviour of the system is shown in the graph of figure 12.10 (overleaf). Initially, INPUT = POSITION, so that ERROR and DRIVE are 0. (Remember that DRIVE = A × ERROR.) Then INPUT is suddenly changed to a new value; it goes from −4.5 V to +4.5 V. ERROR immediately jumps to +9 V and DRIVE (being larger than ERROR) saturates at +13 V. The motor is switched hard on and POSITION proceeds to ramp upwards towards its target of +4.5 V. The speed with which POSITION changes will remain constant while DRIVE is saturated. However, as POSITION ramps steadily upwards, the size of ERROR gets smaller. Eventually ERROR will be small

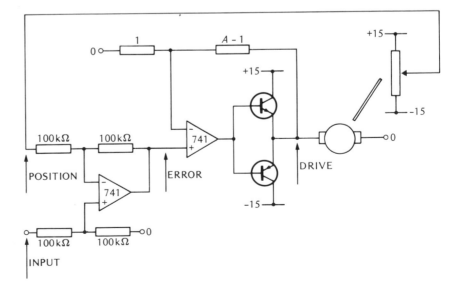

Figure 12.9 A proportional feedback system

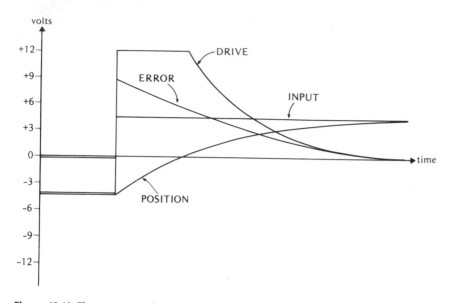

Figure 12.10 The response of a proportional feedback system

enough for the amplifier to stop saturating; this will happen when ERROR = 13/A. Thereafter each decrease in the value of ERROR results in a decrease in the value of DRIVE. So as POSITION gets closer to INPUT, the motor drives the system more and more slowly and the rate of change of POSITION gets smaller and smaller. Eventually DRIVE will become so small that the motor will not provide enough push to overcome friction and the system will stop moving.

So POSITION will end up close to INPUT, but not necessarily exactly equal to it. Since a finite value of DRIVE is needed to get the motor to overcome friction and set the system moving, you may end up with a small static value of ERROR. You can, of course, reduce this uncertainty in the final value of POSITION by increasing the value of A; this reduces the range of values of ERROR which will not set the motor turning. But if you increase the value of A indefinitely, then the system will start to hunt!

You can see that a proportional feedback system is going to respond much more slowly than its on-off counterpart. That is, for a given change of INPUT, the on-off system will be able to change POSITION faster than the proportional system will be able to. Proportional systems do not overshoot, because POSITION will approach its final value smoothly without any hunting. An on-off system, despite its rapid initial response, may have to wait a long time before all the oscillations are damped out.

You can get the best of both worlds by having a proportional feedback system and turning up the open loop gain until the system just starts to hunt. This is the situation illustrated in figure 12.11. A sudden change of INPUT causes ERROR to jump to +9 V. The gain of the amplifier is large, so ERROR drops at a constant rate to start with. Then, when ERROR is small enough, the motor slows down. The inertia of the system tends to keep it going at the same speed, and we have to rely on friction to slow it down when DRIVE is reduced. If ERROR drops too rapidly, the inertia will make the system overshoot a bit. If you get the value of A right, then you can arrange to have the quickest response time with a minimal overshoot.

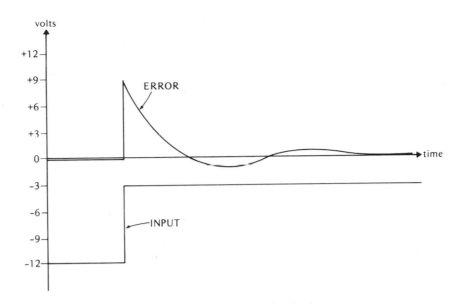

Figure 12.11 The optimum response of a negative feedback system

SPEED CONTROL

Given suitable actuators and sensors, you can use proportional feedback to stabilise any physical quantity you like. We are now going to consider an example in detail to explain how to design this sort of system; you will be introduced to other systems in problems and experiments at the end of this section.

The task is as follows. We need to keep a shaft rotating at a constant speed, regardless of the loading placed upon it. The shaft could be in a lathe or a gramophone turntable; there are many devices where it is important that part of it turns at a known constant speed. In order to apply negative feedback we need a sensor which can give out an electrical signal which depends on the speed of the shaft. That signal is then compared with a reference signal, to generate an error signal. Finally, a signal has to control the actuator (i.e. the motor turning the shaft) so that the error signal is reduced to zero.

A block diagram of a suitable system is shown in figure 12.12. It contains

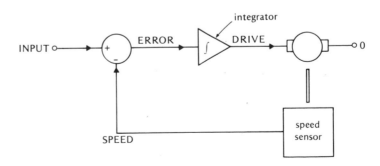

Figure 12.12 The block diagram of a speed control system

an integrator, a device whose output ramps up or down at a rate which is proportional to its input. For example, let us suppose that the motor is going too slowly; perhaps something is loading the shaft and slowing it down. Then SPEED will be smaller than INPUT, so that ERROR is positive. A positive input to the integrator makes its output ramp up. So DRIVE steadily rises, making the motor turn faster. This reduces the size of ERROR, so that DRIVE ramps up at a slower and slower rate. Eventually SPEED matches INPUT exactly and ERROR is zero; thereafter DRIVE does not change, provided that ERROR remains zero.

Clearly, the system that we have just described employs proportional feedback. As ERROR is reduced, its rate of change is also reduced; ERROR will take a long time to get to zero, but it will not overshoot either. The speed with which the system will react depends on the properties of the integrator. By increasing the rate at which DRIVE increases for a given value of ERROR, the response of the whole system can be speeded up. But if the integrator ramps up too rapidly, then the inertia of the system becomes a problem and it will start to hunt.

Figure 12.13 illustrates a speed control system that employs a small DC motor. A large pnp transistor sources current into the motor. An integrator controls the amount of current which flows by pulling down on the end of the 2.2 kilohm resistor; convince yourself that when DRIVE goes down, the current flowing through the motor goes up. (The rate of change of DRIVE = −ERROR × 10, i.e. when ERROR is +0.5 V, DRIVE goes down 5 V every second.) The motor acts as its own speed sensor; providing that it is not spinning too slowly, the voltage across a DC motor is proportional to its speed.

The graph of figure 12.14 shows what happens to the important signals in the system when INPUT is suddenly changed. Notice how ERROR decays exponentially after its sudden rise from zero; this is the hallmark of a true proportional feedback system.

If you want to control a larger system then you will be faced with trying to stabilise the speed of an AC motor. A large system will have plenty of inertia, so its speed will tend to remain constant anyway. You can usually

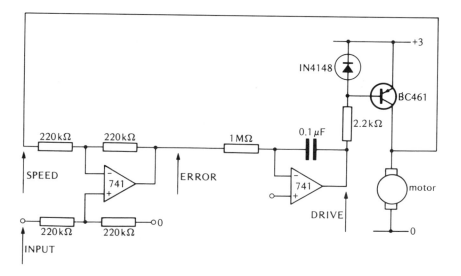

Figure 12.13 A speed control system for a DC motor

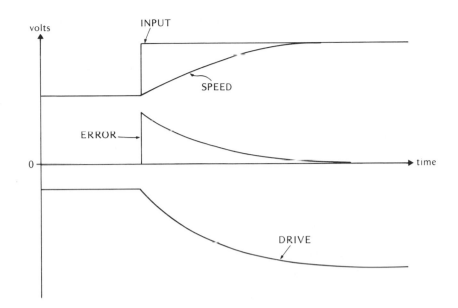

Figure 12.14 The response of the speed control circuit

get away with using an on-off feedback system, feeding bursts of AC current into the motor whenever it gets too slow. AC motors are best controlled by triacs, so you would be able to feed energy into the system in 10 ms bursts (half a cycle of 50 Hz mains). There are various types of speed sensor, but they mostly emit pulses at a rate dictated by the speed of the system. For example, there are magnetic pickups which emit a brief pulse whenever a piece of iron passes in front of them. A suitable analogue signal can be generated by shaping the pulses with a monostable and then feeding them into a treble cut filter.

EXPERIMENT
Examining the behaviour of a system which controls the speed of a small DC motor

The circuit is shown in figure 12.13. Assemble the circuit, and check that the comparator and integrator work properly. Adjust the 3 V supply to match the type of motor that you are using. The base resistor of the transistor limits the maximum collector current to about 500 mA. It would be a wise precaution to put a heat sink onto the transistor so that it does not get too hot! Use the network of figure 12.15 to generate the speed reference signal (INPUT).

Monitor ERROR with a voltmeter. Check that the system always stabilises with ERROR equal to zero whenever you change the value of INPUT. Furthermore, check that the system will compensate for any loading that you place on the motor—try slowing it down with your fingers. **Do not stop the motor with your fingers; under these conditions the transistor may fail**.

Finally, use the network shown in figure 12.16 to generate INPUT. When you press the switch, or release it, INPUT will suddenly change. Use a CRO with the timebase switched off to monitor ERROR (it will respond faster than a voltmeter). Study what happens to ERROR when INPUT is rapidly changed. Try speeding up the integrator by replacing the 1 MΩ resistor with smaller values. Find the optimum value that will give the fastest response without any hunting; use trial and error.

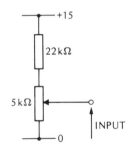

Figure 12.15

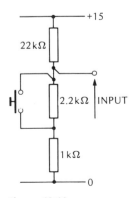

Figure 12.16

EXPERIMENT

Assembling and evaluating an automatic gain control (AGC) system

AGC systems are used to ensure that the mean amplitude of an AC signal emerges with a constant value. For example, suppose that a microphone is sending a signal to a tape recorder. For that signal to be recorded on the tape without distortion its amplitude must always be less than a certain value. On the other hand, if the amplitude is too small then the signal will get lost in the noise when it is recorded on the tape. If an AGC system is placed between the microphone and the tape recorder, the amplitude of the signal recorded on the tape will always have the correct average value.

A block diagram of an AGC system is shown in figure 12.17. IN is the audio signal whose amplitude has to be boosted or cut. The unit labelled VCA is a voltage controlled amplifier. Its gain (OUT/IN) is fixed by the voltage at its third terminal (GAIN). So the size of the amplifier output (OUT) can be controlled by an electronic signal. The amplitude of OUT is monitored by the AC-to-peak unit; this feeds out a signal which is equal to the amplitude of OUT. That signal is compared with REFERENCE, a constant voltage which is equal to the desired mean amplitude of OUT. The resulting error signal is fed into an integrator so that GAIN ramps up or down, boosting or cutting the gain of the amplifier as required.

Figure 12.18 shows the details of each part of the system. Work out how to knit them together so that the whole thing is an AGC system; be very careful to make sure that you arrange for the feedback to be negative! An audio signal fed into the amplifier ought to emerge with an amplitude of 70 mV, regardless of its amplitude when it went in.

Assemble the system, testing that each part of it works properly before you close the loop. Feed in a 1 V amplitude sine wave at 1 kHz and look at the output of the system with a CRO. Find out the maximum and minimum input amplitudes that the system can handle.

Look at ERROR. Note what happens to it when the amplitude of the input signal is suddenly changed (set up a switch so that it shorts out part of a voltage divider?). Speed up the response of the system until it just starts to hunt.

Draw a circuit diagram of the final system. Describe its behaviour, state the range of signals that it can cope with and what its response time is.

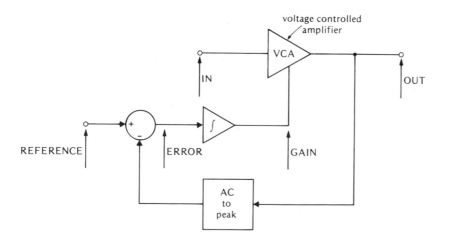

Figure 12.17 An AGC system

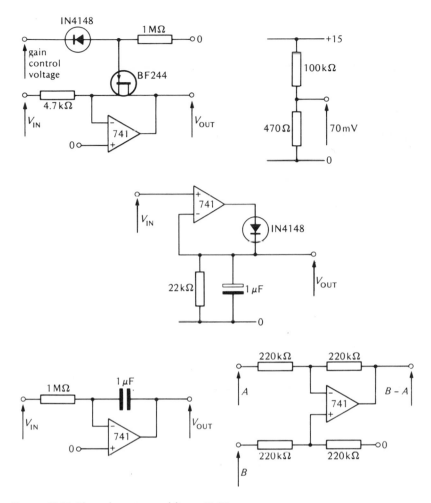

Figure 12.18 The sub-systems of figure 12.17

PROBLEMS

1 A draughtsman's office has its lights controlled by a feedback loop so that the illumination of his desk is at the same level all day long. Draw a block diagram of a suitable system, and state clearly the function of each block.

2 High precision oscillators contain a quartz crystal to fix the frequency of oscillation. The crystal is arranged to act as a capacitor in the circuit and will cause the system to oscillate at its own natural frequency. Unfortunately, the natural frequency of the crystal depends on its temperature. So super-precise oscillators keep their crystals at a constant temperature, in a small enclosure (an oven) held above the temperature of its surroundings.

   Design a circuit which will hold the oven at a fixed temperature, using an on-off feedback system. A resistor is to be used as the heater, a thermistor as the temperature monitor and a potentiometer to fix the temperature. You may assume a split 15 V supply and that the heater only has to dissipate a few watts. Describe how your system works.

Now design a proportional feedback system to control the temperature of the same oven. Describe how it works.

Discuss the relative advantages and disadvantages of the two systems that you have designed.

3 Background music in a pub ought to be loud enough for people at the same table to hear each other, but not people at another table. So the volume of the music depends on the general volume of noise in the room; it needs to be such that the total volume of sound in the room remains constant.

Draw a block diagram of a suitable system to control the volume of the background music. A microphone picks up the sound in the room, and the system must control the music so that the microphone always picks up the same average volume of sound. How are you going to ensure that the music doesn't start hunting whenever someone in the pub coughs?

4 A lathe shaft has a magnet mounted at one end, near a fixed coil of wire. When the shaft spins round, the movement of the magnet induces an AC signal in the coil; as the shaft speeds up both the frequency and the amplitude of the AC signal increases. The shaft is rotated by an AC motor which is run off the mains supply via a power triac.

A potentiometer is used to set the speed of the shaft, and a feedback loop ensures that it runs at a constant speed regardless of the loading on it. Draw the circuit diagram of a suitable on-off feedback system, and explain how the circuit works.

STABILITY

A negative feedback system which has a mechanical element in the loop has to reconcile two factors. The open loop gain has to be as large as possible, so that the output tracks the reference signal faithfully. But the whole loop must not respond too quickly, otherwise the system will become unstable and oscillate.

The inclusion of an integrator solves both problems at once. An integrator has a DC gain of infinity (its output will eventually saturate for any input which is not zero), so the open loop gain of the whole system is sure to be enormous. On the other hand, an integrator can only respond at a finite rate; this slows down the response of the sytem to any changes.

Purely electronic systems which employ negative feedback can also be unstable if they are badly designed. For example, have a look at the system shown in Figure 12.19. It is a misguided attempt to build a treble boost filter with a break frequency of 16 Hz. In practice it is a 3 kHz sine wave generator! The fault lies in the feedback network. Signals which are above the break frequency are fed back from the output of the op-amp to its input with a phase shift. As you will find out as you work through this section, when that phase shift is combined with the phase shift due to the op-amp itself the whole system can appear to have positive feedback at certain frequencies. The circuit starts to oscillate uncontrollably.

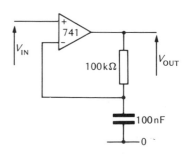

Figure 12.19 An unstable circuit

PHASE SHIFTS

A potential cause of instability in a system with negative feedback is excessive phase shifts of the signal going through the amplifier.

Have a look at the graphs shown in figure 12.20. They show how the open-loop gain and phase shift of a 741 op-amp change with frequency. You will have noticed immediately that the graph of *A* has all the characteristics of a treble cut filter. There is a break frequency of 10 Hz, above which the gain falls at a rate of 20 dB per decade. There is a second break at 3 MHz, with the gain dropping at 40 dB per decade from then on. This characteristic is a consequence of *RC* networks within the IC which also give rise to phase shifts. As you can see from the graph of ϕ, a drop in gain of 20 dB per decade is associated with a phase shift of 90°. Above the second break (i.e. 3 MHz) the phase shift is 180°.

All amplifiers will have similar behaviour. At low frequencies the gain will be flat, with a cut at very low frequencies if they are AC coupled. As the frequency goes up, the gain will start to drop at 20 dB per decade. This will be a consequence of an *RC* network somewhere in the circuit. You do not actually have to include a capacitor; there is plenty of stray capacitance around to provide the treble cut filter. For example, the main source of stray capacitance is usually the junction between the base and the collector of a transistor. It acts like a reverse biased diode, so looks like a pair of

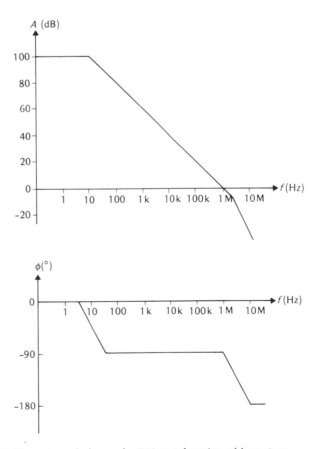

Figure 12.20 The gain and phase of a 741 as a function of frequency

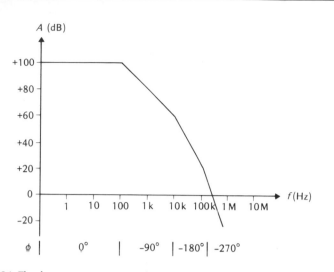

Figure 12.21 The frequency response of a conditionally stable op-amp

conductors separated by an insulator. This capacitor is very small, but it will start to look like a short-circuit for signals which have a high enough frequency!

Each stray capacitor will introduce a break point in the graph of gain against frequency. So the graph will start off flat, drop at 20 dB per decade after the first break, 40 dB per decade after the second break, 60 dB per decade after the third break etc. Furthermore, each break will introduce an extra phase shift of 90°. This is illustrated in the graph of figure 12.21.

Now suppose that we strap resistors around this op-amp, so that we introduce negative feedback. Figure 12.22 shows an arrangement which ought to give a closed loop gain of 100 or 40 dB. That is, we have arranged the beta network so that $B = 0.01$. The closed loop gain will be given by our negative feedback formula:

$$G = \frac{A}{1 + AB}$$

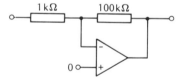

Figure 12.22

Let us consider what happens to G as we alter the frequency of the signal going through the amplifier. At low frequencies A will be much larger than $1/B$ (look at figure 12.23), so the loop gain (AB) will be much greater than 1. G will be almost the same as $1/B$, as planned. As we go up in frequency, A starts to get smaller. This is not disastrous in itself; it just tends to make G somewhat less than $1/B$. Above 10 kHz we really do run into trouble, because of the phase shift of 180°. You will recall from Chapter Eight that a phase shift of 180° changes the sign of a signal. That is, between 10 kHz and 100 kHz our op-amp is going to have a negative value of A. So between those two frequencies we will have to modify our formula for the closed loop gain:

$$G = \frac{-A}{1 - AB} \qquad (10\,\text{kHz} < f < 100\,\text{kHz})$$

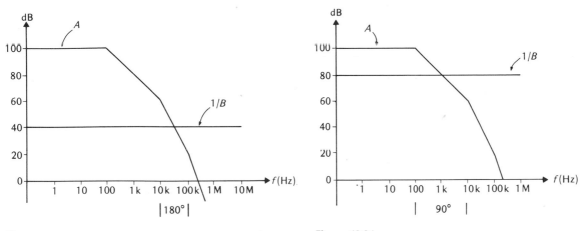

Figure 12.23

Figure 12.24

Disaster strikes when we get to 33 kHz! At that point, $A = 1/B$, and the loop gain is equal to 1. If you look at the formula for G you will see that under these conditions the closed loop gain will be infinite. The tiniest signal at 33 kHz will produce an enormous output at the same frequency; the whole system will become unstable. We have built an oscillator instead of an amplifier.

Had we gone for a higher closed loop gain, we would not have encountered this problem. Figure 12.24 illustrates the situation where $1/B$ is 80 dB or 10 000. The point where $AB = 1$ occurs when the phase shift is only 90°; A is not negative, so there is no possibility of oscillation.

An op-amp with the characteristics we have been discussing is conditionally stable. You have to be careful about the value of B if you are to avoid an unstable amplifier. Op-amps which will always be stable, regardless of the value of B, are said to be unconditionally stable. With this type of op-amp (a 741 for example) you can rest easy that it won't oscillate when you apply negative feedback. An op-amp is made unconditionally stable by deliberately building in a low frequency break point. Then by the time that the other high frequency breaks occur, A is less than 1; this means that the 180° phase shift occurs at a point where AB cannot equal 1. (You cannot make B bigger than 1; think about it.)

If you look at the graph of figure 12.20 (the characteristics of a 741) you will see that the second break point which marks the start of the 180° phase shift occurs at a gain of about −10 dB. You can go for any closed loop gain you like, safe in the knowledge that AB can never equal −1.

Bandwidth

In order to make an op-amp unconditionally stable you deliberately cripple it by forcing its gain to start falling above a fairly low frequency. In the case of a 741, the break frequency is 10 Hz. This limits the range of frequencies which an op-amp amplifier will handle.

Take a look at the graph of figure 12.25. $1/B$ is 40 dB, so the target value of G is 100. Below 10 kHz that target will be achieved because A is much

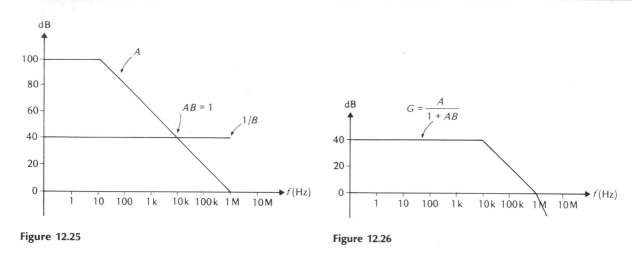

Figure 12.25

Figure 12.26

greater than $1/B$. Above 10 kHz, A is much less than $1/B$. This makes the loop gain AB much less than 1, so that $G = A$:

$$G = \frac{A}{1 + AB} = \frac{A}{1} = A \qquad\qquad (AB \ll 1)$$

So the closed loop gain will drop by 20 dB per decade above 10 kHz.

A graph of the closed loop gain as a function of frequency is shown in figure 12.26. The amplifier has a bandwidth of 10 kHz.

It should be clear that if you want to build an amplifier with a large bandwidth you are going to have to settle for a small closed loop gain. If you are using a 741, a gain of 100 limits the bandwidth to 10 kHz. If you reduce the gain to 10, the bandwidth rises to 100 kHz; look at figure 12.25 and convince yourself of this. In fact, the gain multiplied by the bandwidth always gives the same number for a 741 based amplifier. That number is 10^6 and it is called the **gain-bandwidth product**. Once you know the value of the gain-bandwidth product and of the DC gain of an op-amp, then you can work out the value of the open loop gain at any frequency. The gain-bandwidth product tells you the frequency at which the open-loop gain is 1. You know that the gain will rise by 20 dB per decade below this frequency until it gets to the value of the DC gain. Thereafter the gain remains constant.

PROBLEMS

1 Each of the circuits in figure 12.27 uses a 741 op-amp. Draw a graph to show how the gain of each one varies with frequency. State the values of their closed loop gains and bandwidths.

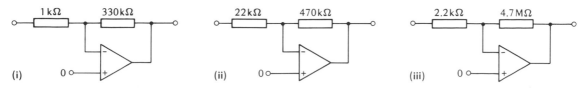

Figure 12.27

2 A cut-price supplier sells op-amps which have DC gains of 10^4 and gain-bandwidth products of 10^5. Draw a graph to show how their open loop gain changes with frequency.

3 You need to build an audio amplifier with a closed loop gain of 10^4, with a bandwidth of 10 kHz. The low frequency break point must be at 10 Hz.

a) Explain why you cannot build the amplifier round a single 741 op-amp.

b) Draw a circuit diagram to show how you could build the system around two 741 op-amps. Show all component values and your reasons for choosing them. Draw a graph to show how its gain depends on frequency.

4 An imaginative, but ignorant, circuit designer decides to put two 741 op-amps in series to make an op-amp with a DC gain of 200 dB. He then puts resistors around the pair of op-amps in an attempt to build an amplifier with a gain of 100 000 over the whole audio frequency range. His arrangement is shown in figure 12.28.

a) Draw a graph of the open loop gain of the op-amp pair as a function of frequency.

b) Explain why the circuit oscillates at a frequency of 3 kHz.

c) Show how the proposed amplifier can be built using 741 op-amps.

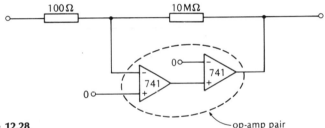

Figure 12.28

POWER SUPPLIES

An important part of any electronic system is the unit which keeps the supply rails at a stable voltage. Despite their obvious importance, we have taken the availability of suitable power supplies very much for granted in this book. We are now going to have a look at how negative feedback can be used to generate supply rails which are well behaved.

AC TO DC

The starting point for most electronic power supplies is the mains supply. This is a 50 Hz AC waveform, which has an rms value of 240 V. The power supply has to convert that signal into a DC voltage with a fixed, known value. For example, if the power supply is to be able to run TTL logic ICs, then it must hold two rails 5 V apart. A very crude system which does that

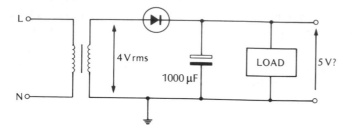

Figure 12.29 A crude 5 V power supply

is shown in figure 12.29. The TTL ICs which the system is supposed to be running are lumped together into the block marked LOAD. We are going to look at that system and find out how bad it really is; later on you will find out how it can be improved with the help of negative feedback.

The first stage in the process of converting 240 V AC to 5 V DC is a transformer. It reduces the voltage of the AC waveform with a minimal loss of electrical energy. It is far more economic than the resistor/diode network which was used in figure 4.11; that system wasted about 3 W of heat in order to provide about 100 mW of power for an IC! A good transformer will convert very little of the electrical energy which passes through it into heat. Transformers are rated according to the rms voltage across their output terminals when mains is fed into the input terminals. We have chosen to use a transformer which gives out 4 V rms. The peak value of the voltage across the secondary will be $1.414 \times 4 = 5.7$ V.

The next stage is to rectify the reduced AC signal with a diode. If you look at figure 12.30 the effect of the diode should be fairly clear. Since one end of the transformer secondary is firmly held at 0 V (look for the earth symbol in figure 12.29) the anode of the diode is continually wobbled between $+5.7$ V and -5.7 V. If we ignore the effect of the capacitor and assume that the load behaves just like a resistor, then current will only flow through the diode when V_{IN} is above $+0.7$ V. That current will flow through the resistor (R), generating a voltage (V_{OUT}) across it. During the positive half-cycles of V_{IN}, V_{OUT} will be 0.7 V below V_{IN}. During the negative half cycles the diode will be reverse biased and the resistor will pull V_{OUT} down to 0 V. The result is the graph shown in figure 12.30; it consists of humps which are 5 V high, about 10 ms long and 20 ms apart. For obvious reasons, the diode-resistor network shown in figure 12.30 is called a half-wave rectifier.

Figure 12.30 A half-wave rectifier

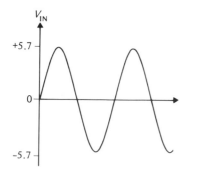

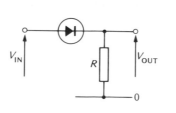

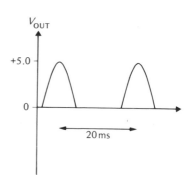

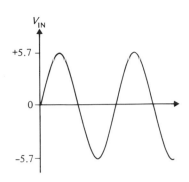

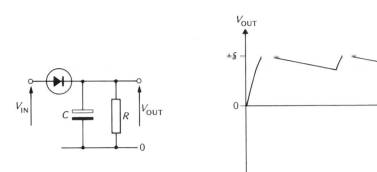

Figure 12.31 The effect of a smoothing capacitor

The final component of the power supply is a smoothing capacitor. As its name implies, it smooths out the humps produced by the half-wave rectifier to give a constant DC voltage. Figure 12.31 illustrates what the capacitor does. Suppose that V_{OUT} starts off at 0 V i.e. we have just switched on the mains supply. As V_{IN} rises it will pull V_{OUT} up, with a drop of 0.7 V across the diode. A lot of charge will flow onto the capacitor in the process. When V_{IN} gets past its peak value of $+5.7$ V, the diode will become reverse biased.

So charge will not be able to flow off the capacitor back to the source of V_{IN}; it will only be able to leak to ground via the load resistor R. So V_{OUT} drops slowly as current flows through R. About 20 milliseconds later V_{IN} will rise above V_{OUT} by 0.7 V and the diode will become forward biased once more. The capacitor will quickly charge up to $+5$ V again, with a burst of current flowing from the transformer through the diode.

Ripple

A major defect of our simple power supply is that it cannot provide a rock steady DC voltage across its output terminals. Every time that the mains supply reaches its peak positive value the diode squirts a quantity of charge onto the capacitor. Between those squirts, that charge leaks away through the load. So the DC output is rapidly pushed up to $+5$ V every 20 ms, and gradually falls towards 0 V for the rest of the time. Figure 12.32 shows you what the output voltage looks like as a function of time. The difference between the maximum and minimum voltage of the DC supply is called the **ripple**; we are now going to obtain a formula for calculating its size.

Look at figure 12.31. During the time that the diode is reverse biased, V_{OUT} will obey the usual RC network formula:

$$V = V_0 e^{-(t/RC)}$$

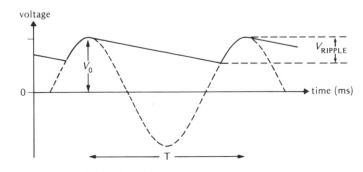

Figure 12.32 A smoothed DC supply

The diode will behave like an open circuit, so the system looks like a capacitor C in parallel with a resistor R. V_0 will be the peak value of the DC signal; if t is the time which elapses before the next lot of charge is dumped on the capacitor, then V will be the minimum value of the DC signal. Since the ripple is the difference between maximum and minimum voltage, we can say that:

$$V_{RIPPLE} = V_0 - V_0 e^{-(t/RC)}$$
$$= V_0(1 - e^{-(t/RC)})$$

Now it is obvious that we want the ripple to be small. This means that we will aim to have t much smaller than RC, so that the half-life of the system is much longer than 20 milliseconds. So we can assume that $t/RC \ll 1$. Using the standard approximation:

$$e^x = 1 + x \qquad\qquad (x \ll 1)$$

therefore $V_{RIPPLE} = V_0[1 - (1 - t/RC)]$
$$= V_0 (t/RC)$$

If the ripple is small, then t is going to be very nearly T, the period of the mains waveform, i.e. the time taken for one cycle. So our final approximate formula for the amount of ripple is:

$$V_{RIPPLE} = \frac{V_0 T}{RC}$$

Let's use this formula to work out how much ripple will be present on the output of our simple 5 V supply. We need to know something about the load namely how much current it draws from the supply. If the load consists of a couple of TTL ICs, it could easily draw 20 mA. The load would then behave like a resistor of value $5/20 = 0.25\,\mathrm{k\Omega}$. We have already established that $V_0 = 5\,\mathrm{V}$ and that $T = 20\,\mathrm{ms}$. Plugging these values into the formula, remembering that we are using a 1 000 µF smoothing capacitor, we get:

$$V_{RIPPLE} = \frac{V_0 T}{RC} = \frac{5 \times 20}{0.25 \times 1\,000} = 0.4\,\mathrm{V}$$

So the load sees the supply line wobble up and down by about 10%, which is not a very satisfactory state of affairs!

You can improve the system by using a larger capacitor. 10 000 µF, the largest you can get at a reasonable price, will reduce the ripple to about 0.04 V. However, once you start trying to run a reasonable number of ICs off the supply, the ripple becomes unacceptably large again. A single seven segment LED can easily draw 30 mA from its supply. A breadboard full of ICs with some LEDs might take 100 mA from its 5 V supply, resulting in a 5% ripple.

Even if you only want to run a single IC off the supply, so that the ripple is negligible, the circuit of figure 12.29 makes a poor power supply. This is because it lacks **regulation**. The size of the DC voltage fed out by the supply depends on the size of the AC voltage fed into it. So any long term alteration in the mains supply voltage is passed straight on to the DC supply. A 10% change in the rms value of the mains voltage results in a 10% change in the DC voltage placed across the load.

REGULATION

The standard way of regulating a power supply is to put a zener diode into the circuit. Figure 12.33 shows how a 5.1 V zener diode can be used to regulate a 5 V supply. You feed a smoothed DC signal in from the left, and a steady 5.1 V emerges at the right. The size of the unregulated DC is not important, provided that it is at least a volt or two above the breakdown voltage of the diode. The size of the resistor has to be chosen carefully if the system is to work properly; we will now calculate how big it has to be for a specific example.

Assume that the load will draw up to 100 mA from the 5 V supply, and that we have a smoothed 8 V DC supply coming in from the left. Then the load current, I_L, can be anything from 0 mA to 100 mA. We will need to have a minimum current of perhaps 10 mA flowing through the zener (I_Z) to keep the voltage across it equal to 5.1 V. (Remember that the breakdown voltage does vary slightly with current.) Neglecting any ripple on the unregulated 8 V supply, we can state that there will be $8 - 5 = 3$ V across the resistor at all times. The current through it has to be $100 + 10 = 110$ mA, so its resistance has to be $3/110 = 0.027$ or $27\,\Omega$.

When the load draws its full 100 mA, a current of only 10 mA flows through the zener. Heat will be generated in it at a rate of $5.1 \times 10 = 51$ mW; no problem there. But if the load ceases to draw its current (i.e. it is removed) what happens then? All of the current which flows through the resistor will have to flow through the zener. Since there is still a 3 V drop across the 27 ohms, the full 110 mA will be diverted into the diode. Heat will be generated at a rate of $5.1 \times 110 = 561$ mW and the temperature of the diode will get uncomfortably high.

The way round this problem of waste heat is to buffer the system with a transistor wired up as an emitter follower. This is shown in figure 12.34. V_U is any DC voltage which is at least a couple of volts larger than V_Z, the breakdown voltage of the zener diode. The resistor R allows a few mA to flow through the diode, holding the transistor base V_Z above the bottom supply rail. The load is therefore held between supply rails that are $V_Z - 0.7$ volts apart. The current which flows through the load is mostly drawn from the source of V_U by the collector of the transistor. If you ignore the base current, it should be obvious that the current through the zener (I_Z) is independent of the load current (I_L) unless there is a lot of ripple on V_U.

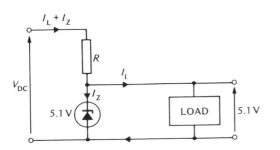

Figure 12.33 A regulated supply using a zener diode

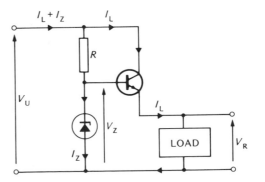

Figure 12.34 Using a pass transistor to buffer the zener diode

A regulated 5 V supply

The full circuit diagram of a regulated 5 V supply is shown in figure 12.35. We are going to discuss its operation briefly, and try to point out some of its remaining defects.

The AC signal coming out of the transformer secondary is 6 V rms. Its peak value is therefore $6 \times 1.414 = 8.5$ V. The smoothed supply rail will therefore sit at about $+8.5 - 0.7 = +7.8$ V above ground. The transistor base will be held 5.8 V above ground by the two diodes; its emitter will therefore hold the regulated supply rail at $+5.1$ V.

How much current can the load draw from the regulated rail without pulling it below $+5.1$ V? Obviously, as I_L increases, so does the amount of ripple on the smoothed supply rail. If the smoothed supply rail goes below $+5.8$ V at any time the zener diode is not going to be able to hold the transistor base steady. So we need to calculate the value of I_L which will place a ripple of $7.8 - 5.8 = 2.0$ V on the smoothed supply rail. Using the approximate ripple formula:

$$V_{\textbf{RIPPLE}} = \frac{V_0 T}{RC}$$

therefore $\qquad 2.0 = \dfrac{7.8 \times 20}{R \times 1000}$

therefore $\qquad R = \dfrac{7.8 \times 20}{2.0 \times 1000}$

$$= 0.078 \text{ k}\Omega$$

So if we linked the smoothed supply rail to ground by a 78 Ω resistor, we would have the desired ripple of 2 V. Now the current which would flow through that resistor would be about $7.8/0.078 = 100$ mA. (In fact the 78 Ω resistor is really the BC441 and the load.) So I_L can go up to 100 mA before

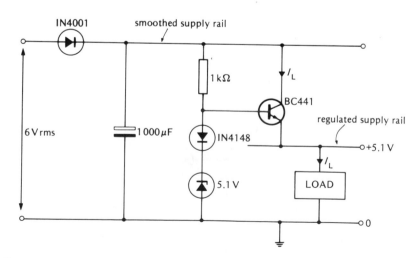

Figure 12.35 A 5 V regulated supply

the regulation is seriously threatened. In practice, it would get a bit soggy at a somewhat lower current; the voltage across the zener diode will stray from 5.1 V if too little current flows through it.

EXPERIMENT

Assembling the circuit of figure 12.35 and evaluating its performance

Use a 220 Ω resistor as the load so that a current of about 23 mA is drawn from the regulated supply rail. Use a CRO to look at the waveforms at the following points in the circuit; the 6 V AC supply, the smoothed supply rail, the transistor base and the regulated supply rail. Draw graphs of each waveform to the same scale.

Use a 10 kΩ resistor as the load. Use the CRO to measure the DC voltage and AC ripple of the smoothed supply rail and the regulated supply rail. Repeat the measurements with 4.7 kΩ, 2.2 kΩ, 1 kΩ, 0.47 kΩ, 0.22 kΩ and 0.1 kΩ resistors as the load. Draw a graph to show how the DC voltage and AC ripple on each of the supply rails depends on the current which flows through the load.

Finally, modify the circuit so that it produces a −5 V regulated supply. When you have got it to work, draw the circuit diagram of the modified power supply.

PROBLEMS

1 A simple power supply, like the one shown in figure 12.29, has a 12 V rms transformer, a 4 700 µF capacitor and a diode.

   a) What is the size of the DC voltage fed out by the supply?

   b) Calculate the ripple on the DC supply when the following resistors are attached to its output: i) 1 kΩ; ii) 100 Ω; iii) 10 Ω.

2 Show how you would modify the circuit of 12.35 so that it produced a +15 V regulated supply. You may use a 15 V rms transformer and a 15 V zener diode. If you use a 1 000 µF smoothing capacitor, how much current can the load draw from the modified circuit before the regulation fails?

3 Design a regulated power supply whose voltage can be varied between +5 V and +15 V by tweaking a potentiometer. You may assume that you have a 15 V rms transformer and a 15 V zener diode. Ensure that the capacitor is large enough for the load to be able to draw up to 500 mA from the regulated supply without upsetting it.

4 Look at figure 12.35. Any ripple on the smoothed supply rail is going to wobble the transistor base up and down slightly. This is because the voltage across the zener diode rises a bit when the current flowing through it increases. In order to keep the current through the zener diode rock steady, an electronic engineer decides to take advantage of the regulated supply and builds the circuit shown in figure 12.36 (overleaf). Will it work? If not, why not? If it does, is it any better than the original of figure 12.35? If it does not, can you modify it so that it does work?

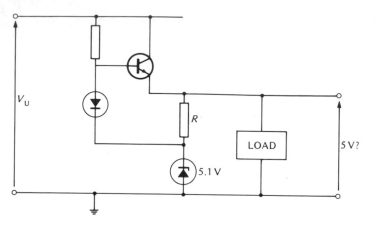

Figure 12.36

5 Look at the split rail power supply of figure 12.37.

a) If each zener diode is rated at 15 V, 500 mW, what is the maximum safe current which can be drawn from the smoothed supply rail?

b) Choose a value of smoothing capacitor which will give a ripple of about 0.2 V on the smoothed supply rails for that current.

c) Calculate a suitable size for the resistor R. How much heat does it have to be able to dissipate?

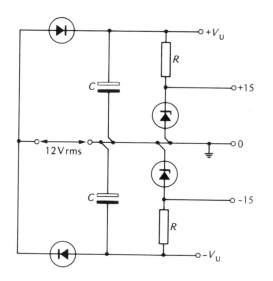

Figure 12.37 A regulated split rail power supply

STABILISATION

The 5 V power supply shown in figure 12.35 is fairly good, but it is not as good as it could be. The voltage of the regulated supply rail will fall slightly as more and more current is drawn from it. This will happen even if you use an enormous smoothing capacitor to wipe out the ripple and keep a constant current flowing through the zener diode. The source of the problem is the transistor. Its value for V_{BE} depends on its collector current. So the regulated supply might drop by up to 300 mV when current is drawn from it. Furthermore, the value of V_{BE} depends on the temperature of the transistor. As most of the supply current is passing through the transistor it is likely to get quite hot and so you cannot rely on V_{BE} remaining static.

In order to compensate for these drifts in the voltage of the regulated supply rail you need to apply negative feedback. You compare the supply voltage with the voltage across a zener diode, and use the resulting error signal to push the supply rail back to where it ought to be. The power supply will be stable, with the supply rail sitting at a constant voltage regardless of changes in load current and temperature.

Figure 12.38 shows the circuit diagram of a 5 V stabilised supply. The feedback connects the non-inverting input of the 741 to the stabilised supply rail. This is compared with the 5.1 V across the zener diode. Suppose that the supply rail is less than 5.1 V. Then the output of the 741 will start to drop towards its lower supply rail (0 V). The base of the transistor will be pulled down, so that more current will flow through its collector. That current will flow through the load and the 22 kΩ resistor, raising the voltage of the supply rail. Obviously, once the supply rail goes above 5.1 V, the output of the 741 will start to rise up towards its upper supply rail (+8 V). Less current will be pulled out of the transistor base, so less current will flow through the load and the supply rail will drop down.

Do not worry about running the op-amp off a 0 and +8 volt supply. A 741 will operate quite happily provided its supply terminals are at least 6 V apart. Its output can only get to within 2 V of its supply rails, and this is the reason why a pnp transistor has been used to supply the load current. Have a look at the circuit of figure 12.39 (overleaf). It shows part of a stabilised supply which uses an npn transistor. The emitter follower acts like a power amplifier for the op-amp, sitting inside the feedback loop. Since the op-amp output cannot get below +2 V, the stabilised supply cannot get below +1.3 V. Furthermore, the op-amp cannot push its output higher than $V_U - 2$, so $V_S \leqslant V_U - 2.7$. The actual value of V_S is, of course, fixed by

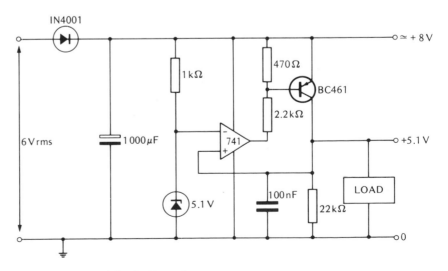

Figure 12.38 A stabilised 5 V supply

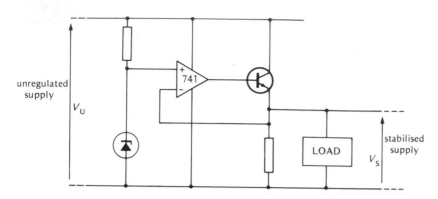

Figure 12.39 Using an npn pass transistor

the zener diode. So with this arrangement you must have a voltage drop of at least 2.7 V between the collector and the emitter of the transistor. As the load current falls through that voltage, heat is going to be generated; a current of 1 A will make heat appear in the transistor at a minimum rate of $2.7 \times 1 = 2.7$ W.

A similar analysis of the circuit of figure 12.38 shows that the pnp transistor need only dissipate a fraction of this heat. In order to minimise the heat wasted in the transistor, we need to make the voltage drop between emitter and collector as small as possible. We can allow the smoothed supply rail to drop down to +6 V without affecting the op-amp; the stabilised supply rail is fixed at +5.1 V of course. The voltage divider at the base of the transistor allows the op-amp output to sit between +2 V and +3 V when the transistor is switched on, so it will have no problem in controlling the collector current. So we can reduce the value of V_{CE} to $6 - 5.1 = 0.9$ V. The rate at which heat is generated in the transistor when the load current is 1 A is now only 0.9 W.

There are two things that you have to watch for in the circuit of figure 12.38. You must make sure that the op-amp can switch off the transistor if it wants to. The voltage divider shown in the circuit will hold the base 0.35 V below the emitter when the 741 output sits 2 V below its top supply rail, so the transistor will definitely be off. Secondly, you have to make sure that the op-amp can sink enough current from the transistor base. For example, when the op-amp output is as low as it can go (+2 V), the current through the 2.2 kΩ resistor will be about 2.3 mA. 1.5 mA of this will have to flow through the 0.47 kΩ to get 0.7 V across it. This leaves 0.8 mA of base current for the transistor. A current gain of 120 means that the load current cannot go above about 100 mA. You can, of course, increase the load current by having smaller resistors in the voltage divider and using a Darlington pair.

EXPERIMENT
Assembling and evaluating a stabilised 5 V supply

Assemble the circuit shown in figure 12.38. Use resistors between 10 kΩ and 0.1 kΩ as the load; check that the stabilised supply rail sits at a constant +5.1 V, with no appreciable ripple. See what happens if you remove the 100 nF capacitor.

Modify the circuit so that the voltage of the stabilised supply can be continually altered between $+2\,V$ and $+5\,V$ by means of a potentiometer. When you have got it to work, draw the circuit diagram.

Finally, alter the circuit so that it holds the stabilised supply rail between $-2\,V$ and $-5\,V$, depending on the setting of the potentiometer. Check that it behaves properly by trying different loads and checking that the output is stable. Draw the circuit diagram of the supply.

PROBLEMS

1 Draw the circuit diagram of a stabilised power supply with the following properties. The stabilised supply rail sits $+10\,V$ above ground, and it can source 250 mA before it starts to drop. Assume that you have a 10 V rms transformer and a 10 V zener diode at your disposal. Show all component values with your justification for them.

2 A good power supply ought to have some sort of current limiting facility. Figure 12.40 shows part of a stabilised supply which cannot, under any circumstances, source more than 1 A from the stabilised supply rail. Explain how the current limiting works.

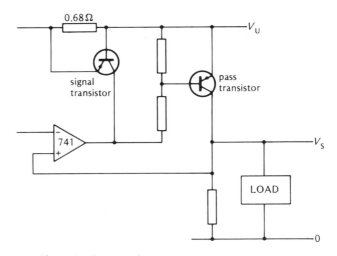

Figure 12.40 Short-circuit protection

3 Another feature of a good power supply is some sort of protection against overheating. Suppose that a thermistor was attached to the transistor. Then if the transistor got too hot, the thermistor could generate a signal which cut down the current flowing through the transistor. Draw the circuit diagram of a 5 V stabilised supply which features this protection. Explain how the circuit works.

4 You need to make a stabilised power supply whose output can be varied from 0 V to $+15\,V$ with a potentiometer. You have to use a 15 V rms transformer and a 5.1 V zener diode. Draw a suitable circuit diagram; you might find that figure 12.1 contains a hint of how to deal with the zener diode problem.

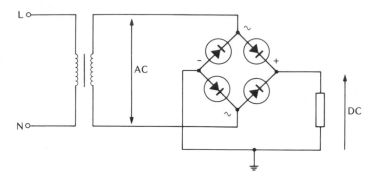

Figure 12.41 A diode bridge

FULL-WAVE RECTIFICATION

A well designed stabilised supply will be rock steady provided that the ripple on the smoothed supply rail does not get too great. You can ensure this by using large smoothing capacitors (bulky and expensive) and by having a large difference between the smoothed supply voltage and the stabilised supply voltage (generating a lot of heat in the transistor). A third way of cutting down the ripple is to use a diode bridge to rectify the AC signal from the transformer.

Figure 12.41 shows what a diode bridge looks like. It consists of four diodes in a ring, with four terminals coming out of the ring. Two of those terminals are connected to the transformer, and the other two to a resistor. If you study the two diagrams of figure 12.42 you should appreciate that the + terminal will always be at a higher voltage than the − terminal. When the − terminal is anchored to 0 V, as in figure 12.41, the voltage across the resistor will be as shown in figure 12.43. Since the current has to flow through two diodes, the peak voltage across the resistor is going to be 1.4 V less than the peak value of the AC which is being rectified.

A 5 V stabilised supply which employs a diode bridge is shown in figure 12.44. Note the way in which the secondary of the transformer is allowed to float. The smoothing capacitor is going to be charged up twice in every mains cycle, so the ripple on the smoothed supply rail is going to be given by:

$$V_{\text{RIPPLE}} = \frac{(6 \times 1.414 - 1.4)}{R \times 1000} \times 10 = \frac{0.07}{R}$$

where *R* is the load drawing current from that rail. Since the smoothed

Figure 12.42 How a diode bridge converts AC to DC

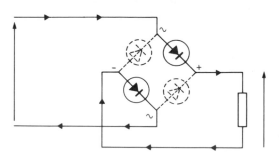

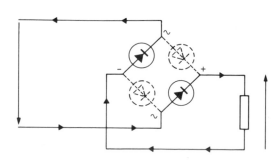

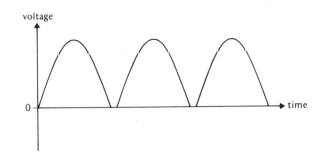

Figure 12.43 The output of a full-wave rectifier

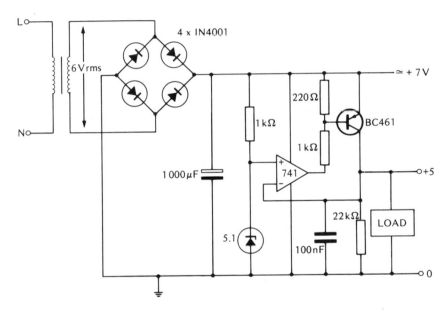

Figure 12.44 A 5 V stabilised supply

supply rail must keep above $+6$ V for the 741 to work properly, we can only afford a ripple of 1 V. So *R* must be 0.07 kΩ or less, i.e. the load current must be less than about $7/0.07 = 100$ mA.

Split supplies

If you combine a diode bridge with a centre-tapped transformer, you can make a full-wave rectifier which has a positive and a negative output. For example, the circuit of figure 12.45 (overleaf) shows how such an arrangement can be used to obtain a pair of smoothed supplies at $+12$ V and -12 V from a 9 V rms transformer. The transformer has two separate, but identical secondaries. These have been connected in series and the join (the centre tap) held at 0 V. Each secondary feeds out 9 V rms, so the AC waveform across the $\sim$ terminals of the diode bridge is 18 V rms. The peak voltage between the $+$ and $-$ terminals must therefore be $18 \times 1.414 - 1.4 = 24$ V. As the centre tap is held at 0 V the two smoothed supply rails will sit at $+12$ V and -12 V.

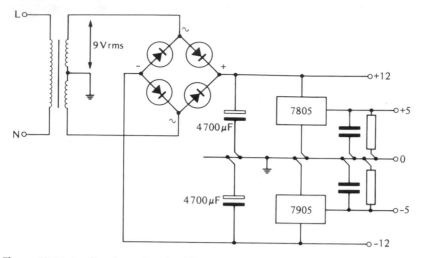

Figure 12.45 A split rail supply using ICs

The two boxes labelled 7805 and 7905 in figure 12.45 represent ICs which contain all of the circuitry necessary to generate the $+5$ V and -5 V stabilised supply rails. These cheap and versatile units are protected against overheating and are short-circuit proof; if they get too hot, or you try to draw more than 1 A from them they simply close down.

EXPERIMENT

Understanding what a diode bridge does

Look at the circuit shown in figure 12.46. The transformer feeds a 6 V rms waveform into the $\sim$ terminals of the bridge. The current which flows through the diodes passes down a pair of resistors; the junction between the resistors is held at 0 V.

Decide what the waveform at each of the four terminals of the bridge will look like. Sketch them, to the same scale, one under the other.

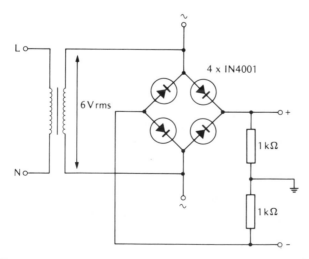

Figure 12.46

Now assemble the circuit. Trigger a CRO on the waveform at one of the terminals and use it to look at the waveforms at each terminal in turn. If necessary, sketch the correct waveforms.

Explain clearly why the waveform at each terminal has the shape it does.

PROBLEMS

1 This problem concerns the circuit shown in figure 12.47.

a) If the transformer has 12 V rms across each of its secondaries, what is the peak value of V_U?

b) What is the range of values of the stabilised voltage V_S?

c) Explain how the stabilisation of each supply rail is done. In particular, explain why they are always kept at equal and opposite voltages.

d) If the system is designed to provide currents of up to 1 A, work out a suitable value for the smoothing capacitors. You will have to make an educated guess at a value for the maximum acceptable ripple on the smoothed supply rail.

Figure 12.47 A variable split rail supply

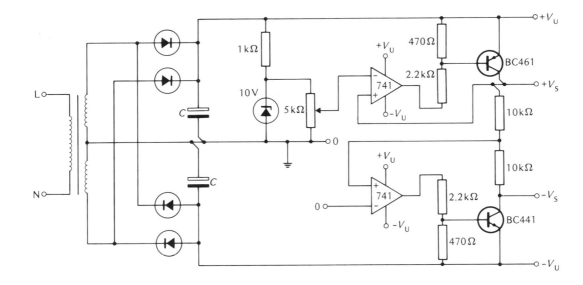

2 This problem concerns the circuit shown in figure 12.45.

a) Each IC can draw up to 1 A from the smoothed supply rail; provided the smoothed supply is greater than 7.5 V, the IC will work correctly. Calculate the ripple on the smoothed supply when the IC pulls 1 A out of it.

b) How small can you make the capacitors before you run the risk of the smoothed supply dipping below 7.5 V when it provides a current of 1 A?

c) If you have to use 4 700 µF capacitors, what is the minimum rms voltage that the transformer has to provide if the supply is to function correctly?

d) What is the maximum rate at which heat will be generated in the ICs in figure 12.45?

e) What do you think the capacitor and the resistor which connect the stabilised supply rails to 0 V are for?

DC TO DC CONVERTERS

Digital circuits, such as logic gates, counters and microprocessors almost invariably run off supply rails at $+5$ V and 0 V. Analogue circuits, generally based on op-amps, tend to run off split supplies. It frequently happens that a system will contain a large number of digital components, the parts that do all of the processing, with a few analogue components to handle the input and output functions. For example, you might have a microprocessor, some RAM, some ROM, a clock, input and output port and a DAC all on one circuit board. Only one component (the last one) will need a split supply.

It does not make sense to go to the trouble of generating $+15$ V and -15 V, routing them round to the circuit board and feeding them into a single IC. Instead, it is common practice to provide only a $+5$ V supply for the board, and to generate a -5 V supply on the board itself if it is needed. The analogue ICs are then run off split supplies of $+5$ V and -5 V. The device which generates the -5 V rail is known as a DC to DC converter. It only has to supply a few mA at -5 V, but its output must be well stabilised.

Level shifters

At the heart of a DC to DC converter is a unit called a level shifter. One of these is shown in figure 12.48. As you can see, it has two input terminals and two output terminals. If you feed a square wave into the system you get the same square wave out, but shifted down; the system adds a negative DC signal to the AC signal fed into it. It does this without the benefit of supply rails.

We will discuss its operation in more detail. Assume that the square wave has been going in for some time, so that the circuit has settled into equilibrium. Forget about any current drawn out of the output as well. The bottom input terminal is held at 0 V and a $+5$ V square wave fed into the top input terminal. The diode on the left, D_1, prevents the top output terminal from going above $+0.7$ V. So when the left-hand plate of C_1 goes from $+5$ V to 0 V, the right-hand plate must go from $+0.7$ V to -4.3 V. The other diode, D_2, will pull the right-hand plate of C_2 down to -3.6 V when the top output terminal goes down to -4.3 V. If no current is drawn from the output terminals then the bottom terminal will be stuck at -3.6 V, as D_2 will always be reverse biased.

So we feed a 5 V square wave and 0 V in. We get a shifted down 5 V square wave and -3.6 V out.

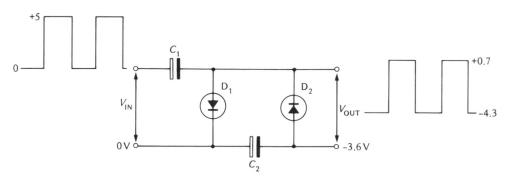

Figure 12.48 A level shifter

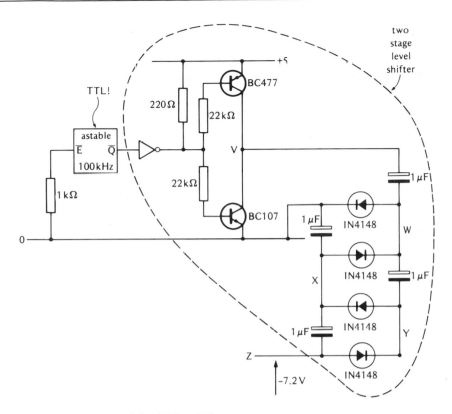

Figure 12.49 An unstabilised DC to DC converter

The beauty of level shifters is that they can be cascaded in series to obtain as low a voltage as you like. You simply feed the output of one level shifter into another; the square wave is shifted down further and you also get a lower DC voltage. If you look at the circuit shown in figure 12.49 you will see that part of it is two level shifters in series. The final DC output is going to be -7.2 V, i.e. 2×-3.6 V.

Of course, once you start drawing current from that output, its voltage will rise and acquire some ripple. So the output needs to be stabilised if it is to act as a supply rail. A method of doing this is shown in figure 12.50. A

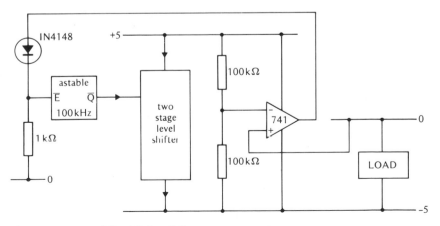

Figure 12.50 A stabilised DC to DC converter

741 looks at the mean voltage of the $+5\,V$ and $-5\,V$ supply rails and compares this with $0\,V$. If the $-5\,V$ rail is not low enough, the op-amp output goes low, pulling $\bar{E}$ down to logical 0. The astable fires a square wave into the two stage level shifter (figure 12.49), and charge is pumped onto the capacitors. So the $-5\,V$ rail is pushed down. Once it gets to exactly $-5\,V$, the op-amp output goes up to $+3\,V$ and $\bar{E}$ is hauled up to $+2.3\,V$. This is a logical 1 if the astable is built out of TTL components. So it stops oscillating and the $-5\,V$ rail is free to drift up again as it sinks current from the load.

The whole system uses on-off feedback to keep the lower supply rail at the right voltage. The astable is pulsed on briefly to push the lower supply rail down whenever the op-amp detects that it has drifted up. The amount of current that the lower supply rail can sink without losing stabilisation depends on several factors. Big capacitors help, and a large power amp to buffer the square wave helps too.

EXPERIMENT
You are going to assemble and test a DC to DC converter.
Then you will design one yourself and try it out.

Assemble the circuit shown in figure 12.49. Look at the waveforms at V, W, X, Y and Z in the circuit with a CRO; trigger the CRO on the waveform at V. Draw the waveforms, one under the other, to the same scale.

By trial and error find the size of resistor which, when placed between Z and $0\,V$, will pull Z up to $-5\,V$.

Now add a 741 so that the system becomes the circuit of figure 12.50. Again, by trial and error, find how much current the load can sink into the $-5\,V$ rail before the ripple becomes unacceptable and the stabilisation fails.

Finally, design a DC to DC converter which will generate a stabilised $+10\,V$ supply from a $+5\,V$ supply. Try out your design. When you have got it to work, find out how much current you can draw from the $+10\,V$ supply rail. Draw a circuit diagram of the system, and explain how it works.

PROBLEM

One way of providing supply rails for the components in a large system is to feed a smoothed $+12\,V$ supply to each board. An IC, such as the 7805 can use this to set up a $+5\,V$ supply for the board. Draw the circuit diagram of a circuit which could use the $+12\,V$ supply to generate the $+9\,V$ and $-9\,V$ needed to supply a few analogue ICs on a board.

REVISION QUESTIONS

1 Draw a block diagram of a negative feedback system. Derive an expression for the closed loop gain of the system as a function of the open loop gain and the feedback fraction (B).

2 Draw the circuit diagram of a simple on-off feedback position control system which uses a DC motor. Explain how the position control works, and how the system reacts to a sudden change in the input signal.

3 What is a proportional feedback position control system? In what way, and why, is its response different to that of an on-off feedback system?

4 Explain what is meant by hunting. Using a speed control system as an example, discuss how hunting can be avoided.

5 Sketch a graph of the gain as a function of frequency of an op-amp which is conditionally stable. Explain why it will oscillate for a range of values of the feedback fraction. Discuss the characteristics of an unconditionally stable op-amp.

6 Explain how the bandwidth of an amplifier which uses an op-amp may be calculated from the feedback fraction, the DC gain and the gain-bandwidth product.

7 Derive an expression for the amount of ripple on a half-wave rectified waveform which has been smoothed by a capacitor.

8 Draw the circuit diagram of a simple smoothed 15 V supply, capable of delivering up to 100 mA with a ripple of less than 5%. Show all component values, including the rms voltage across the transformer secondary. Explain why the supply is unsatisfactory.

9 Why is it a good idea to use a pass transistor to buffer the zener diode of a regulated power supply? Illustrate your answer by considering a supply which uses a 9 V, 500 mW zener diode.

10 Draw the circuit diagram of a +9 V stabilised supply which runs off a 9 V rms transformer. Use a diode bridge, show all component values and justify them; the system has to be able to push at least 500 mA into a load without losing stabilisation. Explain, in detail, how the system works.

11 Show how a centre-tapped transformer can be used to obtain a pair of smoothed split supply rails at +10 V and −10 V. Work out the rms voltage needed across each of the secondary windings.

12 Draw a level shifter which raises the DC level of the square wave fed into it. Draw waveforms for various points within it.

13 Explain why DC to DC converters are useful.

13

Signal transfer

This final chapter is all about communication. It will show how electronics can be used to transfer information over long distances. Signal transfer is a very rich field of study, and we shall only be able to introduce you to some aspects of it. The electronic circuits used in communications are, in general, an interesting mix of analogue and digital devices. It is for this reason that this chapter is near the end of the book; it is going to draw on the knowledge of techniques which you have acquired from previous chapters.

Signal transfer is also a very important discipline in the modern world. Our civilisation depends on the rapid and accurate transfer of information by electronic means. We rely on radio, television, telephone, telex etc. to allow people to communicate with each other, and recent advances in technology will probably improve communications dramatically in the near future.

COMMUNICATION SYSTEMS

There are many ways in which you can transfer information from one place to another by electronic means, but they all share a common foundation. Before we start to describe communication systems which are currently widely used, we shall outline the general operation of all such systems. Look at the block diagram of a communication system shown in figure 13.1. It takes in a signal, encodes it in some way, feeds the encoded signal into a communication link and finally decodes the signal. A perfect communication system will feed out a signal which is a perfect copy of the signal which was fed into it; such systems do exist! The signal that is fed into the communication link may or may not be electrical in nature, and there are many ways in which a particular signal can be encoded for a particular communication link.

Figure 13.1 A communication system

WIRE LINKS

The simplest form of communication link, and the simplest one to use, is a length of wire. Strictly speaking, since an electronic signal involves an electric current flowing in a loop, a wire link requires two lengths of wire. One of the wires is usually grounded and wound around the other to make a coaxial cable; the grounded braid around the outside shields the inner wire from electrical interference. You are probably aware that coaxial cable is used to pipe signals from TV aerials to TV receivers, and it is widely used to transfer signals over short distances.

Wire links can carry signals of frequencies up to about 1 GHz, (i.e. 1000 MHz) but above that they start to get inefficient. At low frequencies a length of cable will have a very low resistance. At high frequencies however, the resistance will start to rise because of something called the **skin effect**. High frequency currents can only flow in the surface of a wire, so only the outer skin of the wire is used to carry the current. Effectively, the thickness of the wire appears to get smaller as the frequency gets higher, making the link appear to have an increased resistance. For example, at 1 MHz the current flowing in a piece of copper is confined to within about 0.05 mm of its surface; at 100 MHz the skin depth is ten times smaller than this.

When you use a wire link to carry high frequency signals you have to be very careful about the input impedance of the circuit into which the link feeds its signal. A length of cable has a characteristic impedance whose value depends on its shape, dimensions and material; for TV coaxial cable it is 75 ohms. A correctly terminated cable feeds all of its signal into the receiver circuit. If the input impedance of the receiver does not equal the characteristic impedance of the cable some of the signal will be reflected, bouncing off the end of the cable and going back down the cable the way it came. Some of that reflected signal will bounce off the other end of the cable and will arrive back at the receiver where some of it will be reflected again. Obviously, these electrical echoes bouncing back and forth along the length of the cable will eventually die away, but they do mean that the same signal is presented to the receiver several times. This is not very important if the signal does not change very rapidly, but it can wreak havoc with high frequency signals. Figure 13.2 shows the signal which emerges from a badly terminated cable when a rising edge is fed into it; the signal which emerges from the cable is far from being a perfect copy of what was fed in.

The amplitude of the signal which travels down a length of cable is attenuated by a factor of $e^{-(R/Z_0)}$ where R is the resistance of the cable and Z_0 its characteristic impedance. The signal dies away as it travels down the cable because the current associated with it generates heat as it flows through the cable resistance. So if you push a signal into a long length of

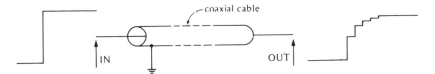

Figure 13.2 The effect of terminating a cable with the wrong impedance

cable you probably lose a lot of it in the cable itself, especially at high frequencies where the cable resistance starts to get high. In order to compensate for this signal loss a long length of cable will have to contain a number of **repeaters**. These are amplifiers which boost the signal as it goes down the cable, so that it arrives at the other end without too much attenuation.

The spacing of the repeaters will depend on two factors namely their gain and the frequency of the signal. For example, suppose that you have to send a 100 MHz signal down a 50 ohm coaxial cable. If it is a good quality (i.e. low loss) cable the signal will be attenuated by approximately 1 dB for each 10 m. Since the logarithm of the attenuation is proportional to the cable resistance, a 20 m length will attenuate by 2 dB, a 100 m length by 10 dB etc. You can get cheap and versatile 100 MHz IC amplifiers which have gains of 40 dB; if you use these as your repeaters they will have to be placed at intervals of 400 m along the cable.

Things are not so bad at lower frequencies. At 10 MHz the attenuation of the cable will be about half that at 100 MHz, so the repeaters can be placed further apart. Nevertheless, the need to use repeaters in long lengths of cable is a considerable nuisance as they have to be supplied with power.

OPTICAL LINKS

An alternative to cable which has recently become widely available is optical fibre. These are thin threads of very pure glass which will transmit red and infra-red light with an attenuation of less than 30 dB per kilometre. You put an LED at one end of the fibre, and use it to fire light towards a photodiode at the other end of the fibre. Figure 13.3 illustrates a system which can send audio signals down an optical fibre by this means. The emitter is an infra-red LED (fibre links work best in the infra-red) connected up to an op-amp such that the current through it obeys this equation:

$$I = (V_{IN} + 15)/R$$

As the amount of light emitted by the LED is proportional to the current flowing through it, the amount of light going down the optical fibre is going to be proportional to the instantaneous value of V_{IN}. The current which flows through the phototransistor is converted to a voltage by another op-amp, so that V_{OUT} varies linearly with the amount of light arriving down the fibre.

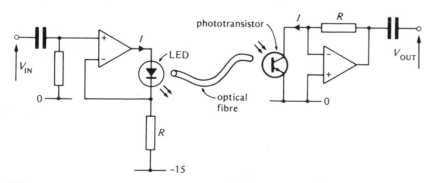

Figure 13.3 An optical fibre transmitting an analogue signal

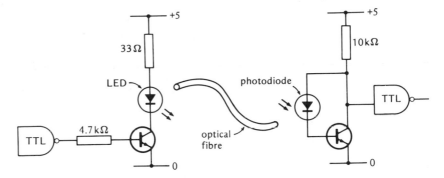

Figure 13.4 Transmitting a digital signal down an optical fibre

Optical fibres are better than cable in several ways. To start with, they are much smaller and lighter, so a single coaxial cable could be replaced by at least 100 optical fibres without taking up any more room. Then optical fibres are very quiet. They do not suffer from crosstalk, electrical interference or mains hum, so they are ideal for carrying signals around in noisy environments such as factories. Finally, they can operate at up to 1 GHz so they are just as good as coaxial cable at high frequencies.

The circuit shown in figure 13.3 shows how an analogue signal can be sent down an optical fibre. The system only functions well for signals that are well below 1 MHz, mainly because of the limited speed of the phototransistor. Photodiodes respond faster, but generate a far smaller current. You can get the best of both worlds (speed and current) by using a fast transistor to amplify the photocurrent. You can get LEDs and photodiodes with response times of a few nanoseconds, matching TTL response times nicely.

Figure 13.4 shows how digital signals can be sent down an optical fibre. A transistor is used to drive a high intensity LED (up to 100 mA perhaps) which feeds infra-red light into an optical fibre. At the receiving end the photocurrent is boosted by a transistor. The system shown can tolerate a signal attenuation of up to 80 dB before the receiver starts to give up, so the fibre could easily be a couple of kilometres long.

RADIO WAVES

A full discussion of the physics of radio waves is beyond the scope of this book. So we are not going to discuss exactly how they are generated, how they propagate or how they induce signals in conductors which they cross. Instead, we shall offer a simple, but useful picture of a radio wave and how it is emitted and absorbed.

Suppose that a wire carries an AC signal. An alternating current will flow down the wire, transporting electrical energy from one end of it to another. That energy leaks out of the wire as a radio wave and rushes away at the speed of light. If part of that radio wave subsequently hits another wire, then some of the wave energy is converted back into electrical energy. An AC signal appears in the wire, with a frequency identical to that of the signal in the radiating wire. So radio waves allow AC signals to be transmitted across empty space from one wire to another.

A radio wave is characterised by its frequency and its wavelength. These are not independent parameters as frequency × wavelength equals the speed of light. The frequency is dictated by the frequency of the AC signal which generated the wave e.g. a 50 MHz signal generates a 50 MHz radio wave. As the speed of light is 3×10^8 m/s, the wavelength of this wave will be $3 \times 10^8 / 5 \times 10^7 = 6$ m. This tells you about the size of aerial needed to convert the radio wave efficiently back into an AC signal.

High frequency radio waves have the advantage of needing small aerials to pick them up. Furthermore, they can be steered in particular directions by relatively small dish reflectors; the dish diameter has to be at least several wavelengths across if it is to beam the wave effectively. The big disadvantage is that high frequency, short wavelength radio waves travel in straight lines and can only be used for "line of sight" communications. At lower frequencies, between 5 MHz and 30 MHz, radio waves are strongly reflected off the ionosphere, so they travel all round the earth and can be used for world-wide communications.

ULTRASONIC LINKS

You can transmit signals over short distances (up to a few metres in air) with ultrasound. This is a high frequency sound wave, well above the audio region. The emitter is usually designed to work at one particular frequency (40 kHz perhaps), with the detector matched to the same frequency. Ultrasonic transducers compete effectively with infra-red ones when signals have to be transmitted short distances without the aid of connecting wires; both methods are widely employed for the remote control of televisions, videos and music centres.

MODULATION

Each type of communication link carries a different sort of signal. Cable carries electrical signals, optical fibre carries light signals, free space carries radio waves and air carries ultrasonic signals. All of these links transfer energy from one point to another, from the emitter to the receiver. However, the information content of a constant signal transmitted down a link is fairly limited. At best you can interpret the presence of the signal as a logical 1 and its absence as a logical 0, or treat the amplitude of the signal as being proportional to a voltage. In either case, a constant signal sent down the link results in a constant signal being fed out by the receiver. If you want to change the information being transferred down the link as time progresses, then you have to change the signal in the communication link. This process is called **modulation**.

There are many different ways in which signals can be modulated, and you will be meeting some of them in the next few sections of this chapter. The type of modulation that you employ depends on the type of communication link being used and what sort of signal you wish to transfer down it. Nevertheless, all of the methods share some common features. We shall use those common features to explain what modulation is, before discussing specific examples.

A different block diagram of a communication system is shown in figure 13.5. The carrier is a constant signal of the type which can be transmitted down the link. So if the link is an optical fibre, the carrier will be a constant intensity beam of infra-red light. The properties of the carrier are altered by the signal as it passes through the modulator. The modulated carrier is then fed into the communication link and carries information about the signal

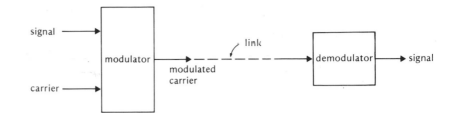

Figure 13.5 Modulation and carriers

towards the receiver. In effect, any change of the modulated carrier is a consequence of a change in the signal. When the modulated carrier arrives at the detector the signal is extracted from the carrier by a demodulator. Naturally, the demodulator in the receiver has to be matched with the modulator in the emitter; the receiver has to know how the signal was coded onto the carrier if it is to successfully decipher it!

An example of modulation in action should make things clearer. Figure 13.6 illustrates the essential parts of a simple, one-way telephone link. It contains just two components, a carbon microphone and a moving-coil speaker. The microphone is basically a resistor made up from little spheres of carbon packed together; when a sound wave hits the microphone, those spheres are squeezed and stretched so that the resistance of the carbon is changed. When the microphone diaphragm moves in, the carbon is squashed and its resistance goes down as the contact between the spheres is improved. An outward movement of the diaphragm moves the spheres apart, increasing the resistance of the microphone. The carrier is the current which flows down the cable. When sound waves hit the microphone the current will be modulated by the changing resistance in the circuit. So the current will increase and decrease as the microphone diaphragm is moved in and out by the sound wave. That change of current will make the speaker at the other end of the cable emit a sound wave; the speaker acts as the demodulator, as it only reacts to AC currents.

The graphs of figure 13.6 illustrate how the signal is coded onto the carrier. This type of coding is called **amplitude modulation**, or just simply **AM**. The signal makes the amplitude of the carrier (a DC current in this case) wobble up and down. The demodulator has to extract that wobble in amplitude to recreate the original signal. One of the optical systems that we looked at earlier (figure 13.3) uses exactly the same technique, except that the carrier is a constant intensity infra-red beam; the capacitor in the detector rejects the DC carrier and transmits the AC wobble on it.

Figure 13.6 A simple telephone link

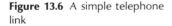

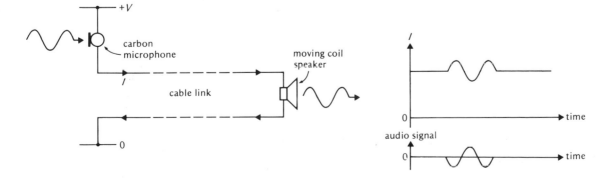

Each of the next few sections of this chapter deal with a modulation technique in detail. Most of the examples used will assume that the signal is modulated onto a carrier which can be sent down a wire link. Some types of link are more suited to a particular type of modulation, but wire links can be used with all of the modulation techniques.

CARRIER KEYING

This is the simplest modulation technique of all. It can only be used to transmit digital signals as the carrier is modulated by being switched on and off. Carrier keying was the technique used by early telegraph systems; a DC carrier was switched on and off by the emitter, usually a simple switch in the form of a morse code key. At the receiving end the signal arriving from the wire link operated a buzzer. A modern circuit to perform the same function is shown in figure 13.7.

Carrier keying can transfer a sequence of digital signals to make up a binary word. Early telegraph systems employed Morse code, with short and long buzzes acting as the equivalent of logical 1 and logical 0. A modern carrier keying system employs the techniques of serial transfer outlined in Chapter Ten to transmit binary words. The presence of the carrier is interpreted as one logic state and its absence as the other logic state.

AC CARRIERS

DC carriers have several disadvantages. To start with, they are very prone to noise, especially low frequency noise such as mains hum and pickup from audio amplifiers. So when the carrier arrives at the receiving end of the cable it will have picked up a lot of rubbish along the way. (Telephone links are a classic example of this.) Then there is the problem of repeaters. A Schmitt trigger would clearly make a suitable repeater for a communication link which transmitted a DC voltage, but each repeater might only provide a DC gain of 14 dB (i.e. thresholds at $+1$ and $+4$ V). So the repeaters would have to be fairly close together. Finally, you can only put one DC carrier onto a link whereas you can put a large number of AC ones onto it. The possibility of multiplexing many carriers onto a single link has enormous economic advantages, and it is for this reason that carrier keying usually involves an AC carrier.

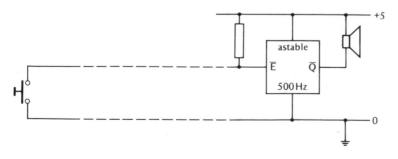

Figure 13.7 A Morse code transmission system

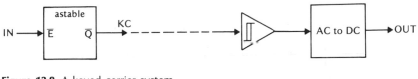

Figure 13.8 A keyed carrier system

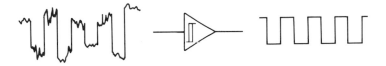

Figure 13.9 A Schmitt trigger cleaning up a dirty waveform

A simple digital transfer system which uses a keyed AC carrier is shown in figure 13.8. The digital signal which is to be transferred controls an astable; a logical 0 feeds a square wave into the cable and a logical 1 feeds a DC signal into it. The receiver feeds out a logical 0 when an AC signal enters it, otherwise it feeds out a logical 1. Figure 13.11 shows how the **keyed carrier** (**KC**) depends on the state of the digital signal being transferred (IN). The insertion of a Schmitt trigger at the receiving end of the link cleans up the keyed carrier, removing the noise and interference which it picked up in transit. (Look at figure 13.9.) This possibility of restoring the modulated carrier to its original form at the end of the communication link allows the system to be virtually error-free. Unless the cable is very noisy, no information about the signal will be lost as it is transferred.

The receiver contains a circuit (called AC-to-DC) which will recognise the presence of an AC signal fed into it. An example of such a circuit is shown in figure 13.10. Study it and convince yourself that it will only feed out a logical 0 when an AC signal of at least about 2 V peak value is fed into it.

Frequency shift keying

We can improve the system which we have just described by using two carriers instead of one. A glance at figure 13.11 shows that the DC level of the keyed carrier, KC, changes by 2.5 V when it is switched on or off. This tends to cause problems when the keyed carrier is fed through a communication link which can only cope with AC signals, such as a telephone line or a link with AC repeaters. Every time that the carrier is keyed the change of DC level will mean that any coupling capacitors in the system have to charge or discharge accordingly. This introduces **transients** (electronic

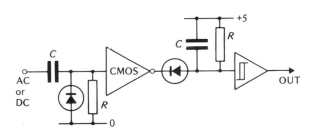

Figure 13.10 An AC-to-DC converter

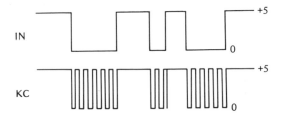

Figure 13.11

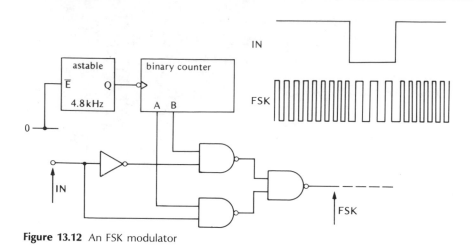

Figure 13.12 An FSK modulator

hiccups) into the carrier which can be inconvenient. So when the keyed carrier is destined to be processed by an AC amplifier you would be advised to try and keep its DC value constant, regardless of the signal modulated onto it.

The standard way of transmitting digital signals via audio links is to use 2.4 kHz to represent logical 1 and 1.2 kHz to represent logical 0. This ensures that an AC signal is being sent down the link at all times, so that the DC level remains steady. This method of changing the frequency of the carrier instead of switching it on and off is called **frequency shift keying** (or **FSK**). It is widely used for transmitting binary words in serial form down telephone lines and for storing computer programs on cassette tape.

A system which can act as an FSK modulator is shown in figure 13.12. The two outputs of the binary counter have frequencies of 2.4 kHz and 1.2 kHz. These are selectively gated onto the communication link by IN; when IN is 1, FSK oscillates at 2.4 kHz and when it is 0, FSK oscillates at 1.2 kHz. A compatible demodulator is shown in figure 13.13. Note the use of a Schmitt trigger to clean up the FSK carrier as it arrives from the link. The restored FSK carrier gates 25.6 kHz pulses into a binary counter and latches the state of D onto a flip-flop every time those pulses are gated off. If D is 0, then the carrier must have a frequency above 25.6/16 = 1.6 kHz.

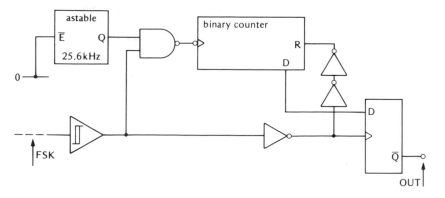

Figure 13.13 An FSK demodulator

Conversely, if the carrier has a frequency between 1.6 kHz and 0.8 kHz then D will be 1 when the flip-flop is clocked. So a 1.2 kHz carrier will reset $\bar{Q}$ to 0 and a 2.4 kHz carrier will set $\bar{Q}$ to 1.

Pulse position modulation

The technique of carrier keying is not confined to the transfer of digital signals. With suitable forms of modulation it can be used to transfer analogue signals as well. One of these techniques, called **pulse position modulation** or **PPM**, is illustrated in figure 13.14. The carrier is keyed on for fixed lengths of time, but the intervals between these pulses is proportional to the analogue signal being modulated onto the carrier. So as the analogue signal goes up the pulses get closer together; as the signal goes down the pulses move apart. One of the nice things about this technique is that the pulses can be cleaned up by a Schmitt trigger when they get to the other end of the communication link, so that it opens up the possibility of virtually error-free transmission of analogue signals. Furthermore, it allows analogue signals to be transferred via communication links which are best suited to digital signals. For example, infra-red and ultrasonic links through free space need to be keyed in order to combat the high level of background noise; direct modulation of an analogue signal onto either of these carriers would not work very well.

A PPM modulator which can encode audio signals is shown in figure 13.15 (overleaf). It is basically an astable with an op-amp controlling the pulse spacing. As V_M goes up, the 6 µs pulses move apart and as V_M goes down the 6 µs pulses move together. Demodulating a PPM carrier is dead easy as you just have to average it. The mean DC level of the PPM carrier goes up and down as the pulses move together and go apart, so the demodulator just has to respond to the average value of the signal emerging from the communication link. Figure 13.16 (overleaf) shows a possible circuit. After being cleaned up with a Schmitt trigger (a CMOS one as an op-amp one is likely to be too slow) the PPM carrier is fed through a simple treble cut filter. The break frequency of the filter is $1/2\pi \times 22 \times 0.001 = 7.2$ kHz, so the op-amp only sees the audio frequency components of the PPM carrier. The higher frequency bits will be attenuated, but you will not hear much of them anyway. Indeed, you can feed a PPM carrier straight into a loudspeaker, via a suitable power amplifier, if you want to; the speaker will respond to the average signal being fed into it as inertia prevents it from reacting quickly. (The transistors in the power amp will run cool as well. They will be either off or saturated; either way, not much heat will be generated in them.)

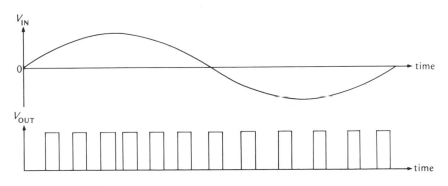

Figure 13.14 PPM

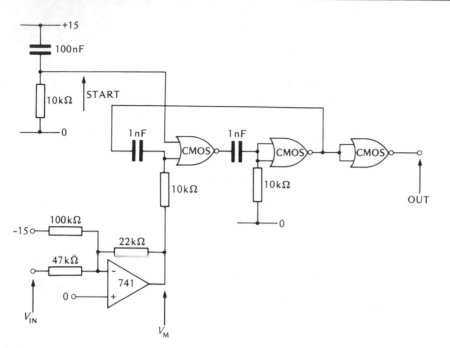

Figure 13.15 A PPM modulator

Self-checking codes

Before we leave carrier keying for other fields we would like to discuss the use of **redundant information** to ensure error-free transmission of data. Techniques which are similar to the ones that we are going to describe allow vast quantities of information to be shuttled around the Earth without degradation. They also enable satellites which are deep in space to give us a close look at the planets of our solar system despite the tiny size of the signals which the satellites transmit back to Earth.

The most precise way to transfer an analogue signal is to convert it into a binary word first. That word is then transmitted in serial form, one bit at a time. If the serial form of the binary word is used to key a carrier (or a pair of carriers if you use FSK) then the signal can be cleaned up before demodulation. Nevertheless, if the noise added to the carrier is particularly severe, some noise may get past the Schmitt trigger and cause the serial-to-parallel converter to assemble the wrong binary word. In order for the receiver to recognise that this has happened, it is customary to end each serial transmission of a word with a **parity bit**. If you look at figure 13.17

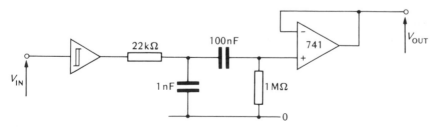

Figure 13.16 A PPM demodulator

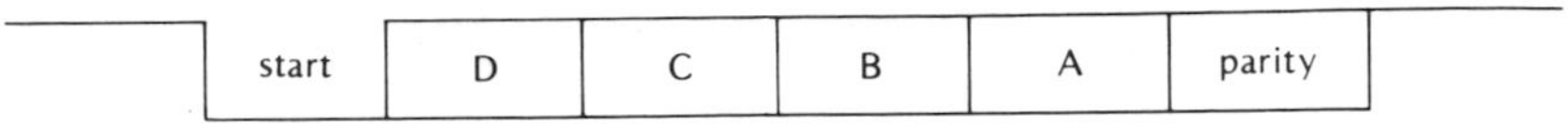

	start	D	C	B	A	parity

Figure 13.17 A nibble in serial form with start and parity bits

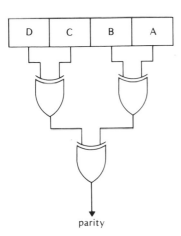

Figure 13.18 An odd parity generator

you will see the format of a typical serial nibble. The start bit comes first, followed by the four bits of the nibble and ending with the parity bit. The state of a parity bit is decided by the number of 1's in the nibble; if that number is odd, then the parity bit will be set. (A circuit which generates this odd parity signal is shown in figure 13.18.) Now if the receiver checks that the parity of each word matches its parity bit then it can tell if it has received the word correctly. Since a well designed system should be virtually error-free anyway, the chances of two bits of the word going wrong in transmission are fairly slim, so if the parity matches the parity bit all is going to be well.

Suppose that the receiver does detect an error. What can it do about it? If you transmit each word twice then the receiver can have two goes at decoding it. So if the first word of the pair is correctly transmitted the second one is ignored. When an error is detected in the first word, the second word is read in and, if it is error-free, used instead of the first one. Of course this strategy does mean that twice as much information is transferred through the link, taking twice as long, but that is the price you have to pay for ensuring no loss in transmission.

EXPERIMENT

Investigating a PPM system

Have a go at getting the PPM system of figures 13.15 and 13.16 to work. Run the CMOS gates off $+15\,$V and $0\,$V. Use an audio signal (e.g. 1 V peak to peak at 1 kHz) to modulate the carrier. Evaluate the performance of the system and try to improve it.

PROBLEMS

1 Explain how the circuit of figure 13.10 works. A few sketches of waveforms at various places in the circuit may help your argument along. Work out the minimum size of AC signal that the system will detect and in what frequency range it has to lie.

2 Explain how the FSK demodulator of figure 13.13 works. Draw timing diagrams for the system at each of its two frequencies.

3 An alternative method of modulating an analogue signal onto a keyed carrier is known as pulse width modulation, or PWM. The carrier is pulsed on at fixed intervals of time, but the duration of each pulse (its width) is dictated by the value of the analogue signal. Draw the circuit diagram of a suitable PWM modulator, with the pulses emerging once every 10 µs. Explain how the system works.

4 Draw the circuit diagram for a PWM demodulator, assuming that the signal coded onto the carrier lies in the audio frequency range.

FREQUENCY MULTIPLEXING

So far we have only considered the case of one carrier going down one communication link. By using a number of carriers, all at different frequencies, it is possible to send many signals down a single link at the same time. This technique is known as **frequency multiplexing**. Figure 13.19 illustrates how three keyed carriers can be multiplexed onto a single link and demultiplexed at the other end. A, B and C are digital signals which control three astables; each astable runs at a different frequency. The output of each astable is cleaned up by feeding it through a filter which converts the square wave into a sine wave at the same frequency (look at figure 13.20). All three carriers are sent down the link simultaneously in the form of three sine waves of frequencies f_0, f_1 and f_2. At the end of the link three filters guide each carrier towards one of the AC-to-DC converters. So if only C is logical 0, a sine wave of frequency f_2 is sent down the link and through the f_2 pass filter to activate the top AC-to-DC unit on the right.

The frequency spectrum shown in figure 13.21 shows how the multiplexing works. Each filter will only transmit a narrow range of frequencies.

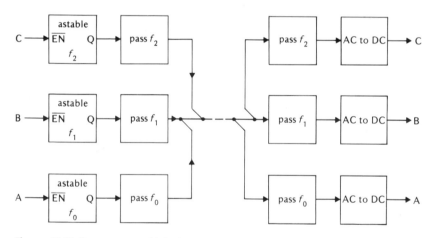

Figure 13.19 Frequency multiplexing

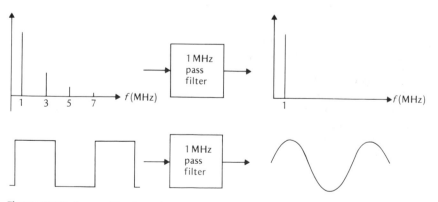

Figure 13.20 A pass filter in action

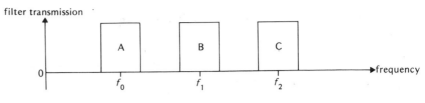

Figure 13.21

Provided that the filters do not transmit any of the frequencies which pass through the other filters the carriers at f_0, f_1 and f_2 will be correctly routed when they arrive at the receiver. The three frequencies can go down the link independently so there will be no crosstalk between the signals. A, B and C will be independently and simultaneously transmitted down the communication link.

Tuned circuits

It should be clear that successful frequency multiplexing places fairly stringent demands on the filters that are used. They must only pass a limited range of frequencies and must totally reject all of the others. If the filters are not selective enough there will be crosstalk between the various signals being transmitted. So before we do any more work on frequency multiplexing, it would be wise to explore the properties of filters which can be built with today's technology.

Pass filters with suitable **selectivity** are almost invariably built around **tuned circuits**. These systems are assembled from passive components (i.e. they need no power supply) in such a way that they only respond to a narrow range of frequencies. This behaviour is summarised in figure 13.22. Although the signal going in has a wide range of frequencies the output consists of a single frequency. The tuned circuit rejects all of the frequency components of the signal except those grouped around the **resonant frequency** of the circuit. As you will find out, both the resonant frequency and the spread of frequencies transmitted depends on the values of the components which go to make up the tuned circuit. Obviously a tuned circuit has all the characteristics of a pass filter like the one shown in figure 13.20.

A tuned circuit contains three components. A capacitor, a resistor and an inductor. You know about the first two of these, but you need to be introduced to the third. The construction of an **inductor** is simplicity itself; it is no more than a length of wire wound into a coil around a piece of magnetic material. An ideal inductor has no resistance at all and in practice its resistance will be small, a few ohms perhaps. The reactance of an inductor is proportional to the frequency of the current which flows through it.

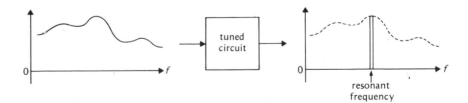

Figure 13.22 The transmission of a tuned circuit

That is, the voltage across the inductor depends on the current and its frequency. More precisely:

$$\frac{V}{I} = X = 2\pi f L$$

L is the inductance of the inductor. If L is measured in μH (microhenries) and the frequency f is in kilohertz then X will be in kilohms. (The standard units are henries, hertz and ohms) So an inductor behaves like a resistor whose value rises with the frequency of the current flowing through it. Figure 13.23 illustrates how the reactance of a 220 μH inductor varies with frequency. It also shows the circuit symbol of an inductor.

Series *LC* network

Suppose that you put an inductor in series with a capacitor, as shown in figure 13.24. How is the impedance of this network going to vary as the frequency is changed? At very low frequencies the inductor is going to look like a short circuit, but the capacitor will behave like an open circuit. So we can expect the low frequency impedance to be high. Similarly at high frequencies the impedance will also be high; this time it is the inductor which looks like an open circuit. The network behaves like an inductor at high frequencies and like a capacitor at low frequencies.

We can write down an approximate expression for the impedance of the network. It is going to be approximate because it neglects the phase shifts introduced by the inductor and the capacitor; a more complete treatment of the problem is set out in Appendix A. It also neglects the resistance of the inductor.

$$Z = 2\pi f L + \frac{1}{2\pi f C}$$

A glance at the graph of figure 13.24 should convince you that the

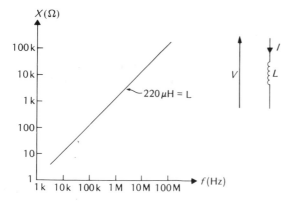

Figure 13.23 The reactance of a 220 μH inductor as a function of frequency

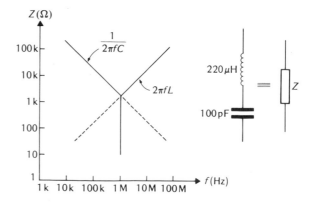

Figure 13.24 A series *LC* network

minimum value of Z ought to occur when the reactances of the inductor and the capacitor are equal. This will happen at a frequency f_0 given by:

$$2\pi f_0 L = \frac{1}{2\pi f_0 C}$$

therefore $\quad f_0 = \dfrac{1}{2\pi\sqrt{LC}}$

We would expect the impedance to be given by:

$$Z = 2 \times (2\pi f_0 L) = 4 \left(\frac{1}{2\sqrt{LC}}\right) L = 2\sqrt{\frac{L}{C}}$$

In practice, the impedance suddenly dives to a very low value at this frequency. Our expression for the impedance takes no account of the phase shifts which make the capacitor and inductor cancel each other out. At the resonant frequency f_0 the impedance of the series LC network almost vanishes completely.

The graph of figure 13.24 shows how the impedance of $220\,\mu H$ in series with $100\,pF$ depends on frequency. The resonant frequency is given by the formula:

$$f_0 = \frac{1}{2\pi\sqrt{LC}} = \frac{1}{2\pi \times \sqrt{220 \times 10^{-6} \times 100 \times 10^{-12}}} = 10^6$$

So at 1 MHz the impedance suddenly drops from about 1 kilohm to 10 ohms, the resistance of the inductor. Below and above the resonant frequency the network has an impedance of at least 1 kilohm, increasing steadily as the frequency gets further away from 1 MHz.

Figure 13.25 shows how the addition of a resistor converts the network into a filter which will only transmit signals at one frequency. Provided that R is chosen so that it is much larger than the resistance of the inductor and

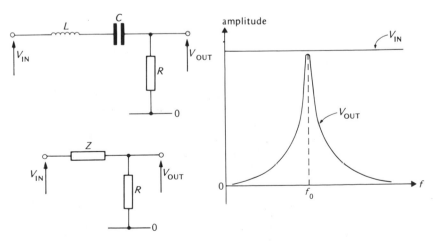

Figure 13.25 A tuned circuit based on a series LC network

also much smaller than $2(L/C)^{1/2}$, then the circuit will have the transfer characteristics shown in figure 13.25. The series *LC* network and the resistor behave like a voltage divider. So V_{OUT} can only equal V_{IN} when Z is very much smaller than R. When Z is much larger than R, V_{OUT} will be zero. Since Z is only smaller than R at the resonant frequency, only signals at or near that frequency are transmitted by the system.

The bandwidth of the filter is dictated by the **Q** of the inductor. Signals which have frequencies between $f_0(1 + 1/2Q)$ and $f_0(1 - 1/2Q)$ will be attenuated by less than 3 dB by the filter. This is a useful definition of the **bandwidth** of the filter, i.e. the range of frequencies which it will transmit. The bandwidth is therefore f_0/Q.

We will now look at a practical example. Suppose that we want to incorporate our series network of 220 µH and 100 pF into a 1 MHz filter. We must obviously choose a value of 100 Ω for the resistor, so that it is greater than the resistance of the inductor (10 Ω) and less than the impedance of the network close to resonance (about 1 kΩ). The Q of the inductor might well be 100 at a frequency of 1 MHz, so the bandwidth of the filter would be 1000/100 = 10 kHz. It would transmit signals within 5 kHz of 1 MHz without appreciable attenuation.

The Q of an inductor is a useful parameter. Its value is given by the formula:

$$Q = \frac{2\pi f_0 L}{R_0}$$

where R_0 is the resistance of the inductor. The Q of an inductor is usually specified at a particular frequency, so you have to calculate what it will be at any other frequency. Since a large Q results in a small bandwidth when the inductor is part of a tuned circuit, it obviously pays to use a large inductor at all times. Furthermore, narrow bandwidths are going to be easier to generate at high frequencies than at low ones. Indeed, tuned circuits tend to be fairly useless as narrow bandwidth filters much below 1 MHz.

Parallel *LC* network

If you put an inductor in series with a capacitor you have a network whose impedance suddenly drops at a particular frequency. Conversely, if you put the inductor in parallel with the capacitor the impedance will shoot up dramatically at one particular frequency. The parallel *LC* network has exactly the opposite behaviour of the series *LC* network. The detailed algebra is laid out in Appendix A.

Look at the graph shown in figure 13.26. At low frequencies the inductor will have a much lower reactance than the capacitor, so the whole network behaves like the inductor ($2\pi fL \ll 1/2\pi fC$ at low frequency.) On the other hand, at high frequencies the inductor is going to have a very large reactance and the capacitor a very small reactance. So all of the current will flow through the capacitor and we can forget that the inductor is there. But when the two reactances are the same the phase shifts in the network push the impedance up to a value of $Q^2 R_0$. This is usually much larger than the expected value of $\frac{1}{2}(2\pi f_0 L)$ or $\frac{1}{2}(1/2\pi f_0 C)$, about 0.5 kΩ for 220 µH in parallel with 100 pF. If we assume that the values of R_0 and Q are still 10 Ω and 100 for the inductor, the impedance will suddenly shoot up to 100 kΩ at 1 MHz.

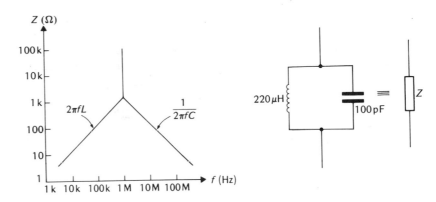

Figure 13.26 A parallel *LC* network

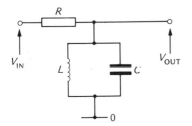

Figure 13.27 A tuned circuit based on a parallel *LC* network

Figure 13.27 shows how this behaviour can be exploited to make a pass filter for a frequency of $1/2\pi\sqrt{LC}$. R has to be chosen so that it is much larger than the impedance of the network away from the resonant frequency and yet much smaller than its impedance at the resonant frequency itself. If the *LC* network is the one shown in figure 13.26, then you can see that 10 kilohms would be a good choice for R.

Routing frequencies

You will have noticed that the two types of tuned circuit have similar behaviour. Given the same values of L and C, both the series and parallel networks have the same resonant frequency. So which one do you use in practice?

It all depends on which end of the communication link you place the filter. At the transmitting end you want the filter to have a low output impedance (you will remember that it is always best to feed a signal out of a low output impedance). The series *LC* arrangement has a very low impedance at its resonant frequency, so you put the filter based on it at the transmitting end, as shown in figure 13.28.

The filter based on the parallel *LC* network presents a very high impedance at its resonant frequency, so it is used at the receiving end of the link. The link is then driven out of a low impedance (the series *LC* network) into a high impedance (the parallel *LC* network).

If you study the circuit shown in figure 13.29 (overleaf), you should appreciate how tuned circuits can be used to route different frequencies different ways in a circuit. For example, consider the signal at frequency f_1 which enters at the top left hand of the diagram. It sees L_1 and C_1 in series as a short circuit if $f_1 = 1/2\pi\sqrt{L_1 C_1}$, so it goes straight onto the link and

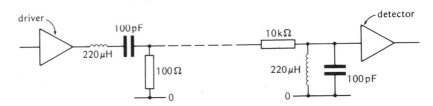

Figure 13.28 Tuned circuits in a transmission system

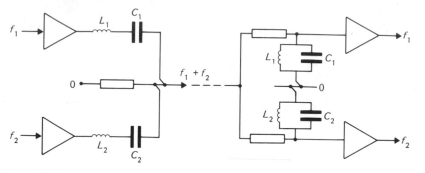

Figure 13.29

sets off towards the receiver. When it gets there it will see L_2 in parallel with C_2 as a short circuit, so no signal at f_1 gets through to the amplifier at the bottom right hand corner. Because the signal sees L_1 in parallel with C_1 as a large resistance the top right hand amplifier sees almost all of it.

EXPERIMENT
*Assembling a system which transmits a digital signal
down a wire link as a sine wave*

Start off by assembling the 1 MHz square wave oscillator shown in figure 13.30. Put it at the far left of your breadboard. Check that it oscillates when the switch is pressed.

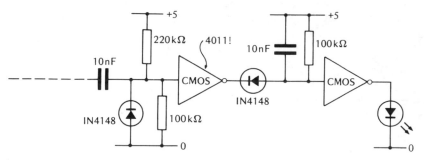

Figure 13.30

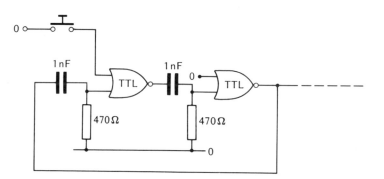

Figure 13.31

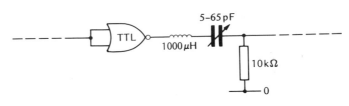

Figure 13.32

Then construct the AC detector shown in figure 13.31, putting it at the far right-hand end of the breadboard. Link the two systems that you have built and press the switch; the LED should light up. If it does not work it is possible that your CMOS IC cannot cope with 1 MHz signals. The gates on a 4011 IC appear to cope with high frequency signals better than those on a 4069 IC. If necessary, make your oscillator run at 500 kHz by replacing the 470 Ω resistors with 1 kΩ ones.

Use a CRO to look at the inputs and outputs of both gates in the receiver circuit when the oscillator is on and off. Sketch the waveforms that you see, and use them to explain how the receiver works.

Convert the output of the astable to a sine wave by inserting the filter shown in figure 13.32 in front of it. You will have to tweak the capacitor until the maximum sine wave emerges from the filter. When you have got the resonant frequency of the *LC* network matched with the frequency of the astable, only the 1 MHz component of the square wave comes through the filter. Link the output of the filter to the receiver circuit and check that the system still works.

Now assemble the system shown in figure 13.33. There are two astables, one operating at 500 kHz and the other at 1 MHz. They feed their signals into a common transmission line via a pair of tuned circuits. If you get the resonant frequency of the filter matched to either 500 kHz or 1 MHz, by tweaking the capacitors (you need two to get the frequency range), you should be able to get the receiver to respond to either signal A or signal B but not both.

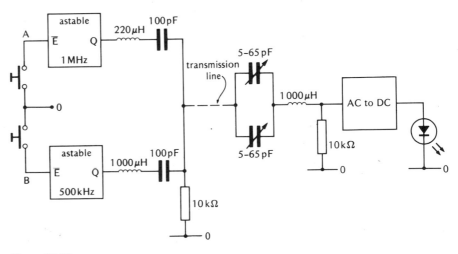

Figure 13.33

BANDWIDTH

Frequency multiplexing allows us to send several streams of information down a single link. What is the limit to the number of streams that we can have? Well, it depends on two things. The bandwidth of the filters which are used to select the carriers at the receiving end and the rate at which the carriers are modulated. The limit imposed by the filters is easy to work out; if the filters have bandwidths of 10 kHz, then the carriers have to be spaced apart by at least 10 kHz if there is to be no crosstalk between channels. The limit imposed by the rate of carrier modulation is more subtle; we shall need some algebra to help explain it.

Sidebands

Suppose that a carrier of frequency f_c is keyed on and off by a square wave of frequency f_m. The result is a chopped sine wave as shown in figure 13.34. We are going to obtain an approximate expression for its frequency spectrum; this will show that as the carrier is keyed at a faster rate, the resultant signal spreads over a wider range of frequencies.

The square wave which keys the carrier can be represented as a sum of sine waves:

$$M = 1 + \frac{4}{\pi}\left[\cos(2\pi f_m t) - \frac{\cos(2\pi 3 f_m t)}{3} + \cdots\right]$$

The keying of the carrier is equivalent to multiplying the carrier by the square wave M. So the keyed carrier is given by the expression:

$$\cos(2\pi f_c t)\left[1 + \frac{4}{\pi}\cos(2\pi f_m t) - \frac{4}{3\pi}\cos(2\pi 3 f_m t) \cdots\right]$$

Multiplying this out, we are going to get an enormous number of terms, as we need an infinite number of sine waves to represent a square wave. Here are the first few of them:

$$\cos(2\pi f_c t) + \frac{4}{\pi}\cos(2\pi f_c t)\cos(2\pi f_m t)$$

$$-\frac{4}{3\pi}\cos(2\pi f_c t)\cos(2\pi 3 f_m t)$$

$$+\frac{4}{5\pi}\cos(2\pi f_c t)\cos(2\pi 5 f_m t) - \cdots$$

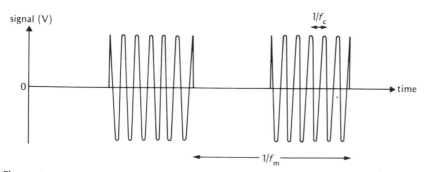

signal (V)

Figure 13.34 A carrier of frequency f_c keyed with a square wave of frequency f_m

Using the identity

$$\cos(x)\cos(y) = \tfrac{1}{2}\cos(x+y) + \tfrac{1}{2}\cos(x-y)$$

we can replace all the products of cosines with sums of cosines. Our expression for the keyed carrier now becomes:

$$\cos(2\pi f_c t) + \frac{2}{\pi}\cos(2\pi(f_c + f_m)t) + \frac{2}{\pi}\cos(2\pi(f_c - f_m)t)$$

$$-\frac{2}{3\pi}\cos(2\pi(f_c + 3f_m)t) - \frac{2}{3\pi}\cos(2\pi(f_c - 3f_m)t)$$

$$+\frac{2}{5\pi}\cos(2\pi(f_c + 5f_m)t) + \frac{2}{5\pi}\cos(2\pi(f_c - 5f_m)t) + \cdots$$

The frequency spectrum of the keyed carrier is now clear. If you look at figure 13.35 you will see that it consists of the carrier at f_c, with a number of smaller spikes (called **sidebands**) at $f_c \pm f_m$, $f_c \pm 3f_m$, $f_c \pm 5f_m$ etc. Although the number of sidebands is infinite, only a few of them contain much energy. As the energy carried by a signal is given by the square of its amplitude $(W = V^2/R?)$ we need only worry about the sidebands that are well above 10% of the carrier amplitude; they will have at least 1% of the energy of the carrier. So we are going to forget the existence of the sidebands which lie more than $3f_m$ from the carrier.

Where has this algebra led us? We have found that when the carrier is keyed at a rate f_m (a quantity we shall call the **modulation frequency**) its frequency spectrum spreads out to at least $3f_m$ on either side of f_c. Furthermore, the keyed carrier will need a bandwidth of at least $2f_m$ if it is to transfer the information without fault. This is an important conclusion, with far-reaching practical consequences. Here is an example.

Telephonic bit rates

Imagine that you are going to use a standard FSK modulation scheme to send streams of digital data down a telephone link. How fast can you transmit data with this sort of system?

Consider the worst case, that you are sending a stream of alternate 1's and 0's down the link. In practice, of course, you would use the standard serial format with a start bit and a parity bit sandwiching a serial byte between them. But if we work out the maximum bit rate for the worst case, then we will certainly be safe in practice. This will become clear later on.

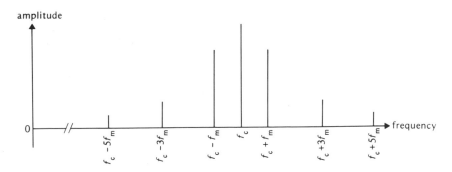

Figure 13.35 The frequency spectrum of a sine wave keyed by a square wave

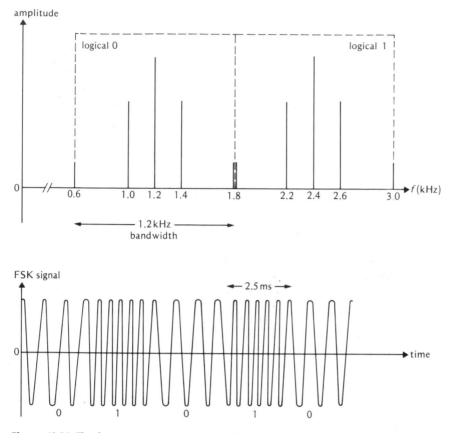

Figure 13.36 The frequency spectrum of an FSK carrier

Anyway, the stream of 1's and 0's will cause the two carriers (2.4 kHz and 1.2 kHz) to be alternately fed down the link. Each carrier will be modulated by a square wave (on, off, on, etc.) so its frequency spectrum will acquire sidebands above and below the carrier frequency.

We dare not let the sidebands of one carrier overlap with those of the other. If we did that the receiver could well get 1's and 0's confused. For instance, a strong sideband of the 1.2 kHz carrier which had a frequency near to 2.4 kHz could have disastrous consequences! So, since the carriers are 1.2 kHz apart, we must not allow either of them to spread more than 0.6 kHz away from their nominal frequency. The modulation frequency (f_m) is therefore limited to less than 0.2 kHz for a stream of 1's and 0's. That is, a 1 or 0 must be transmitted for at least 2.5 ms; look at figure 13.36 if you are not sure about this.

We can use this minimum bit length to work out the maximum bit rate which can be achieved in practice. A logical 0 or 1 must exist for at least 2.5 ms. A serial byte will require at least 10 lots of 2.5 ms, 8 lots for the byte, 1 for the start bit and 1 for a 1 in front of the start bit. So a byte will take at least 25 ms to transmit. This is equivalent to 40 bytes per second i.e. 320 bits per second (not counting start and stop bits).

Channel number

We are now in a position to work out the maximum number of channels which can be independently sent down a single link. Consider the example of a wire link which can cope with signals in the range 10 MHz to 100 MHz. How many channels can you fit into it? It all depends on the type of modulation being used and the modulation technique being employed; you will be seeing several examples later on. For the present example we shall assume that each channel uses a keyed carrier to transmit digital data at a rate of 100 bits per second. Each bit will therefore be transmitted for 10 ms. If we consider a worst case of a stream of alternate 1's and 0's, each carrier will be modulated by a 50 Hz square wave.(1 for 10 ms, 0 for 10 ms, 1 for 10 ms etc.) So each carrier will acquire appreciable sidebands up to 150 Hz on either side of it; each carrier will require a bandwidth of at least 300 Hz. If we space the carriers at intervals of 500 Hz, leaving a gap of 200 Hz between each carrier's territory, we can fit 180 000 of them between 10 MHz and 100 MHz!

PROBLEMS

1 For each of the tuned circuits shown in figure 13.37

   a) calculate the resonant frequency

   b) describe, as fully as possible, what sort of filtering action it provides.

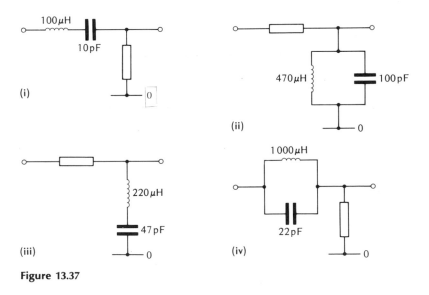

Figure 13.37

2 Suppose that you want to transmit serial digital data down a telephone link. The telephone system will handle audio signals whose frequencies are between 300 Hz and 3.4 kHz. You want to get the data down the telephone line as fast as possible using a keyed carrier. Explain what carrier frequency you would employ and calculate the maximum rate at which bytes could be transmitted in standard serial form.

3 A high quality cassette tape can record signals up to 12 kHz without too much difficulty. If you wanted to store bytes of data on such a tape using FSK techniques so that they could be dumped on the tape and read off it at maximum speed, what two carrier frequencies would you use? What would the maximum byte rate be?

4 An ultrasonic transmission link is used to transfer digital data from a hand-held control module to a TV receiver. The control module sends out binary words which control the action of the TV. So one word changes the channel, another changes the volume and so on. The ultrasonic emitter and detector only work at 40 kHz $\pm$ 4 kHz. They act like tuned circuits, with resonant frequencies of 40 kHz and bandwidths of 8 kHz.

a) Draw the frequency spectrum of the carrier when it is keyed on and off by a 2 kHz square wave.

b) What is the minimum time that it would take the system to transfer a byte from the control module to the receiver? Assume that a keyed carrier technique is being used.

c) How would you arrange to use an FSK carrier technique with the system? What carrier frequencies would you choose? What would the maximum byte rate be?

5 A telephone link is used to transmit a photograph by the following method. The photo is scanned by a photodiode which can only see an area which is 0.1 mm square. The photodiode measures the amount of light coming off the square and transmits a 1 if it is bright and a 0 if it is dark. That signal is used to key a carrier at 1.5 kHz; the keyed carrier is then sent down the line to the receiver. There it is decoded and the digital signal is used to fire a burst of light at 0.1 mm squares on a sheet of photo-sensitive paper. By scanning the light source over the paper in the same way that the photodiode scans the photograph, a facsimile of the photograph will be built up on the paper. A telephone link can handle signals between 300 Hz and 3.4 kHz.

a) What is the maximum rate at which bits can be sent down the telephone link in this fashion?

b) How long would it take to transmit a 9 cm $\times$ 13 cm photograph by this method?

c) If you had to transmit a photograph of this size in just three minutes, what resolution would you have to put up with? In other words, how big would the smallest detail have to be? (The resolution in (b) was 0.1 mm.)

6 A standard TV screen which is controlled by a computer has 625 lines painted on it by an electron beam. Each line is cut into 833 **pixels**; the computer can make each pixel light or dark as the electron beam sweeps past it. The electron beam scans rapidly across the screen from left to right before flying back to the left, rather like the spot on a CRO screen. At the same time the beam is slowly moved down the screen so that it scans the lines one after the other. The electron beam scans all 625 lines (i.e. gets from the top to the bottom of the screen) in 40 milliseconds.

a) If a logical 1 turns a pixel white and a logical 0 leaves it dark, what is the rate at which the computer has to transmit bits to the TV?

b) If those bits are transmitted by keying a 100 MHz carrier and feeding it down a coaxial cable, what sort of bandwidth will that keyed carrier take up?

c) Suppose that you wanted to use this sort of system to send TV pictures via a telephone link. The computer has a block of memory which it fills with 1's and 0's. It then treats that memory like a look-up table, feeding it out in serial form to the TV 25 times a second. As bits arrive down the telephone link they are fed into memory to replace the bits already there. Estimate, giving reasons, how long it would take to fill the memory with 1's and 0's by this method.

d) In practice, the computer will carve the TV screen into a grid which is 128 units wide and 96 units deep; each pixel is then one unit square. How long would it take to paint a new picture on the screen by this method if the bits were sent down a telephone line?

e) When the TV screen is to be painted with numbers and characters, the information is sent down the telephone link in bytes. Each of the 256 bytes, which are possible, corresponds to a different character. The screen is carved up into cells, and one cell can hold one character; there are 32 cells across the screen and 16 down it. How long would it take to fill the screen with characters if the bytes were transmitted down a telephone link using standard FSK?

AMPLITUDE MODULATION

The only technique for modulating analogue signals onto a carrier which we have discussed is pulse position modulation, or PPM. Its big disadvantage is that it does not easily lend itself to frequency multiplexing many analogue signals onto a single communication link. Each PPM carrier takes up too much room in bandwidth terms. For example, suppose that you wish to transmit the full range of audio frequencies i.e. up to 16 kHz. The pulse frequency must be at least twice this value (look at figure 13.14), so the carrier will be keyed on and off at approximately 32 kHz. The bandwidth needed to transmit more or less faithfully that modulated carrier must be at least six times that frequency, i.e. 192 kHz, just under 200 kHz. This is no great problem if you only want to send one signal down a link, provided you can send a suitable carrier down it. But if you want to send more than one signal down the line, then the carriers will have to be spaced at 200 kHz intervals.

The technique of amplitude modulation can be used to transmit a 16 kHz signal with a bandwidth of only 32 kHz. So you can transmit six amplitude modulated carriers down a link for each PPM one. This economy of use of available frequencies is particularly important when radio waves are the communication link. We are now going to describe what amplitude modulation is and how it can be put into practice; the subject of radio receivers will be left to the end of the chapter, after we have discussed the other important analogue modulation technique, frequency modulation.

AM MODULATORS

Take a look at figure 13.38 (overleaf). The box in the middle modulates the signal onto the carrier to produce an **amplitude modulated (AM) carrier**. As you can see, the size of the AM carrier wobbles up and down with the signal. Expressed in algebraic terms, the AM carrier is given by:

$$(1 + m\cos(2\pi f_m t))\cos(2\pi f_c t)$$

f_c and f_m are the frequencies of the carrier and the signal respectively. The amplitude of the signal is contained in m, the **modulation index**. You can

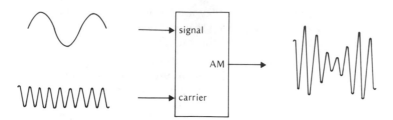

Figure 13.38 Amplitude modulation

think of the modulation index as being equal to the signal amplitude divided by the carrier amplitude. If we multiply out the expression and use our standard identity to replace the product of two cosines with the sum of two other cosines, we get:

$$\cos(2\pi f_c\, t) + \tfrac{1}{2}m \cos(2\pi(f_c + f_m)t) + \tfrac{1}{2}m \cos(2\pi(f_c - f_m)t)$$

The frequency spectrum of the AM carrier is shown in figure 13.39. It contains just three spikes. The ones at $f_c + f_m$ and $f_c - f_m$ (the sidebands) contain the information about the signal; each is proportional to m. The central spike at f_c is the carrier. It contains no information at all (it has a constant amplitude regardless of the signal). It is clear that the AM carrier will need a bandwidth of $2f_m$ when it is transmitted down a link.

Amplitude modulators are quite easy to build. The usual technique is to use the signal to wobble the supply rails of an oscillator up and down. Provided that the frequency of the oscillator is independent of its supply voltage, the result will be a sine wave with a varying amplitude. Figure 13.40 shows a system which can be used to amplitude modulate audio signals onto a 1 MHz carrier. The audio signal is fed into an AC follower, so that its DC level is raised from 0 V to +4 V. The follower (the BC441 in emitter follower mode) drives the modulated supply rail up and down. Since the size of the signal fed out by the oscillator is a function of its supply voltage, the amplitude of the 1 MHz carrier will vary according to the amplitude of the audio signal. Finally, a bass cut filter (100 pF and 47 kΩ) filters out any of the audio signal added to the carrier by this process.

An oscillator which could be used to generate the 1 MHz carrier is shown in figure 13.41. It is usually called a Colpitts oscillator, and relies on a tuned circuit to fix the frequency of oscillation. The transistor is used to fire bursts of current into the parallel *LC* network attached to its collector. Remembering that the *LC* network will have a small impedance at all frequencies

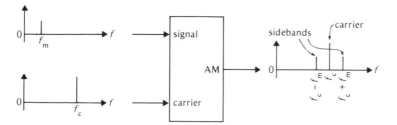

Figure 13.39 The frequency spectrum of an AM carrier

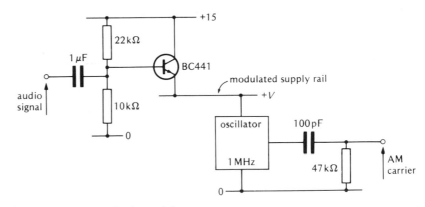

Figure 13.40 An amplitude modulator

except one, the voltage across it will only be significant when the current bursts arrive at its resonant frequency. The feedback from the tuned circuit to the emitter ensures that this is the case i.e. once the collector voltage starts to wobble up and down at 1 MHz, the emitter is only pushed 0.7 V below the base when the collector gets to the bottom of its waveform. (Do not worry if you have difficulty in working out how the circuit works. It is only one out of many possible oscillators which are built around tuned circuits. They are all, to a certain extent, magic; you optimise their performance by trial and error rather than with a calculator.) The 1 nF capacitor in the tuned circuit will always have a reactance which is ten times smaller than that of the 100 pF capacitor. So when we calculate the resonant frequency we can, as a first approximation, forget that the 1 nF capacitor is there at all. The circuit will therefore ring at a frequency of roughly $1/2\pi(220 \times 10^{-6} \times 100 \times 10^{-12})^{1/2} = 1$ MHz.

The behaviour of an amplitude modulator which incorporates the circuits of figures 13.40 and 13.41 is summarised in the graphs of figure 13.42 (overleaf). Notice how the bass cut filter removes the low frequency signals present in the oscillator output, leaving a carrier and two sidebands around 1 MHz.

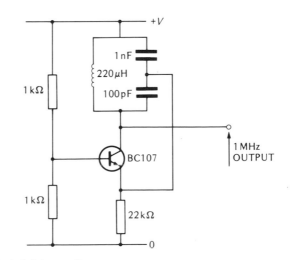

Figure 13.41 A Colpitts oscillator

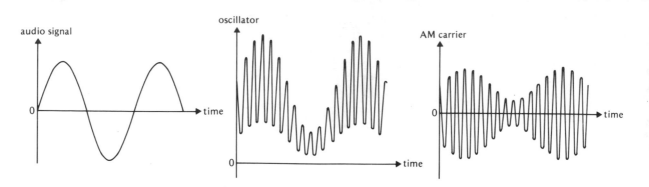

Figure 13.42 Modulating the amplitude of a carrier

AM demodulation

A simple AM demodulator is illustrated in figure 13.43. It is no more than a smoothed half-wave rectifier. Every time that the AM carrier goes positive, charge flows through the diode onto the capacitor. While the AM carrier is negative the capacitor discharges through the resistor. A little thought should convince you that V_{OUT} will be equal to the peak value of V_{IN} provided that we neglect the voltage drop across the diode and that RC is much longer than one cycle of V_{IN}. It is standard practice to use germanium point-contact diodes in AM demodulators. They have a negligible voltage drop when forward biased, and a very small capacitance. The latter property is very important when high frequency carriers are used; the capacitance between anode and cathode would act like a short-circuit across the diode.

How do we choose the values of the resistors and capacitors in the circuit? It pays to use a large resistor because the demodulator is usually fed from a tuned circuit acting as a filter. If you draw too much current from such a filter you degrade its performance; this will become evident when you look at radio receivers. So we have chosen to use 100 kilohms.

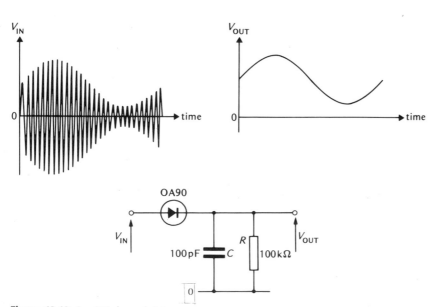

Figure 13.43 An AM demodulator

Now if V_{OUT} is to follow the peak value of V_{IN} then RC must not be too long; V_{OUT} will always rise as fast as V_{IN} does, but it can only fall at a rate dictated by the value of RC. On the other hand, if RC is too short, there will be a lot of ripple on V_{OUT}. So we have to choose a capacitor such that the value of RC is neither too long nor too short. Since the time between the peaks of V_{IN} is $1/f_c$ and the time between peaks of the audio signal is $1/f_m$, we can express this condition of RC in algebraic terms:

$$1/f_c < RC < 1/f_m$$

If $RC > 1/f_c$ then $100 \times C > 1/1000$. So $C > 1/100\,000\,\mu F = 10\,pF$. (We have assumed that $f_c = 1\,MHz$.) Similarly, if $RC < 1/f_m$ then $100 \times C < 1/16$. So $C < 1/1600\,\mu F \simeq 10\,nF$, assuming that f_m is $16\,kHz$ i.e. the highest frequency of the audio spectrum. C must therefore lie between $10\,pF$ and $10\,nF$; our value of $100\,pF$ will do.

It follows that the value of the capacitor used in the demodulator depends on both the carrier frequency and the range of signal frequencies which will be modulated on to it. It also follows that its value is not particularly critical, provided that it falls well inside the limits imposed by those two frequencies.

The output of the circuit of figure 13.43 contains a DC signal which was not present in the original signal. A simple bass cut filter will remove this. Figure 13.44 shows a suitable circuit; the op-amp follower ensures that any subsequent audio amplifying stages will not draw too much current from the demodulator.

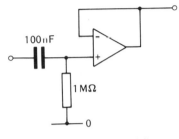

Figure 13.44 An op-amp follower with a bass cut filter

100 nF   1 MΩ   0

EXPERIMENT

Assembling an AM transmission system and evaluating its performance

Start off by building the oscillator shown in figure 13.41. Alter the inductor so that it oscillates at $500\,kHz$ or thereabouts, and alter the value of the emitter resistor until you get a good, stable sine wave. Make $V = 15\,V$ to start with.

Then tack on the rest of the modulator shown in figure 13.40. Feed in an audio signal (a few kHz) and check that this moves the modulated supply rail up and down. Then look at the modulated carrier and note what happens to it as you change the amplitude and frequency of the audio signal.

Finally, assemble the demodulator system shown in figures 13.43 and 13.44. Check that the output of the whole system is a good copy of the audio signal fed into it.

Make the audio signal a $1\,V$ peak value sine wave at $1\,kHz$ and trigger a CRO on it. Look at the waveforms on the modulated supply rail, the AM carrier, the demodulator output and the final output. Draw them one under the other and to the same scale, with the audio input signal.

Frequency multiplexing with AM

AM techniques are generally employed for transmitting audio signals. Although the full audio range is $16\,Hz$ to $16\,kHz$, you can get an acceptable signal transmitted if you ignore the frequencies above $5\,kHz$. So most of the information in an audio signal can be transmitted in a bandwidth of $10\,kHz$ if it is amplitude modulated onto a carrier. You would therefore expect to be able to feed AM carriers onto a single communication link spaced out by as little as $10\,kHz$. The method of putting all of the signals onto one line

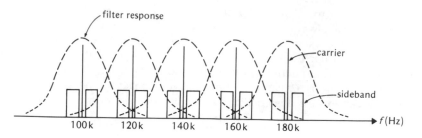

Figure 13.45 Frequency multiplexing AM carriers

and separating them out at the receiving end is no different from the one used in figure 13.29. You use a series of tuned circuits, with their resonant frequencies matched to one of the carriers, to route each AM carrier to its demodulator.

Unfortunately, the need for each filter to pass the sidebands as well as the carrier places limitations on how closely the carriers can be spaced. This is illustrated in figure 13.45. It shows the frequency spectrum of five carriers which have been amplitude modulated with audio signals between 50 Hz and 5 kHz. Since each signal will contain a number of frequencies within those limits, we have represented each sideband with a band rather than a series of lines. The carriers are 20 kHz apart, so there should be a gap of 10 kHz between the end of one set of sidebands and the start of another.

The frequency response of each of the five tuned circuits in the receiver is shown in the diagram. The resonant frequency of each filter coincides with one of the carriers, and the Q has been chosen so that the filter bandwidth is 10 kHz. So each filter passes one carrier and its sidebands with hardly any attenuation, but it also transmits some of the sidebands from adjoining carriers. As it stands, the system will have some crosstalk between channels; to reduce that crosstalk the carriers will have to be moved further apart.

Noise

By using AM techniques you can send the largest number of analogue signals down a single link. That is, it is the technique which requires the smallest bandwidth. The price which has to be paid for this is that AM signals are the ones that are most susceptible to noise and interference.

When an AM carrier goes down a link it will pick up extraneous signals. They could come from noise in the link itself or interference from outside the link. Whatever their source, those unwanted signals will be essentially random in nature, so we can represent their frequency spectrum with a band. Figure 13.46 shows the frequency spectrum of an AM carrier before

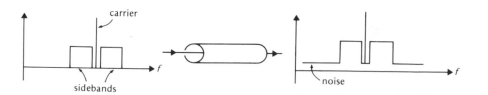

Figure 13.46 An AM carrier picking up noise in transmission

and after transmission down a length of cable. Some of the noise will have frequency components which lie within 5 kHz of the carrier. Those extra signals will pass through the filter which transmits the carrier and sidebands and reach the demodulator. There they will be treated like the sidebands and end up as noise in the audio output signal. There is no way of avoiding this other than ensuring that the amplitude of the sidebands is well above that of the noise in the link.

FREQUENCY MODULATION

By making the frequency instead of the amplitude of the carrier wobble around, you can substantially improve the noise immunity of an analogue transmission system. As you might expect, you have to pay the price of an increased bandwidth for the modulated carrier, as well as having to employ more complicated circuitry.

Figure 13.47 attempts to illustrate a frequency modulated (FM) carrier. The amplitude of the audio signal at any instant fixes the frequency of the carrier at that instant. So as the audio signal goes up in voltage, the carrier goes up in frequency. The amplitude of the carrier carries no information at all.

Frequency modulators

For reasons which will become obvious later on, most FM transmission systems employ very high frequency carriers, typically 100 MHz. Oscillators which are built around tuned circuits that have resonant frequencies in this region tend to have small value capacitors in the tuned circuit. By using a special diode, called a **varactor**, instead of the capacitor, the resonant frequency can be programmed by an electronic signal. The varactor is just a reverse biased diode; its capacitance depends on the voltage across it. By using the audio signal to wobble that voltage up and down, the frequency of oscillation will be forced to wobble up and down as well.

Figure 13.48 (overleaf) illustrates an FM modulator which operates at more accessible frequencies. It employs a CMOS astable multivibrator as a voltage-to-frequency converter; a tuned circuit converts the square wave output to a sine wave. The audio signal to be modulated onto the 150 kHz carrier drives a 741 buffer which, in turn, speeds up and slows down the astable multivibrator. The deviation of the carrier from its normal frequency

Figure 13.47 A frequency modulated carrier

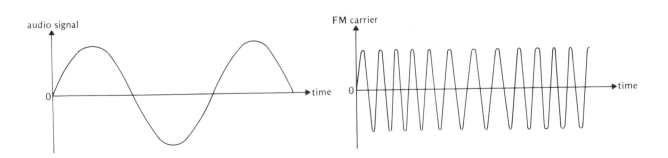

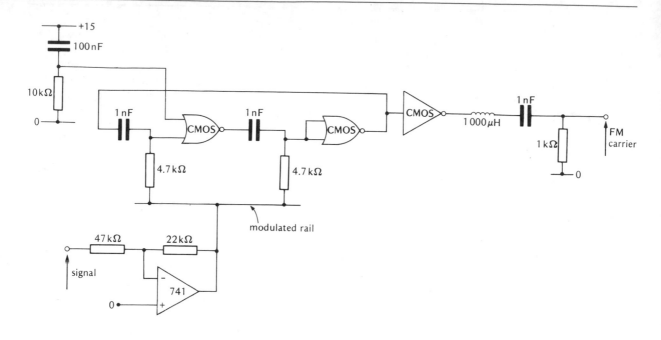

will only be proportional to the amplitude of the audio signal if the deviation is small. So you have to be careful not to overmodulate the carrier and make the system non-linear.

Figure 13.48 An FM modulator

Frequency demodulators

The system which we are about to describe is not the best way of demodulating an FM carrier. Its virtues are simplicity of operation and ease of construction. The system is shown in figure 13.49, and is usually known as a **slope detector**. The reason for this name will become clear in a moment. The FM carrier is first converted into a square wave by a Schmitt trigger, so that a constant amplitude signal is fed into the next stage. This is a tuned circuit whose resonant frequency is a bit different from the unmodulated carrier frequency. If you study the graph of figure 13.50, it should become clear that any change of frequency of the carrier going into the circuit results in a change of amplitude of the signal coming out of it. In particular,

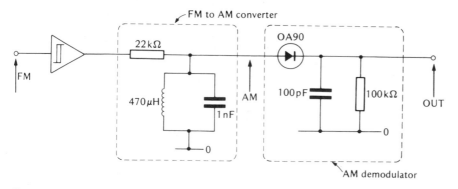

Figure 13.49 A slope detector

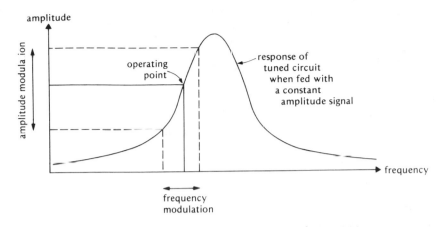

amplitude

amplitude modula ion

operating
point

response of
tuned circuit
when fed with
a constant
amplitude signal

frequency

frequency
modulation

Figure 13.50 The frequency response of the tuned circuit, figure 13.49

if the quiescent operating point is set about halfway down the response curve, any small frequency modulation is proportional to the resulting amplitude modulation. The tuned circuit acts as an FM to AM converter; the AM signal can then be demodulated with a standard half-wave rectifier.

Noise immunity

FM transmission systems are much quieter than AM ones. The source of this increased immunity to noise is the Schmitt trigger at the front end of the receiver. It converts the FM carrier, together with any noise it has picked up, into a square wave. So all information about the amplitude of the FM carrier is deliberately thrown away. If you look at figure 13.9 it should be obvious that if the FM carrier rises and falls sharply enough all of the noise is thrown away too, provided that the noise is not excessively large. In practice, of course, the FM carrier will be a sine wave. Nevertheless, we can make the carrier fall sharply from one threshold of the trigger to the other by making it sufficiently large. So as we increase the amplitude of the FM carrier the noise which gets through the Schmitt trigger drops rapidly. It is for this reason that FM systems are quieter than AM ones. If the noise is small enough the FM system can ignore it almost completely, whereas the AM one can do nothing about it!

Bandwidth

Before we leave our discussion of FM systems we need to find out the bandwidth which they require. It turns out that an FM carrier needs five times the bandwidth of an AM one transmitting the same analogue signal.

Since the deviation of the carrier from its normal frequency is dictated by the amplitude of the signal which is modulating it, we need to specify the maximum amplitude signal allowable. More specifically, we have to state the range of frequencies which can be covered by modulating the carrier. If the normal frequency of the carrier is f_c, then we shall allow it to be deviated up to $\frac{1}{2}f_d$ on either side of f_c. f_d is known as the deviation frequency.

Now suppose that we are modulating an audio signal onto the carrier. We want to calculate the bandwidth required to transmit the signal with no appreciable loss of information, even when the signal has the maximum allowed amplitude. To do the calculation properly requires the use of Bessel functions, so we shall content ourselves with an approximate answer. An audio signal will contain frequencies up to 15 kHz, so the simplest way of finding the bandwidth of the modulated carrier is to obtain an expression for its frequency spectrum. Instead, we are going to sort out the bandwidth taken up by a 5 kHz square wave. This is the acid test of an audio system; providing it can deal with a 5 kHz square wave without distorting it, all is well, as humans cannot tell the difference between a sine wave and a square wave above 5 kHz.

So we modulate the carrier with a 5 kHz square wave which has the largest amplitude allowed. The carrier is therefore pushed from $f_c + \frac{1}{2}f_d$ to $f_c - \frac{1}{2}f_d$ and back again 5000 times a second, spending 100 μs at each frequency. Each of the carrier frequencies ($f_c + \frac{1}{2}f_d$ and $f_c - \frac{1}{2}f_d$) is keyed on and off at a rate of 5 kHz. Now if we key a carrier at a rate of f_s, that carrier acquires sidebands at $\pm f_s$, $\pm 3f_s$, $\pm 5f_s$, $\pm 7f_s$, etc. So the frequency spectrum of the FM carrier will look like the one shown in figure 13.51. We have to arbitrarily decide how many of these sidebands are important; we are going to ignore the ones more than $7f_s$ away from the two carriers. Since f_s is 5 kHz, this means that the upper and lower sidebands will each spread out over a bandwidth of 70 kHz. The sidebands are f_d apart, so the total bandwidth of the FM carrier is $f_d + 14f_s$.

How do you choose the optimum size of f_d? Clearly, the larger you make f_d, the easier it is going to be to demodulate the FM carrier in the receiver; figure 13.50 shows that a large degree of frequency modulation leads to a large degree of amplitude modulation in a slope detector. On the other hand, if f_d is very large, you will be forced to accept more noise than you need to. The optimum situation is where f_d is equal to the bandwidth taken up by the upper and lower sidebands i.e. $f_d = 14f_s$. It might help to look at figure 13.51. If f_d is much larger than $14f_s$, there will be a gap between the two sets of sidebands. That gap will contain noise and no useful signal. So the bandwidth required to transmit a maximum amplitude 5 kHz square wave has to be about $28 \times 5 = 140$ kHz. In other words, the bandwidth is about ten times the highest modulation frequency.

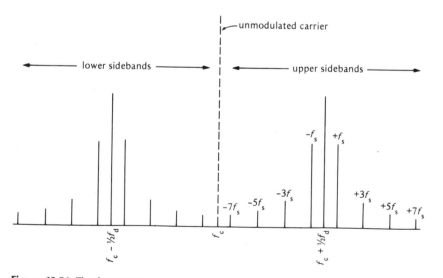

Figure 13.51 The frequency spectrum of a carrier frequency modulated by a square wave

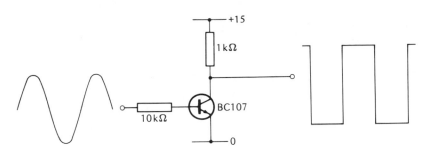

Figure 13.52

EXPERIMENT

Building an FM transmission system

Assemble the circuit shown in figure 13.48; use CMOS gates run off $+15\,V$ and $0\,V$. Check that when the signal input is wobbled up and down the frequency of the output waveform (called FM carrier in the diagram) changes accordingly.

Then put together the slope detector shown in figure 13.49, using the transistor NOT gate of figure 13.52 instead of the Schmitt trigger. Look at the AM signal with a CRO; check that its amplitude wobbles up and down as the signal input of the modulator is raised and lowered.

Feed a $1\,V$ peak-to-peak sine wave at $1\,kHz$ into the signal input of the FM modulator. Use the CRO to look at the waveforms called FM, AM and OUT in figure 13.49. Sketch them.

By trial and error, find the range of signal amplitudes and frequencies which the system will transmit without distortion.

PROBLEMS

1 An audio signal, with frequencies ranging from $16\,Hz$ to $16\,kHz$ is amplitude modulated onto a $250\,kHz$ carrier and sent down a wire link. The receiver at the end of the link contains a tuned circuit which allows the carrier and sidebands through to a half-wave rectifier and bass cut filter. Draw a circuit diagram of a suitable receiver, showing all of the component values and your justification for them. Assume that you have to use a $1\,mH$ inductor.

2 An optical fibre transmission system can cope with frequencies of up to $500\,kHz$. You want to send as many telephone channels down it as you can. If you restrict the audio signals to between $300\,Hz$ and $3.4\,kHz$ (the usual telephone range), roughly how many channels can you have if you use

a) FM

b) AM

c) PPM?

3 FM radio broadcasts use radio waves whose frequencies lie in the range $88\,MHz$ to $108\,MHz$. Each station is allocated a bandwidth of $150\,kHz$. What range of audio frequencies will each station be able to transmit effectively?

4 If each FM radio station decides to transmit audio signals up to 20 kHz, what bandwidth will each one require? How many stations could broadcast simultaneously on the FM band in a particular area?

5 AM radio broadcasts on the MW (medium wave) band use radio frequencies from 540 kHz to 1.6 MHz. If each station transmits audio signals whose frequencies do not exceed 5 kHz, how many stations can broadcast on the MW band simultaneously. (Assume that you have access to perfect filters!)

6 By both frequency modulating and amplitude modulating a carrier it is possible to send two independent audio signals via a single carrier. Draw a circuit diagram of a suitable modulator and explain how it works; use a 500 kHz carrier. Then draw a circuit diagram of a suitable demodulator and explain how it works.

7 Draw a circuit diagram of a system which uses amplitude modulated carriers to permit simultaneous two-way voice transmission down a single wire link.

8 When transmitting an audio signal using AM, why must both the amplitude and frequency of the signal not exceed certain limits? What happens if they do exceed them?

RADIO RECEIVERS

The transmission of signals by radio waves is a very important and time-honoured field of electronics. For many years it was virtually the only major area of application for electronics. Indeed, at one time the word "electronics" was almost synonymous with "wireless". We shall only be able to scratch the surface of the subject in this section, as many of the techniques employed in radio receivers are esoteric and complex. You will find out how to build simple MW receivers, how to improve upon their shortcomings, why superhets work and why TRF's do not.

A MW CRYSTAL RADIO

In the early days of wireless telegraphy a device known as the cat's-whisker receiver was popular among electronics amateurs. The simple radio receiver shown in figure 13.53 is its modern equivalent; it is sometimes called a **crystal receiver**. As you can see, it has no power source other than the aerial and only contains four components. There is a tuned circuit, a point-contact diode (the equivalent of the crystal in the old sets) and a pair of headphones. After we have discussed how the circuit works, we will go on to deal with more realistic MW receivers.

The aerial is no more than a long length of wire. It needs to be draped so that it intercepts radio waves; outdoors works better than indoors. The radio waves induce alternating currents in the wire if one end of it is firmly earthed and a substantial portion of it is held vertically. Since radio waves of all frequencies are always present, an enormous range of different alternating currents will be present in the aerial. However, those currents have to flow through a tuned circuit to get to earth. That parallel *LC* network will

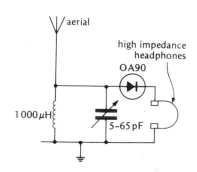

Figure 13.53 A crystal radio receiver

present a large impedance to the AC currents whose frequencies are close to $1/2\pi(LC)^{1/2}$. Only these currents will develop an appreciable voltage as they pass through the network.

If the capacitor has been set so that the resonant frequency of the *LC* network matches that of an incident radio wave, the diode will be fed with an AM carrier. The size of the signal fed into the diode will depend on the proximity and power of the emitter of the radio waves. It rarely gets above 100 mV. If the diode is a germanium one it will be able to rectify the signal, small though it is. The demodulated signal is finally used to run a set of headphones. No smoothing capacitor is needed as the headphones will ignore the high frequency components of the demodulated signal.

The circuit will only work if the headphones have a high impedance. The tuned circuit will only be able to have a large Q, i.e. a large impedance at its resonant frequency, if it drives a very large resistance. In the crystal radio, the tuned circuit has to drive the headphones via the diode. If the headphones contain piezoelectric speakers, they will present a large impedance to the tuned circuit. So the Q will be high and the current flowing in the aerial will be able to generate an appreciable signal across the tuned circuit. On the other hand, if the headphones are of the moving-coil type they will appear to be a low resistance in parallel with the tuned circuit. So the currents flowing in the aerial will only see a low impedance as they flow through the circuit to earth and the signal presented to the diode will be small.

Crystal radios are also deficient in many other respects. Their audio output is very weak, even under favourable conditions, as the whole circuit has to be powered from the energy picked up by the aerial. They lack **sensitivity**, i.e. they can only pick up very strong radio signals. Finally, they are not very **selective**; even when they are tuned to pick up one station, other stations can be demodulated as well. The reason for this lack of selectivity is shown in figure 13.54. It shows how the response of the tuned circuit depends on frequency. Even when the curve is centred on one carrier and its sidebands, an appreciable amount of signal from neighbouring stations gets through as well.

Figure 13.54 The frequency response of a loaded tuned circuit

PROBLEMS

1 For the circuit shown in figure 13.53, what range of frequencies can the radio be tuned for?

2 Sketch the waveform of the signal across

   a) the tuned circuit

   b) the headphones

   of the circuit in figure 13.53, when it is tuned to a radio station.

A SIMPLE MW RECEIVER

Study the circuit shown in figure 13.55 (overleaf). It is a crystal radio set with some of the deficiencies removed. The audio amplifier boosts the demodulated signal before feeding it into a speaker for conversion to a sound wave; the system no longer relies totally on the energy picked up by the aerial. The sensitivity has been improved, because weak stations are amplified once they have been demodulated. The circuit is still not very sensitive because you need to have at least a few mV across the diode for

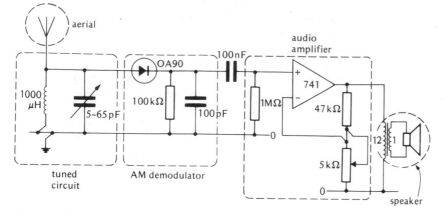

Figure 13.55 A simple MW radio receiver

it to function. Finally, the system is more selective. The tuned circuit only has to drive a light load (100 kΩ), so it can have a fairly high Q; this means that the response curve of the tuned circuit will rise and fall more sharply than before, attenuating neighbouring stations more.

In general, the circuit works better than the crystal radio does, but there is still a lot of room for improvement. We shall discuss the esoteric delights of superhets and TRFs after you have had a chance to explore a simple MW receiver for yourself.

EXPERIMENT
Assembling the MW radio of figure 13.55

Put the aerial and tuned circuit together. The aerial need only be a couple of metres long; try pinning it to the ceiling! Make sure that the bottom end of the tuned circuit is earthed, not merely held at 0 V. Look at the signal across the tuned circuit with a CRO. Tweak the capacitor until you find the strongest AM carrier. Make a rough sketch of it, showing the modulation clearly.

Now insert the AM demodulator, and look at its output with the CRO. Sketch the output. Then sketch the output with the 100 pF capacitor removed.

Finally add on the audio amplifier. You can adjust its gain by altering the 5 kilohm potentiometer; make sure that you do not amplify the audio signal so much that it starts to bang against the supply rails.

Find out what stations you can tune your radio to. Try inserting other values of inductor (470 μH and 220 μH) to get other stations. Experiment with the aerial to try and improve reception.

EXPERIMENT
Modifying the circuit of 13.55 to improve its performance

The first thing you can try is to buffer the tuned circuit, i.e. make its load as light as possible. The ideal device for this purpose is an FET. So try inserting the source follower shown in figure 13.56. Since the gate of the FET draws no current from the tuned circuit, the latter will behave as though it were

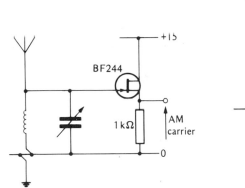

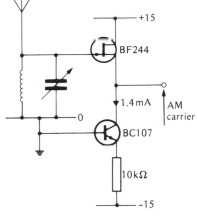

Figure 13.56 Isolating the tuned circuit with an FET follower

Figure 13.57 A better source follower

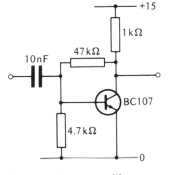

Figure 13.58 An amplifier

completely isolated from the rest of the circuit. You should find that the AM carrier is larger than it was before, and that the tuning is more selective. The capacitor will have to be carefully tweaked to get the resonant frequency matched to the carrier.

If you feed the AM carrier from the source follower into the AM demodulator and audio amplifier, you should be able to see that the audio waveform is not symmetrical. The source follower shown in figure 13.57 should wipe out this distortion.

You can increase the sensitivity of the system by amplifying the AM carrier before demodulating it. Put a transistor amplifier between the source follower and the demodulator, and see if it allows you to pick up weak stations more easily. A suitable amplifier is shown in figure 13.58.

TRFs

TRF is short for **tuned radio-frequency**. The term is used to describe one technique which is used to improve the selectivity and sensitivity of a radio receiver. The gain in sensitivity is provided by amplifying the modulated carrier before presenting it to the demodulator. The way in which the selectivity is improved is rather more subtle. It concerns the way in which extra filters are added to the system, to knock out signals from neighbouring stations.

A block diagram of a TRF system is shown in figure 13.59 (overleaf). The signal across the first tuned circuit is amplified and filtered by a second tuned circuit. That signal is further amplified and filtered in the two subsequent stages. Only then is it demodulated and fed into the audio amplifier. You need not worry about having too much gain with three stages of amplification of the RF (radio-frequency) signal; you are doing quite well if an RF amplifier gives you 20 dB of gain!

The three tuned filters have to be physically ganged together so that they can be tuned by rotating a single shaft. Furthermore, their resonant frequencies have to be carefully adjusted so that they are a few kilohertz apart. This staggering of the three filters can give a bandwidth of 10 kHz, so that an AM carrier and its sidebands get through unscathed, combined with

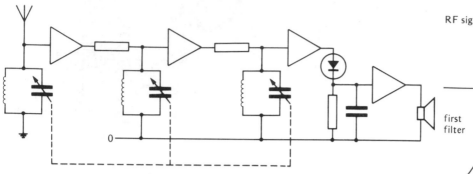

Figure 13.59 A TRF system

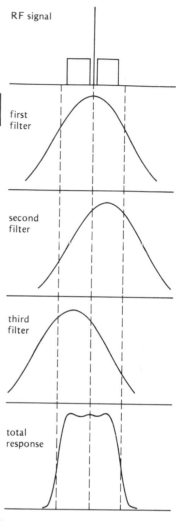

Figure 13.60 The frequency response of the filters in a TRF receiver

a very strong rejection of all other frequencies. Have a look at figure 13.60; the total response curve falls much more sharply than each of the individual filters does. Provided that the carrier and its sidebands are properly centred within that response curve, there will be virtually no crosstalk from adjacent stations.

It sounds too good to be true! You amplify the RF signal and filter it at the same time. So you get better selectivity by knocking out signals that are more than 5 kHz away from the carrier to which you have tuned. You also get better sensitivity because of all the RF amplification. The flaws become evident when you want to tune into another station. All three capacitors must change their values by exactly the same amount, to within at least 0.5 kHz. If the three filters get out of line as they are tuned, the overall response curve will lose its sharp sides. So TRF receivers have to be very carefully engineered if they are to maintain their good selectivity whilst being tuned from one station to another.

SUPERHETS

A block diagram of a **superheterodyne receiver** is shown in figure 13.61. It can give excellent selectivity and sensitivity by doing most of the filtering and amplifying at a fixed frequency of 455 kHz. The trick employed is to copy the sidebands of the RF carrier and stick them to a 455 kHz carrier at the front end of the receiver.

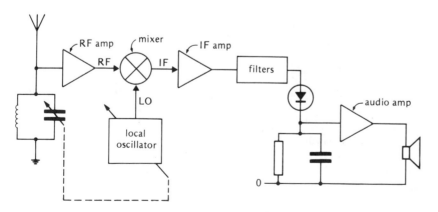

Figure 13.61 A superhet receiver

The signal from the aerial is fed into the standard parallel *LC* network before being amplified. It is then combined with the output of the local oscillator in a mixer. The local oscillator is ganged to the tuned circuit at the end of the aerial, so that both are tuned together. It does not matter if two circuits are not very precisely ganged; the important requirement is that the local oscillator (LO) frequency is about 455 kHz above resonant frequency of the *LC* network.

You can think of the mixer as a very non-linear amplifier, one that has been designed to maximise inter-modulation distortion. Non-linear amplifiers generate spurious signals at frequencies which are sums and differences of the frequencies fed in. So if you add the RF signal to the LO signal and feed them through a non-linear amplifier you will end up with signals at all sorts of frequencies. In particular, there will be one signal at the difference between RF and LO. That difference is 455 kHz. This signal is called the intermediate frequency (IF) signal, and it contains all of the information present in the RF signal. More precisely, IF = LO − RF, where we are talking about frequencies rather than amplitudes. Every frequency component of the RF signal (i.e. the AM carrier and its sidebands) is shifted down by the LO frequency. The IF signal is therefore a copy of the RF one, but centred on 455 kHz rather than LO − 455 kHz.

The operation of the mixer is illustrated in figure 13.62. A radio signal, consisting of a 1.200 MHz carrier and its sidebands, is mixed with an LO signal of 1.655 MHz. This results in a large number of frequencies, not all of which are shown. In particular, the mixer feeds out the IF signal which has the radio signal sidebands centred on a 0.455 MHz carrier.

Returning to the block diagram of figure 13.61, you will see that the IF signal is amplified before being fed into some filters. The amplification further improves the sensitivity of the receiver, and the filters make it selective. By staggering the resonant frequencies of the tuned circuits, the block of filters can be engineered to transmit signals between 450 kHz and 460 kHz and reject all others. Finally, the filtered IF signal is demodulated etc. etc.

The superhet receiver is better than the TRF receiver in two ways. It does not have the tracking problems of the TRF system because the filters which make it selective do not have to be tuned from one station to another. In fact, in the superhet receiver the station is tuned to match the fixed characteristics of the filters! (Imagine what happens to the frequency spectra shown in figure 13.62 as the LO frequency is changed.) Furthermore, much

Figure 13.62 What a mixer does

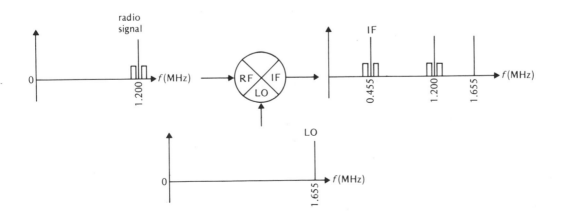

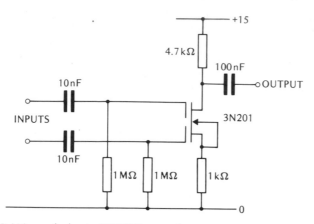

Figure 13.63 Using a dual gate MOSFET as a mixer

of the amplification of the signal takes place at 455 kHz, rather than at the radio frequency itself. RF amplifiers are difficult to build, and you can get much higher gain at lower frequencies. The only RF components of a super-het receiver are the local oscillator, the mixer and the RF amp which buffers the first tuned circuit. High frequency oscillators are relatively easy to build; indeed, it is difficult to stop a high frequency amplifier from oscillating anyway! The buffer need only be a source follower of the type shown in figure 13.57. Finally, the mixer can be built around a dual gate MOSFET, as shown in figure 13.63; you will recall that FET amplifiers are inherently non-linear ($I_D \propto (V_{GS} - V_P)^2$?) so they make efficient mixers.

EXPERIMENT
Investigating a mixer

Have a look at figure 13.64. There are two oscillators, one whose frequency is fixed at about 45 kHz and the other whose frequency can be altered with a potentiometer. The two signals are fed into the mixer of figure 13.63. The output of the mixer is fed into a treble cut filter which knocks out components above 3 kHz. The low frequency components of the mixer output are fed into a loudspeaker.

Assemble the oscillators first. The diodes clamp the 741 outputs to +0.7 V and −0.7 V; this is to get around the problem of the low slew rate of 0.5 V/µs. You should be able to change the frequency of one oscillator smoothly by tweaking the potentiometer. If you suspect that the oscillators get locked together when their frequencies get close, you will have to de-spike the supply rails near each op-amp with a pair of capacitors, as shown. If the two are getting locked, you should be able to see the fixed oscillator change frequency slightly as the variable one is swept past 45 kHz.

Then put in the mixer. Feed in one input only, and check that it emerges amplified. Then feed in the other input only and check that all is well. Finally, feed both signals in and look at the output with a CRO. You should be able to see interesting things happen as you tweak the variable oscillator.

Now insert the treble cut filter to eliminate the unwanted signals at 45 kHz and over. Listen to the difference frequency with the help of the speaker. Note what happens to it as you sweep the variable oscillator past 45 kHz.

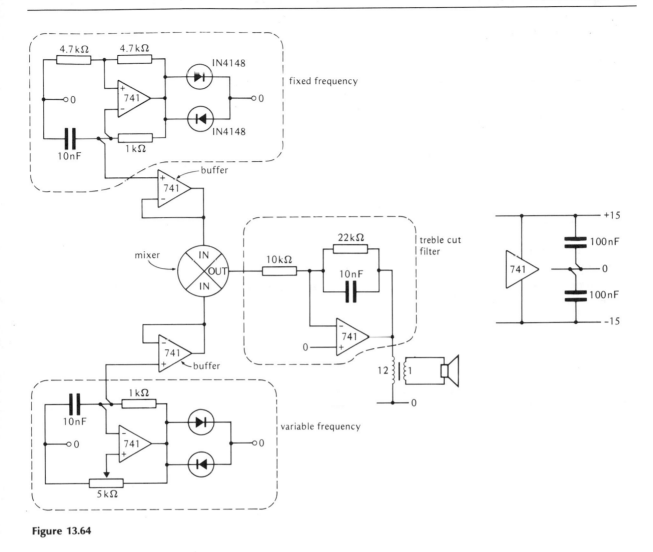

Figure 13.64

Finally, try modifying the circuit so that it becomes a metal detector. You will have to replace the fixed frequency oscillator with one which uses a tuned circuit. A simple Colpitts oscillator with 1000 μH and 10 nF might do. Tweak the variable oscillator until the note coming out of the speaker is very low. You should find that the note changes when you put a screwdriver near to the inductor!

PROBLEMS

1 Design a simple LW radio receiver. The LW band stretches from 150 kHz to 260 kHz and employs AM with the sidebands restricted to within 5 kHz of the carrier. Assume that you have a capacitor which can be tuned from 50 pF to 300 pF. Draw a circuit diagram, show all of the component values and calculations. Explain carefully the function of each part of your circuit.

2 TV stations broadcast using carriers in the region of 1 GHz. The instantaneous brightness of the spot on the TV screen is amplitude modulated onto the carrier. A lot of carrier gives a bright spot, a little carrier gives a dark spot, with grey tones in between. A standard TV receiver has 625 lines, with each line divided up into 833 pixels. As the spot scans along each line, its brightness is varied to make each of the pixels light or dark as required. The spot scans each pixel twenty five times a second. (See problem 6 on p. 406 for more details.)

a) Suppose that a TV station broadcasts a signal which paints every even numbered line black and the odd numbered ones white. How wide a bandwidth would this signal require?

b) The worst signal that the station can broadcast as far as bandwidth is concerned is one which paints one pixel white and its neighbour black. In other words, each line contains 416 bright pixels alternating with 417 dark ones. Calculate the frequency of sine wave that would have to be amplitude modulated onto the TV carrier to give this pattern.

c) In order to save on bandwidth, TV signals are doctored so that the lower sideband is suppressed. The AM carrier looks like a spike with one sideband, rather than the normal two. How much bandwidth does the TV signal need to take up?

REVISION QUESTIONS

1 Discuss the properties of these communication links:
  a) wire links,
  b) optical fibre links,
  c) ultrasonic links,
  d) radio links.

In each case, state and explain what types of modulation can be used.

2 What is a repeater? Why is it necessary? If a repeater has a gain of 30 dB and a cable has a loss of 1 dB for each 20 m length, how far apart can the repeaters be placed?

3 Draw circuit diagrams to show how an optical fibre can be used to carry analogue and digital signals. In what ways are optical fibres superior to wire links?

4 What is meant by the term modulation? Explain the principles behind FSK, PPM, AM and FM, giving suitable circuits for encoding and decoding the carrier.

5 State the range of carrier frequencies usually used in:
  a) telephone links,
  b) MW radio,
  c) TV broadcasts,
  d) standard FSK.

6 Explain why transmission methods which employ digital signals are far less noisy than methods which send analogue signals down the link.

7 Discuss the use of parity bits to ensure that bytes are transmitted down a link without any error.

8 Explain what is meant by frequency multiplexing.

9 State the properties of a series *LC* network; illustrate your answer with a graph of the impedance of the network as a function of frequency. Show how the network can be used to make a filter which only passes a narrow range of frequencies.

10 You have access to 470 µH inductors. Show how you would use them to build

a) a pass filter at 1 MHz,

b) a stop filter at 500 kHz.

11 Draw the frequency spectrum of a 100 kHz carrier which is keyed by a 5 kHz square wave. If the keyed carrier is sent down a link which can support signals between 90 kHz and 110 kHz, work out the maximum rate at which bytes can be transmitted.

12 Go through the calculation of the maximum possible byte rate for standard FSK; support your argument with a drawing of the frequency spectrum of the FSK carrier.

13 Derive an expression for the bandwidth required to transmit an FM carrier.

14 Draw the circuit diagram of a simple crystal receiver. If the headphones have an impedance of 10 kΩ and the system is to be tuned to a radio wave at 810 kHz, calculate the values of all the components. Explain, with the help of some waveforms, how the circuit works. Discuss its limitations.

15 Draw the circuit diagram of a simple MW receiver, and explain the function of each part. Discuss how it can be improved i.e. how it can be made more sensitive and more selective by inserting buffers and amplifiers.

16 Draw a block diagram of a superhet radio receiver. Explain how it works. Discuss why it works better than the TRF radio receiver.

Appendix A
Resistors, capacitors and inductors

This appendix has been written to cater for the interests of the mathematically minded reader. At several places in the book we have quoted formulae or stated rules without any attempt to justify them. If you know how to differentiate and can manipulate algebra patiently, then the next few pages will explain how these formulae have been reached.

Exponential decay in *RC* networks

Consider the circuit shown in figure A.1. V_U and V_L are fixed voltages of any value; V_M is the voltage at the junction of the resistor (R) and the capacitor (C). We are going to derive an expression for the voltage across the resistor as a function of time i.e. we are going to prove that

$$V = V_0 \, e^{-(t/RC)}$$

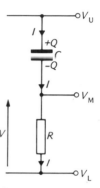

Figure A.1

For the resistor:

$$V_M - V_L = IR \tag{1}$$

For the capacitor:

$$V_U - V_M = \frac{Q}{C} \tag{2}$$

Now, the current $I = dQ/dt$, which is the rate at which charge is dumped on the capacitor. This allows us to combine (1) and (2) and eliminate Q and I. First we differentiate (2):

$$\frac{dQ}{dt} = C \frac{d}{dt} (V_U - V_M) = C \left(\frac{dV_U}{dt} - \frac{dV_M}{dt} \right)$$

But V_U is fixed, so it does not change with time:

$$\frac{dV_U}{dt} = 0$$

So

$$\frac{dQ}{dt} = -C \frac{dV_M}{dt} \tag{3}$$

Moving (1) around a bit we get another expression for dQ/dt

$$I = \frac{dQ}{dt} = \frac{V_M - V_L}{R} \tag{4}$$

Setting (3) and (4) equal to each other we get:

$$\frac{V_M - V_L}{R} = -C \frac{dV_M}{dt} \tag{5}$$

Equation (5) is a differential equation. You solve it by making a guess at its solution. By putting the guess into the differential equation you find out if your guess was correct. If it was not, you have another guess and keep on guessing until you find a solution which works! Obviously, we do not want to go through pages of wrong guesses, so we will only go through the correct guess. But first, we shall introduce an extra variable which will make the equation easier to handle.

Let $V = V_M - V_L$. If you look at figure A.1, you can see that V represents the voltage across the resistor. Since V_L is fixed, it follows that

$$\frac{dV}{dt} = \frac{dV_M}{dt}.$$

Changing the variables of (5) from V_M and V_L to V we get:

$$\frac{V}{R} = -C\frac{dV}{dt} \tag{6}$$

Now for our guess. We are going to try $V = A e^{-Bt}$ as our solution to (6), where A and B are constants. Differentiating this expression we get:

$$\frac{dV}{dt} = A(-B)e^{-Bt} \tag{7}$$

Inserting (7) into (6) we find that our solution works if we choose appropriate values for A and B:

$$\frac{V}{R} = -C(A(-B)e^{-Bt}) = \frac{A}{R}e^{-Bt}$$

So, provided that $CAB = A/R$ our solution works. Therefore $B = 1/RC$. What about the value of A? It is fixed by the value of V when $t = 0$. Since $e^{-0} = 1$, $V = A$ at $t = 0$. If we say that the value of V at $t = 0$ is V_0, then $A = V_0$. The final solution is now:

$$V = V_0 e^{-(t/RC)}$$

This equation describes how the voltage across the resistor changes with time. Provided that V_U and V_M remain fixed (at any value) V gets smaller and smaller as time progresses, regardless of its sign or initial value.

The treble cut filter

In Chapter Eight we introduced the idea of reactance in order to explain the properties of simple filters. In particular, we suggested that you would never go far wrong if you assumed that a capacitor could be replaced by a resistor of value $1/2\pi fC$ for calculating filter characteristics. Thus the voltage divider of figure A.2 should, with this approximation, obey the equation:

$$V_{OUT} = V_{IN}\frac{1}{1 + f_R}$$

where
$$f_R = \frac{f}{f_0} \quad \text{and} \quad f_0 = \frac{1}{2\pi RC}$$

(Work it out for yourself.)

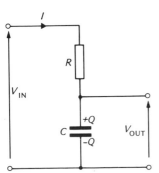

Figure A.2

If you do the algebra accurately, you find that the equation should actually be:

$$V_{OUT} = V_{IN} \left(\frac{1}{1 + f_R^2} \right)^{1/2}$$

where V_{OUT} and V_{IN} refer to the peak values of the output and input waveforms. Before we discuss how this formula is different from the approximate one, we shall show how the formula can be obtained.

Let $V_{IN} = V_0 \cos \omega t$. The angular frequency, $\omega = 2\pi f$. Write down the equations which describe the resistor and the capacitor:

$$V_{IN} - V_{OUT} = IR \tag{1}$$

$$V_{OUT} = \frac{Q}{C} \tag{2}$$

Now differentiate (2) to get an expression for $I = dQ/dt$.

$$\frac{dV_{OUT}}{dt} = \frac{dQ}{dt}\frac{1}{C} = \frac{I}{C}$$

therefore
$$I = C\frac{dV_{OUT}}{dt} \tag{3}$$

Now combine (3) and (1) to eliminate I:

$$I = C\frac{dV_{OUT}}{dt} = \frac{V_{IN} - V_{OUT}}{R}$$

therefore
$$RC\frac{dV_{OUT}}{dt} + V_{OUT} = V_{IN} = V_0 \cos \omega t \tag{4}$$

Now equation (4) has to be solved. We guess a form of solution for V_{OUT} and plug it into the differential equation to see if it works. (It will, of course!) Let's try $V_{OUT} = A \cos \omega t + B \sin \omega t$ where A and B are constants that we do not yet know.

$$V_{OUT} = A \cos \omega t + B \sin \omega t \tag{5}$$

therefore
$$\frac{dV_{OUT}}{dt} = -A\omega \sin \omega t + B\omega \cos \omega t \tag{6}$$

Substitute (5) and (6) into the differential equation (4):

$$RC(-A\omega \sin \omega t + B\omega \cos \omega t) + A \cos \omega t + B \sin \omega t = V_0 \cos \omega t$$

therefore
$$(RCB\omega + A - V_0) \cos \omega t = (-B + RCA\omega) \sin \omega t \tag{7}$$

Now if (7) is to be true for all values of t, then the two brackets must contain zero.
So:

$$B = RCA\omega \tag{8}$$

$$A = V_0 - RCB\omega \tag{9}$$

After a bit of manipulation we obtain these two expressions for A and B:

$$A = \frac{V_0}{(1 + f_R^2)}$$

$$B = \frac{V_0 f_R}{(1 + f_R^2)}$$

where $\qquad\qquad f_R = \omega RC = 2\pi fRC$

So the solution of the differential equation is:

$$V_{OUT} = V_0 \frac{1}{1 + f_R^2} \cos \omega t + V_0 \frac{f_R}{1 + f_R^2} \sin \omega t$$

The solution would be easier to interpret if we could put it into the form $V_{OUT} = V \cos(\omega t + \phi)$. The peak value would then be just V, and ϕ would represent the phase difference between the input and output waveforms. Re-jigging the form of our solution is straightforward if we use the identity:

$$\cos(x + y) = \cos x \cos y - \sin x \sin y$$

therefore $\quad V \cos(\omega t + \phi) = V \cos \phi \cos \omega t - V \sin \phi \sin \omega t$

If you compare the above equation with our solution, you can see that:

$$V \cos \phi = V_0 \frac{1}{1 + f_R^2} \qquad (10)$$

$$V \sin \phi = -V_0 \frac{f_R}{1 + f_R^2} \qquad (11)$$

You will recall that $\sin^2 \phi + \cos^2 \phi = 1$. If we use that identity, we can easily obtain an expression for V.

$$V^2 \sin^2 \phi + V^2 \cos^2 \phi = V_0^2 \frac{(f_R^2 + 1)}{(1 + f_R^2)^2} = \frac{V_0^2}{1 + f_R^2} = V^2$$

therefore $\quad \boxed{V = V_0 \left(\frac{1}{1 + f_R^2} \right)^{1/2}}$

Similarly, we can obtain an expression for the phase as a function of frequency:

$$\frac{V \sin \phi}{V \cos \phi} = \frac{-V_0 f_R/(1 + f_R^2)}{V_0/(1 + f_R^2)} = -f_R = \tan \phi$$

therefore $\quad \boxed{\phi = \tan^{-1}(-f_R)}$

Figure A.3 (overleaf) shows how V and ϕ change with frequency. You may already have realised that f_R is a measure of how the input frequency compares with the break frequency $f_0 = 1/2\pi RC$. In fact, $f_R = f/f_0$. Notice how the graph of V as a function of f_R is identical with the one that you would draw using the two straight line approximation except near to $f_R = 1$.

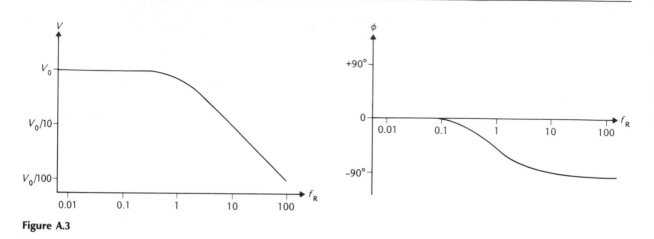

Figure A.3

The bass cut filter

Now we shall find an expression for the signal which gets through the simple bass cut filter shown in figure A.4. It is not going to take very long as a bass cut is just a treble cut upside down. A treble cut filter feeds out the voltage across the capacitor whereas the bass cut feeds out the voltage across the resistor; otherwise the arrangement is the same. We already have an expression for the voltage across the capacitor from the last section; if we take it away from the voltage across the whole system we must be left with the voltage across the resistor.

So, for the bass cut filter:

$$V_{OUT} = V_0 \cos \omega t - V_0 \frac{1}{1 + f_R^2} \cos \omega t - V_0 \frac{f_R}{1 + f_R^2} \sin \omega t$$

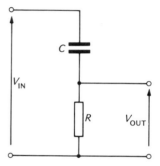

Figure A.4

therefore $\quad V_{OUT} = V_0 \dfrac{f_R^2}{1 + f_R^2} \cos \omega t - V_0 \dfrac{f_R}{1 + f_R^2} \sin \omega t$

As before, we need to convert this expression into a more understandable form. If $V_{OUT} = V \cos \omega(t + \phi)$, then:

$$V = V_0 \left(\frac{f_R}{1 + f_R^2} \right)^{1/2}$$

$$\phi = \tan^{-1}(1/f_R)$$

(Do the rearrangement for yourself.) Figure A.5 shows how V and ϕ depend

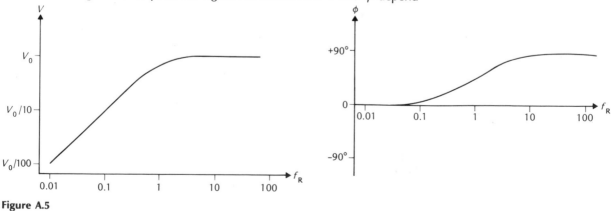

Figure A.5

on f_R. Note how the graph for V is similar to the one which you would draw using the two straight lines approximation.

The series *LC* network

In Chapter Thirteen we glibly stated that an inductor had a reactance given by $2\pi fL$, and that a series *LC* network would appear to have a very low resistance at the resonant frequency $f_0 = 1/2\pi(LC)^{1/2}$. We shall now try to convince you that both statements are more or less correct.

The voltage across an inductor is proportional to the rate of change of the current flowing through it. More specifically:

$$V = L\frac{dI}{dt}$$

If the voltage across the inductor is a sine wave, $V_0 \cos \omega t$, then:

$$\frac{dI}{dt} = \frac{V_0}{L}\cos \omega t$$

To find an expression for the current (I) we have to integrate:

$$\int dI = \frac{V_0}{L}\int \cos \omega t\, dt$$

therefore $\qquad I = \frac{V_0}{L}\frac{\sin \omega t}{\omega}$

therefore $\qquad \dfrac{\text{peak voltage}}{\text{peak current}} = \text{reactance} = \omega L = 2\pi fL$

So if we neglect phase shifts, an inductor appears to have a resistance of $2\pi fL$.

Figure A.6 shows a series *LC* network. We are now going to get an expression for the current flowing through when a sine wave is fed into it.

Our starting point is the two expressions for the voltage across the inductor and the capacitor:

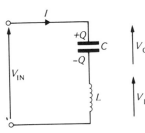

Figure A.6

$$V_L = L\frac{dI}{dt} \tag{1}$$

$$V_C = \frac{Q}{C} \tag{2}$$

Obviously, $V_{IN} = V_L + V_C$. If we let $V_{IN} = V_0 \cos \omega t$, then we get:

$$V_0 \cos \omega t = L\frac{dI}{dt} + \frac{Q}{C} \tag{3}$$

Expression (3) needs to be differentiated so that we can make use of the relationship $I = dQ/dt$.

$$-V_0 \omega \sin \omega t = L\frac{d^2I}{dt^2} + \frac{I}{C} \tag{4}$$

As usual, we are going to have to solve our differential equation (4) with a guess. Suppose that $I = A \sin \omega t$, where A is a constant. Then:

$$I = A \sin \omega t \tag{5}$$

$$\frac{dI}{dt} = A\omega \cos \omega t$$

$$\frac{d^2 I}{dt^2} = -A\omega^2 \sin \omega t \tag{6}$$

Now substitute (5) and (6) into (4) to see if our guess is correct.

$$-V_0 \omega \sin \omega t = -LA\omega^2 \sin \omega t + \frac{A}{C} \sin \omega t$$

therefore $\qquad -V_0 \omega = A(1/C - L\omega^2) \tag{7}$

So our guess is correct if A fits equation (7). If we move (7) around a bit, we get:

$$A = -V_0 \sqrt{\frac{C}{L}} \frac{f_R}{1 - f_R^2}$$

where $f_R = f2\pi\sqrt{LC}$

therefore $\quad I = -V_0 \sqrt{\frac{C}{L}} \frac{f_R}{1 - f_R^2} \sin \omega t$

f_R is, of course, the frequency relative to the resonant frequency $1/2\pi(LC)^{1/2}$. Let us see what happens to the peak value of I when f_R is very large, one, and very small.
So:

$$V_0 \sqrt{\frac{C}{L}} \frac{f_R}{1 - f_R^2} = V_0 \sqrt{\frac{C}{L}} \frac{f_R}{f_R^2} = V_0 \sqrt{\frac{C}{L}} \frac{1}{f2\pi\sqrt{LC}} = \frac{V_0}{2\pi fL} \quad \text{when } f_R \gg 1$$

So the network behaves like the inductor on its own.
Then:

$$V_0 \sqrt{\frac{C}{L}} \frac{f_R}{1 - f_R^2} = \infty \quad \text{when } f_R = 1$$

So the network has no resistance when $f = 1/2\pi(LC)^{1/2}$.
Finally:

$$V_0 \sqrt{\frac{C}{L}} \frac{f_R}{1 - f_R^2} = V_0 \sqrt{\frac{C}{L}} f_R = V_0 \sqrt{\frac{C}{L}} f2\pi\sqrt{LC} = V_0 2\pi fC \quad \text{when } f_R \ll 1$$

The network behaves like the capacitor on its own.

The parallel LC network

This final section will be concerned with showing how the voltage across a parallel LC network depends on the frequency of the current pumped through it.

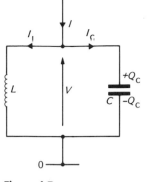

Figure A.7

Referring to figure A.7, you can see that:

$$I = I_L + I_C$$

therefore　$$\frac{dI}{dt} = \frac{dI_L}{dt} + \frac{dI_C}{dt} \qquad (1)$$

For each of the two components:

$$\frac{dI_L}{dt} = \frac{V}{L} \qquad (2)$$

$$Q_C = CV$$

therefore　$$I_C = \frac{dQ_C}{dt} = C\frac{dV}{dt}$$

therefore　$$\frac{dI_C}{dt} = C\frac{d^2V}{dt^2} \qquad (3)$$

If we now substitute (2) and (3) into (1) we get a differential equation for V, the voltage developed across the network:

$$\frac{dI}{dt} = -I_0 \omega \sin \omega t = \frac{V}{L} + C\frac{d^2V}{dt^2} \qquad (4)$$

where we have let $I = I_0 \cos \omega t$.

By now you can probably see that a wise guess for the solution of (4) is going to be $V = A \sin \omega t$.

$$V = A \sin \omega t \qquad (5)$$

therefore　$$\frac{dV}{dt} = A\omega \cos \omega t$$

therefore　$$\frac{d^2V}{dt^2} = -A\omega^2 \sin \omega t \qquad (6)$$

If you substitute (5) and (6) into (4), you will find, eventually, that A can be given by:

$$A = -I_0 \sqrt{\frac{L}{C}} \frac{f_R}{1 - f_R^2}$$

where $f_R = f 2\pi \sqrt{LC}$

So the signal which appears across the network is:

$$V = -I_0 \sqrt{\frac{L}{C}} \frac{f_R}{1 - f_R^2} \sin \omega t$$

It follows that when $f_R = 1$, (i.e. $f = 1/2\pi(LC)^{1/2}$), V is going to be infinite. In practice, V will not be infinite because of the small resistance of the inductor, and because of the effect of the input impedance of the circuit which the network is driving. If you study figure A.8, you can see that the maximum signal across the load resistor is going to be $I_0 R$; this will happen at $f_R = 1$, when the LC network will appear to have a much larger resistance than R. As f_R departs from 1, the apparent resistance of the LC network will fall until it gets to R; at that point, the signal across the load

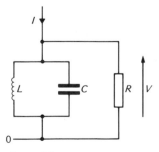

Figure A.8

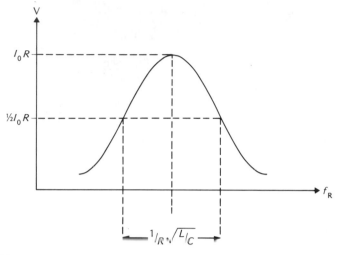

Figure A.9

will be half of its maximum value. The graph of figure A.9 shows how the signal across the load depends on the value of f_R; the size of the band-width, df_R can be roughly calculated as follows.

$$R = \sqrt{\frac{L}{C}} \frac{f_R + \frac{1}{2}df_R}{1 - (f_R + \frac{1}{2}df_R)^2} = \sqrt{\frac{L}{C}} \frac{1}{df_R}$$

if $f_R = 1$ and df_R is small compared with f_R. If we reorganise this expression we can see how the bandwidth depends on the load resistance:

$$df_R = \frac{1}{R} \sqrt{\frac{L}{C}}$$

So the bandwidth increases as the size of R decreases. If you want a narrow bandwidth (so that it is selective), you need to have a high value for R.

Appendix B
Components

This appendix brings together all the information about components that you need to assemble the circuits described in this book. Details of how to assemble them have been left to the next appendix on breadboard.

Resistors

Use metal oxide, $\frac{1}{2}$W types. You will need the values shown in table B.1. There is no point in keeping a vast range of resistors in stock, when you can build any value you need out of a limited number which span the whole range. For example, if you need 1.5 kΩ, then put 1 kΩ and 470 Ω in series (figure B.1). Similarly, 3.3 kΩ is more or less the same as 2.2 kΩ in series with 1 kΩ. Putting resistors in series makes larger resistances; putting them in parallel makes them smaller. For example, you can manufacture 9 kΩ by putting 10 kΩ in parallel with 100 kΩ (figure B.2). Table B.2 shows how you can make integral resistance values from 9 to 1 within 10% out of no more than two resistors.

Resistors are colour coded. If you look at a resistor you will see that one end has a silver or gold band. Hold the resistor as shown in figure B.3, with the gold or silver band at the right. The three coloured bands on the left tell you the value of the resistor. Each colour represents a number; take a look at table B.3. The first two bands on the left give you a number

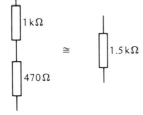

Figure B.1

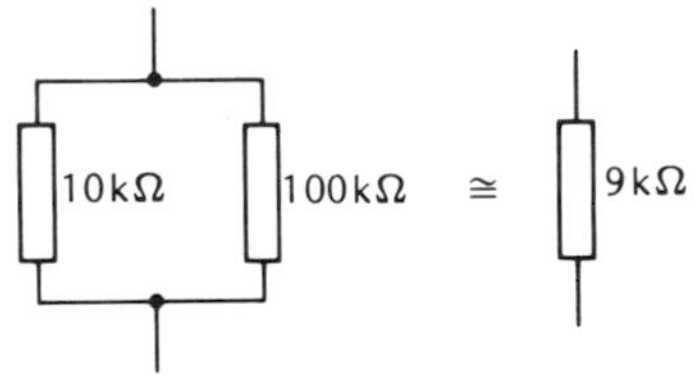

Figure B.2

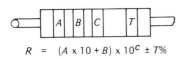

$$R = (A \times 10 + B) \times 10^C \pm T\%$$

Figure B.3

Table B.1

		Brown	Red	Orange	Yellow	Green	
Brown	Black	100	1 k	10 k	100 k	1 M	silver
Red	Red	220	2.2 k	22 k	220 k	2.2 M	or
Yellow	Purple	470	4.7 k	47 k	470 k	4.7 M	gold

Table B.2

Total resistance	Resistor connection
9	10 parallel 100
8	10 parallel 47
7	4.7 series 2.2
6	4.7 series 1.0
5	4.7
4	4.7 parallel 47
3	2.2 series 1.0
2	2.2
1	1.0

Table B.3

Colour	Value
black	0
brown	1
red	2
orange	3
yellow	4
green	5
blue	6
purple	7
grey	8
white	9

between 10 and 99; the third band tells you how many noughts to put on the end of the number. The whole number is the resistance in ohms. A silver band means that the tolerance is 10%. This means that the actual value of the resistor will be within 10% of its supposed value. If the tolerance is 5%, the band on the right will be gold.

Table B.1 shows the colour coding for the fifteen resistors you will be using. It assumes that you are holding the resistor as shown in figure B.3. So if the coloured bands read red, red, orange, gold from the left, then the resistance is 22 kΩ.

Capacitors

You will need two classes of capacitor. The following will be electrolytic; 1000 μF, 100 μF, 10 μF and 1 μF. Use single ended types if you can get them. Ensure that they can cope with up to 30 V. You will also need these values of non-polarised capacitor; 1 μF, 100 nF, 10 nF, 1 nF and 100 pF. Guard against getting ones that are very small physically; they can be very fiddly on breadboard.

If you want other values of capacitance you will have to use pairs of capacitors in series or parallel. You get a bigger capacitance if you put two in parallel, and a smaller one if you put them in series. Table B.4 shows how to get some values between 1 μF and 10 μF.

Inductors

1000 μH, 470 μH, 220 μH and 100 μH. They should have Q values of close to 100 near 1 MHz.

Trimmers

You can get small adjustable capacitors which slot easily into breadboard. They can be adjusted from 5 pF to 65 pF with the aid of a screwdriver.

Potentiometers

You need a 5 kΩ potentiometer. Get the type which sits horizontally on breadboard and is adjusted with a screwdriver. The wiper will invariably be the middle pin.

Switches

You need an array of four SPST switches in a DIP format (i.e. same format as an IC). Use a pencil to open and close the contacts; lever contacts are easier to manage than slide ones.

Single push switches designed to slot into breadboard tend to have fragile pins, but they are useful. Go for miniature ones rather than standard keyboard size.

SPDT switches are available in DIP format, but they are a bit fiddly. You might do better with a microswitch on the end of a three-stranded length of ribbon cable.

Relays

If you are going to do the relay experiments of Chapter One, you are going to need a relay which fits into breadboard. A DIP format relay is available which has a coil resistance of 560 Ω and is designed to be operated by TTL gates. Its pin-out is shown in figure B.4. (See the IC section to find out how

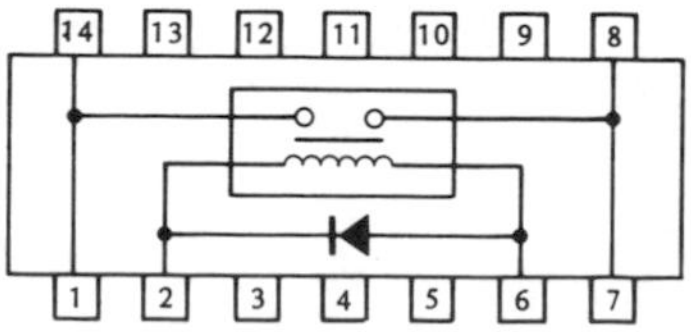

Table B.4

Total capacitance	Connection
10 μF	10 μF
5 μF	10 μF series 10 μF
2 μF	1 μF parallel 1 μF
1 μF	1 μF

Figure B.4

to identify pin 1.) Before using the relay for the experiments in Chapter One you will have to blow the diode by holding pin 6 at +5 V and pin 2 at 0 V.

LEDs

Use 0.2″ red LEDs with the legs cut down so that the body of the LED sits close to the breadboard. Figure B.5 shows how to identify the cathode of the LED; the package has a flat edge near the cathode leg. Most LEDs will not survive if they are put the wrong way round in a circuit. Furthermore, you must ensure that the current never exceeds about 25 mA. Yellow and green LEDs are pretty, but not as bright as red ones.

LDRs

Use an ORP12. If it is going to be part of a voltage divider, make sure that the other resistor is at least 1 kΩ. Otherwise the LDR may overheat when a lot of light falls on it.

TTL logic ICs

Figure B.7 shows the pin-out of all the TTL logic gate ICs which you will need to use. All run off +5 V and 0 V, with unattached inputs floating up to logical 1. Outputs can sink up to about 16 mA at logical 0 and you have to sink about 1 mA from an input to pull it down to logical 0. Signals below +1 V are definitely logical 0.

How do you identify pin 1? Look at figure B.6. You place the IC so that the labelling on its top reads from left to right. There will either be a dot in the package next to pin 1, or there will be a semicircular depression at the left hand end.

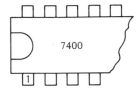

Figure B.5

Figure B.6

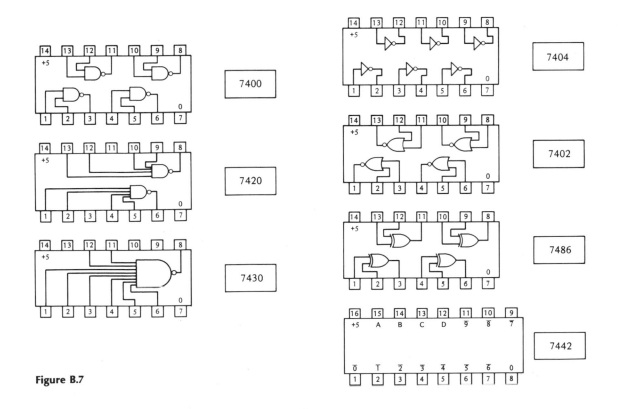

Figure B.7

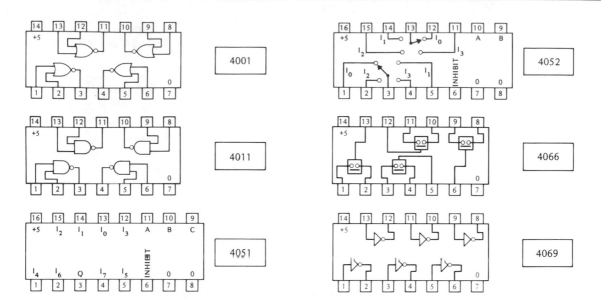

CMOS ICs

Figure B.8

These are shown in figure B.8. All of them can be run off 0 V and +5 V or +15 V. The 1/0 threshold will be halfway between the supply rails. The two multiplexers and the analogue switch ICs can handle analogue signals which lie between the two supply rails. Unused inputs of a gate must be anchored to one of the supply rails.

Be careful when handling CMOS ICs. They are easily damaged by static electricity. So avoid wearing nylon underwear and get into the habit of touching a grounded terminal before handling a CMOS device.

Sequential ICs

Four ICs are shown in figure B.9. The first three have all the characteristics of the TTL family. The last has its own idiosyncracies; assume that it only recognises +5 V as logical 1 and 0 V as logical 0 to be safe!

The two D flip-flops in the 7474 are completely independent. It sometimes pays to hold the set and reset terminals at logical 1 rather than let them drift up.

You can use the A counter separately from the BCD counter in the 7493 if you wish to. Normally, you convert it into a four bit counter (ABCD) by connecting pin 12 to pin 1 and feeding pulses in at pin 14. Note the odd power supply arrangements.

Figure B.9

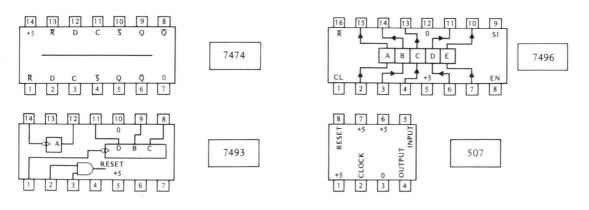

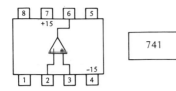

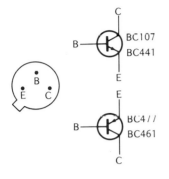

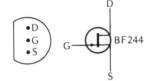

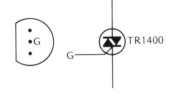

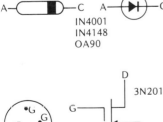

Figure B.10

Op-amps

The pin-out of a 741 is shown in figure B.10. Most op-amps housed in an 8 pin DIP format have exactly the same pin out. The supply can go down to +3V and −3V before the 741 stops working. You can take the inputs anywhere you like between the supply rails; the output can only get to within about 2V of either supply rail. If you need an op-amp whose output goes right up to the supply rails, use a CMOS one. The 741 output has an internal current limit of about 13mA; the IC will self-destruct if the output is directly attached to one of the two supply rails.

Discrete semiconductors

The pin-out of the various discrete semiconductors is shown in figure B.11. It is always assumed that you are looking at the package from the pin side.

The BC107 and BC477 are complementary signal transistors. Their current gains can range from 100 to 400, usually being about 200. The collector current must be less than 100mA and they will fry if the heat dissipation goes above 360mW.

The BC441 and BC461 are complementary power transistors. Current gains are typically 100, but quite variable from specimen to specimen. They can take collector currents of up to 1A and can cope with up to 4W of heat dissipation. If you clip on a heat sink they can cope with much more.

All four transistors can cope with their collectors being held 30V away from their emitters. If the base-emitter junction is reverse-biased by more than about 5V there will be zener breakdown.

The BF244 is a signal n-channel JUGFET. Its parameters vary quite a lot. V_P is about 1.5V and I_{DSS} is about 3mA.

The TRI400 is designed to be used with mains voltages. It can take currents of up to 350mA and is capable of running a 60W mains light bulb quite happily.

You will need four types of diode. A 1N4148 for general purpose work (maximum current 75mA), a 1N4001 for supply rectifiers (maximum current 1A), an 0A90 for AM demodulation and a 5.1V zener (maximum heat dissipation 500mW). All should have a bar on their package to show the cathode end of the diode.

Seven segment LED display

You will need a common anode seven segment LED as shown in figure B.12. It will not be in DIP format, but it will sit facing the right way on your

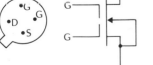

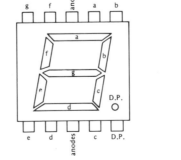

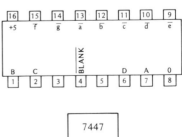

Figure B.11

Figure B.12

breadboard because of that. The two anode pins are connected inside; you can use either or both. It is dangerous to have a limiting resistor of less than 220 Ω.

Loudspeaker

For many experiments a 500 mW 8 ohm speaker will do; the larger it is the better it will sound. Some experiments require the use of a 12 : 1 matching transformer which can cope with audio signals. This will be a bit bulky for breadboard, so it may pay to mount it separately. The best arrangement is to have it in an enclosure with the loudspeaker.

Voltmeter

A voltmeter is invaluable for fault finding, as well as being necessary for measuring the properties of a device or system. A centre-zero voltmeter which can read from +15 V to −15 V is most useful for analogue circuits, whereas a range of 0 V to +5 V is sufficient for digital circuits. Aim to have an input impedance of at least 100 kΩ.

The µP system

The chapter on microprocessors has been written around a particular CPU (the Z80) embedded in a particular support system. A microprocessor system which is virtually the same as the µP system described in Chapter Eleven is available from Educational Electronics, 30 Lake Street, Leighton Buzzard, Bedfordshire.

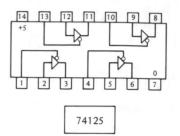

Figure B.13

Appendix C
Breadboard

The purpose of this appendix is twofold. The first part extols the virtues of solderless breadboard and tells you how best to use it. The second part is a collection of fault finding tips.

Solderless breadboard

A solderless breadboard is a flat plate of plastic punctured by holes on a 0.1 inch grid. The pins of electronic components, such as ICs, transistors and resistors, can be pushed through the holes to make contact with metal strips underneath. Connections between strips can be made by inserting short lengths of single strand wire in the appropriate holes.

Figure C.1 shows how the strips run underneath the surface of the breadboard. The two long strips at the top and the bottom are designed to be used as power supply rails. Figure C.2 illustrates the layout for the third experiment in Chapter One; note how the resistors have been mounted vertically so that they take up less space.

Solderless breadboard is a very useful tool for assembling and testing circuits. A vast range of components are available which can plug into it without any trouble. Connections can be changed quickly, and when you have finished, all of the components can be removed and used again.

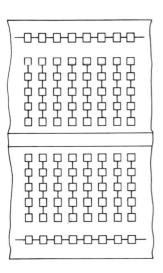

Figure C.1

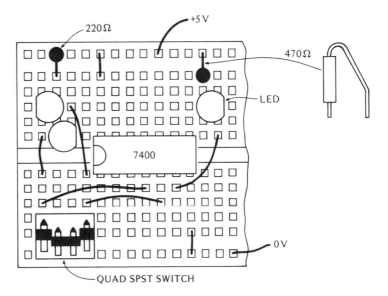

Figure C.2

In order to use breadboard successfully, you must obey the following commandments.

1) **Make your layout as much like your circuit diagram as possible**. So always have the top long strip as $+5\,V$ and the bottom long strip as $0\,V$ for logic systems. Alternatively, use the top strip for $+15\,V$ and the bottom one for $-15\,V$ (or $0\,V$) when building analogue circuits. Furthermore, have the information flow from left to right, i.e. inputs at the left going towards outputs at the right.

2) **Use colour coded wires**. We suggest that all wires connected to the top supply rail are red, ground wires are green and negative supply rail wires are black. Then use blue for connections within a system and yellow for connections from one system to another. You might even use blue for one system, grey for another, orange for a third and so on. A colour coded board makes fault finding much easier. Keep wires short and neat.

3) **Keep a record of what you have built**. When using ICs it is not always possible to lay out the circuit exactly as it is shown in the circuit diagram. So draw a diagram of the positions of all the gates on your breadboard, number them, and mark the numbers on your circuit diagram. This makes it much easier to sort out which gate is doing what on the breadboard.

4) **Do not have the supply connected when you are making or breaking connections**.

5) **Assemble the inputs first and the outputs last**.

6) **Test each functional system as you assemble it**. When you have assembled a system (such as an oscillator or a binary counter), check that it works. This may mean that you have to insert a potentiometer or voltmeter temporarily, but it is most important that you establish that a system is working perfectly before you connect it to another system.

7) **Check the supply rails carefully before you connect them up to your power source**. It is worthwhile applying the flame test to a circuit when you have just switched it on. Run your fingers over the circuit; if any components appear to be getting unreasonably hot switch off immediately! If you act quickly, i.e. within a few seconds, you may be able to save the hot components from irreversible melt-down.

Here is a list of the most common mistakes, in order of popularity:

1) getting the supply lines round the wrong way;
2) forgetting to connect the supply rails to the ICs;
3) inserting a wire into the strip one place to the left or right of the correct strip;
4) putting both pins of a resistor into the same strip;
5) forgetting the external link on a 7493;
6) forgetting to hold at least one of the RESET pins of a 7493 at logical 0;
7) omitting to hold the INHIBIT pin of a 4051 at logical 0;
8) leaving the EN pin of a 7496 to float up to logical 1;
9) leaving inputs of a CMOS gate to float;
10) using the wrong pin-out diagram (e.g. getting a 4001 confused with a 7402);

11) inserting an LED round the wrong way;

12) getting confused over the legs of a transistor;

13) letting the can of a transistor touch another bit of metal;

14) using the $-15\,V$ supply rail as logical 0 for a CMOS gate run off $+15\,V$ and $0\,V$;

15) shorting the output of a 741 to its $+15\,V$ pin with a screwdriver;

16) inserting a polarised capacitor round the wrong way;

17) using $220\,k\Omega$ instead of $220\,\Omega$;

18) connecting a component to a strip which is attached to an unused IC pin.

Fault finding

You assemble a system, plug it into the power supply and it does not work properly. At best, your circuit will lie there inert and lifeless. At worst, an integrated circuit or capacitor will spectacularly explode. What do you do now?

Fault finding is an art that you can only acquire by doing lots of it. All we can do is to give you some guidance about how to approach fault finding, so that you find your fault with the minimum of fuss. The first thing that you do is take a deep breath and calm down. A circuit which has taken ten minutes to assemble and then refuses to work can get you a bit annoyed. Then reach for your circuit diagram, your collection of pinout diagrams and a voltmeter or CRO.

Start off by checking the obvious things. Is each IC connected to its power supply rails? Have you inserted every wire? Are the wires in the right place? Are all the components such as diodes, LEDs and transistors the right way round? In other words, compare your circuit on breadboard with its circuit diagram. Do not do anything silly like connecting the power supply and swapping wires at random. Just be patient and spend a little time ensuring that the circuit diagram has been faithfully transferred onto the breadboard.

Now connect the supply again and use your voltmeter or CRO to find out what, if anything, is going on in your circuit. A high impedance probe is useful because it does not usually load up the circuit very much. A device which measures voltage can be used to probe points in the circuit without the need to alter connections within it. You will however have to use a CRO if the circuit processes or generates AC signals.

Start off at the input end of the circuit and work your way through to the output. Every time that you move your probe to a new place, check that the new reading is what it ought to be. It goes without saying that it helps if you have a good understanding of how your circuit is supposed to work. If the circuit is not one of a chain of interconnected systems, you may have to temporarily insert a potentiometer or a set of switches to set up some input signals to test the circuit.

At some stage you may suspect that one of your components is broken or faulty. The cause of most faults is missing or wrongly connected wires; very few faults are caused by bad components. Logic ICs very rarely fail, provided they have not been abused by wrong connection to the power supply. However, if you suspect that a component is behaving badly, do not be hasty. First of all, check that each pin of the component has the same voltage as the strip that it is plugged into; sometimes the clips on the strip get prised too far apart and fail to make contact with the pin. Then, without altering the wiring in any way, extract the offending component. If it is an IC or a diode, replace it straight away with another one. Should the circuit now function correctly, throw the faulty component away. If the fault persists, then you must look elsewhere for it; the chances of your picking up two faulty components in succession are very small.

Of course, it is always possible that you are damaging the component as soon as you connect the power supply. If you appear to insert two faulty components in succession, that may be exactly what you are doing. Transistors are especially fragile, so you must be absolutely sure about which leg is which. Check that each of the three strips carries the voltage that it ought to before you replace a transistor; you might have to check the integrity of the transistor by inserting it in a simple test circuit like that of figure 5.11.

Finally, binary counters have their own special idiosyncracies. Should a counter appear unwilling to count the pulses fed into it, it may be receiving pulses through the supply lines as well. This can be cured by putting a 100 nF capacitor across the supply lines close to the oscillator which is generating the pulses.

Solutions to problems

1 LOGIC SYSTEMS

Problems on page 6

1 i) Q is 1 when either or both switches are closed; 5 mA
ii) Q is 1 only when both switches are closed; 1 mA

2

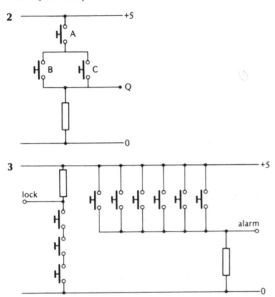

3

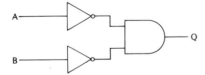

Problems on page 9

1 Look at figure 1.18 upside down!

2 Look at figure 1.19 upside down!

3 Q = (NOT A) AND (NOT B)

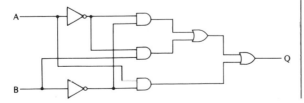

4 Q = [(NOT A) AND (NOT B)] OR [(NOT A) AND B]
    OR [A AND (NOT B)]

5

A	B	Q
0	0	1
0	1	1
1	0	1
1	1	0

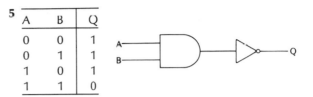

Problems on page 16

1

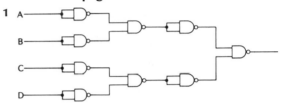

2 Figure 1.31 (iii)

3 i)

A	B	Q
0	0	0
0	1	0
1	0	1
1	1	1

ii)

C	D	R
0	0	1
0	1	0
1	0	0
1	1	0

iii)

E	F	S
0	0	1
0	1	0
1	0	0
1	1	1

4 a) 500 Ω   **b)** approximately 6 mA

Problems on page 21

1 a) $\bar{A} + \bar{B} \cdot C$ **b)** $\bar{A} \cdot \bar{B} + \bar{A} \cdot \bar{C}$
c) $A \cdot B \cdot C + A \cdot B \cdot \bar{D}$ **d)** $A + B + \bar{C} \cdot D$

Problems on page 24

1a)

B	A	Q
0	0	1
0	1	0
1	0	0
1	1	1

b)

B	A	Q
0	0	1
0	1	1
1	0	0
1	1	1

c)

B	A	Q
0	0	1
0	1	0
1	0	1
1	1	0

d)

B	A	Q
0	0	0
0	1	1
1	0	1
1	1	0

2 i) $Q = \bar{A} \cdot \bar{B} + A \cdot \bar{B} + \bar{A} \cdot B$ **ii)** $Q = A \cdot B$
  iii) $Q = A \cdot B + A \cdot \bar{B} + \bar{A} \cdot B$

3 i) $Q = A \cdot B$ **ii)** $R = D \cdot \bar{C}$ **iii)** $S = \bar{E} + \bar{F} + \bar{G}$
  iv) $T = \bar{J} + I \cdot \bar{H}$

Problems on page 34

1

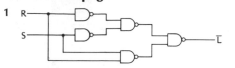

2

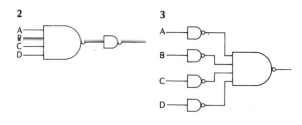

3

4

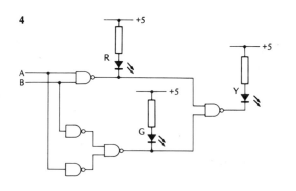

5

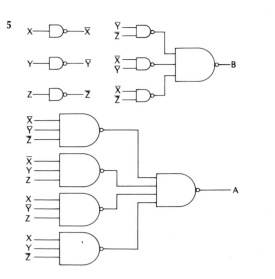

6

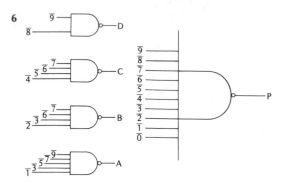

7

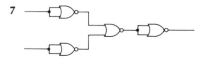

8 A B

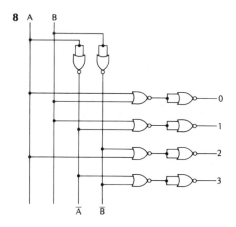

9

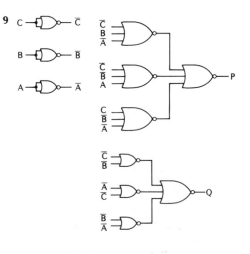

Problems on page 49

1
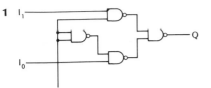

2
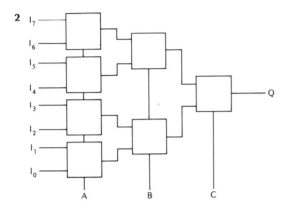

3
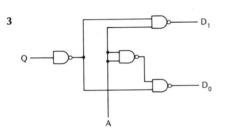

4 Look at the solution to question 2 in a mirror!

5
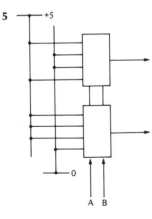

Problems on page 58

1 a) 528 ms   **b)** 0.33 ms   **c)** 17 ms

2 a) +7.6 V   **b)** −1.35 V   **c)** +3.6 V

3 a) 282 µF   **b)** 116 µF   **c)** 4314 µF

5 Q goes to 0 approximately 7 s after the switch is pressed. It returns to 1 approximately 4 s after the switch is released.

6
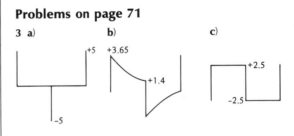

Problems on page 65

1 10.4 kHz

3 a)   **b)**   **c)**
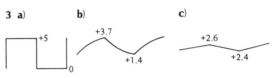

Problems on page 71

3 a)   **b)**   **c)**
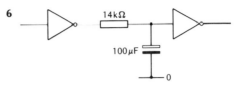

Problems on page 76

1
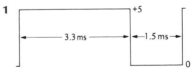

2 An amalgam of the graphs of figures 2.35 and 2.36 with +5 V swapped with 0 V and viewed upside down!

Problems on page 80

1 a) 3 V   **b)** 30 mA maximum   **c)** 45 mW
d) 7.5 mA maximum   **e)** 1.6 kΩ

2 a) V^2/n^2R **b)** V/n^2R **c)** n^2R

3 25 : 1

Problems on page 91

1 a) 0.68 mA   **b)** 6.8 V

3 i) +1.6 V   **ii)** −4.1 V   **iii)** +5.4 V   **iv)** −1.4 V

4 a) 185 mA   **b)** 66 ohms   **c)** 15 mW

5 i)   **ii)**   **iii)**

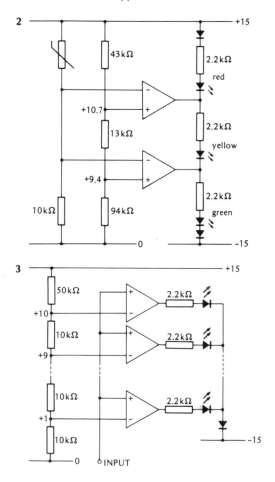

6 i) 4.7 V to 0 V   **ii)** +7.7 V to −7.7 V
iii) +5 V to +4.1 V

Problems on page 101

1 a) The LED is initially off and comes on suddenly
b) +5.6 V   **c)** approx 10 kΩ

Problems on page 107

1 i) +5.8 V, −3.4 V   **ii)** +2.3 V, −2.3 V
iii) +6.1 V, −6.1 V   **iv)** +12 V, −0.2 V

3 Period of oscillation 29 ms, trigger thresholds +4.2 V
and −4.2 V.

4

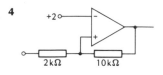

Problems on page 115

1 i) +3 V   **ii)** −0.5 V   **iii)** −11.4 V   **iv)** +5.2 V

2 a)   **b)**

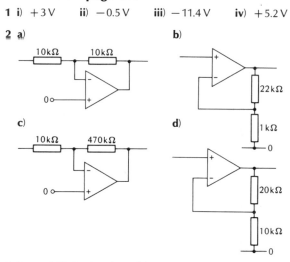

c)   **d)**

4 Figure 3.74 feeding into this:

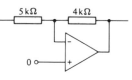

5 All resistors have the same value, except one

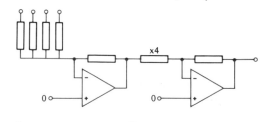

6 i) +2.1 V, −4.4 V   **ii)** −1.4 V, +8.5 V

Problems on page 123

1 i) −1.8 V/ms   **ii)** +0.42 V/ms   **iii)** −0.065 V/s
   iv) +0.32 V/ms

2

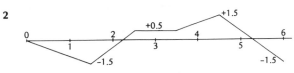

3 The output drops by 165 mV for each rising edge which enters from the left.

Problem on page 136

2

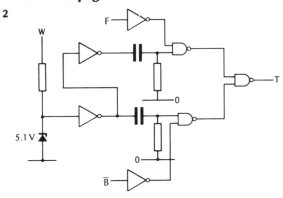

Problems on page 142

1   **i) a)** 9.5 V   **b)** 9.5 mA   **c)** 0.048 mA   **d)** 52 mW
   ii) a) 1.8 V   **b)** 8.2 mA   **c)** 0.082 mA   **d)** 26 mW
   iii) a) 4.4 V   **b)** 44 mA   **c)** 0.88 mA   **d)** 246 mW
   iv) a) 19 V   **b)** 40 mA   **c)** 0.27 mA   **d)** 200 mW

2 a) 244 mA   **b)** 0.0244 mA   **c)** 1.32 W, 24 mW

Problems on page 144

1   **i) a)** −8.3 V   **b)** 38 mA   **c)** 0.25 mA   **d)** 255 mW
   ii) a) +3.7 V   **b)** 11.3 mA   **c)** 0.23 mA   **d)** 211 mW
   iii) a) +4.7 V   **b)** 103 mA   **c)** 0.52 mA   **d)** 484 mW
   iv) a) −6.3 V   **b)** 13.4 mA   **c)** 0.13 mA   **d)** 117 mW

Problem on page 151

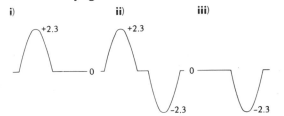

Problems on page 158

1 i) 3.2 V   **ii)** 2.8 V   **iii)** 4.6 V   **iv)** 4.0 V

2   **i) a)** 0.043 mA   **b)** 3.6 mA   **c)** 85
   ii) a) 0.024 mA   **b)** 6.4 mA   **c)** 266

3 logical 0 less than 0.7 V, logical 1 above 2.9 V, fanout of 21

4 a) i) NOT   **ii)** NAND
   b) logical 1 above 4.3 V, logical 0 below 2.0 V

Problems on page 169

1 a) 2 kΩ   **b)** 3.75 mA   **c)** 0.375 V   **d)** 2.9 kΩ and 37 l
   e) 47 μF, approximately

2 F is exactly the same as A, and E is exactly the opposite of A

Problems on page 186

1 0.94 V, 2.05 V, 2.79 V

2

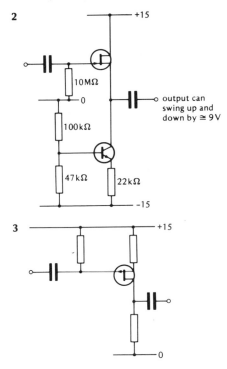

output can swing up and down by ≅ 9 V

3

5 $I_{DSS} = 1.8$ mA, $V_P = 5$ V, $V_S = 1.1$ V, $V_D = 19$ V

Problems on page 191

1

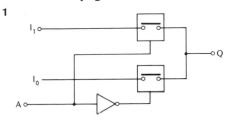

2 a) 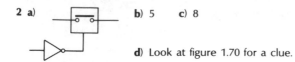   **b)** 5   **c)** 8

d) Look at figure 1.70 for a clue.

3 The frequency, in kHz, is equal to the decimal value of the binary number CBA; i.e. $f = ((A + 2B + 4C)/10) \times 1/(2 \times 0.047)$.

4 a) i) Ramps normally   **ii)** V_{OUT} clamped to 0 V
b) Use a Schmitt trigger to connect V_{OUT} to $V_{CONTROL}$; keep V_{IN} negative.

Problems on page 195

1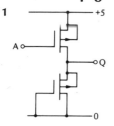

3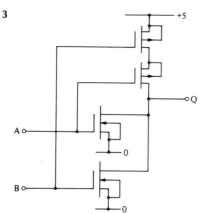

4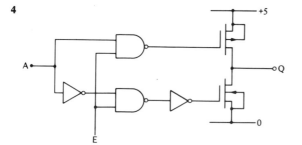

Problems on page 202

1 a) 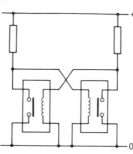   **b)**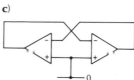

c)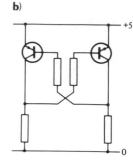

2 Figures 7.4 and 7.5 upside down

3

$\bar{S}$	$\bar{R}$	Q	$\bar{Q}$
1	1	0	1
0	1	1	0
1	1	1	0
1	0	0	1
1	1	0	1

4 Place a switch *either* between the base of each transistor and ground, *or* between the collector of each transistor and the top supply rail.

5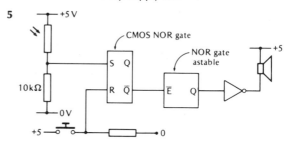

Problems on page 208

1 a) $+4$ V   **b) i)** 2.6 ms   **ii)** 1.6 ms   **iii)** 1.2 ms

2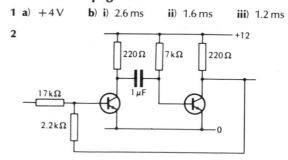

Problems on page 213

1

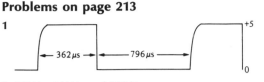

2 862 Hz, 911 Hz and 936 Hz

3 a) 2.1 V   **b)** 2.26 kHz at +30 V, 317 Hz at +2.1 V

5 Look at figure 7.27 upside down.

Problems on page 233

2

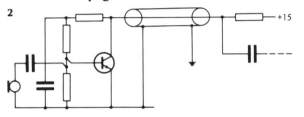

3 Elimination of crosstalk between high current output stage and sensitive input stage.

Problems on page 241

1 i) Bass cut, break at 3.4 kHz, high frequency gain 10.
ii) Treble cut, break at 328 Hz, low frequency gain 0.1.
iii) Bass boost, maximum gain 1000, break at 338 Hz, high frequency gain 1.
iv) Treble boost, maximum gain 1, low frequency gain 0.01, break frequency 318 Hz.

2 i)

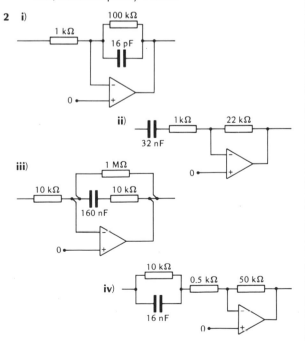

3 Break frequencies at 339 Hz and 3.39 kHz.

4

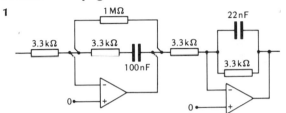

Problems on page 244

1

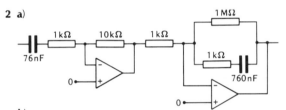

2 a)

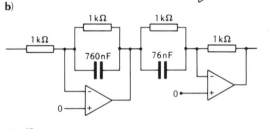

b)

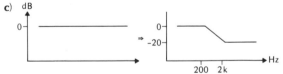

c)

Problems on page 253

1

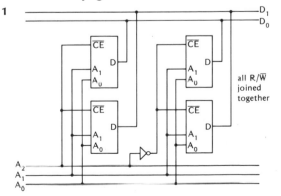

all R/$\overline{W}$ joined together

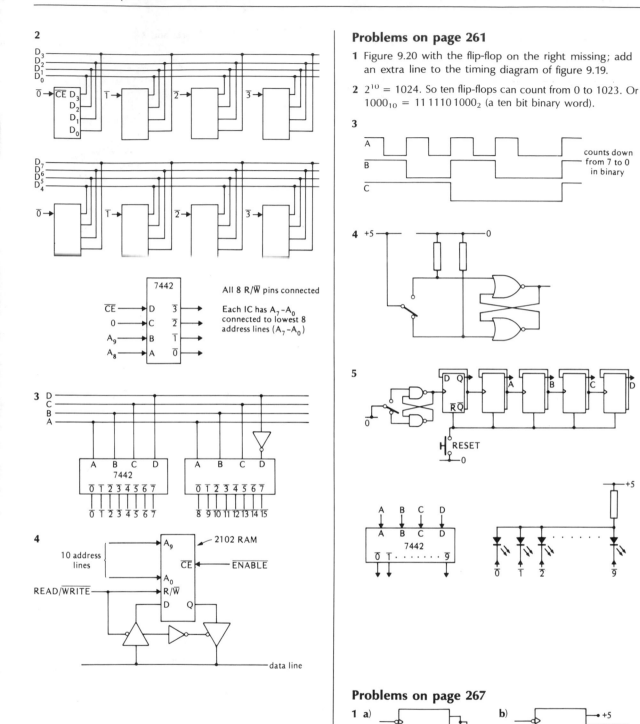

2

D_3
D_2
D_1
D_0

$\bar{0} \rightarrow$ CE D_3
D_2
D_1
D_0

$\bar{1} \rightarrow$ $\bar{2} \rightarrow$ $\bar{3} \rightarrow$

D_7
D_6
D_5
D_4

$\bar{0} \rightarrow$ $\bar{1} \rightarrow$ $\bar{2} \rightarrow$ $\bar{3} \rightarrow$

7442

$\overline{CE} \rightarrow$ D $\bar{3} \rightarrow$
$0 \rightarrow$ C $\bar{2} \rightarrow$
$A_9 \rightarrow$ B $\bar{1} \rightarrow$
$A_8 \rightarrow$ A $\bar{0} \rightarrow$

All 8 R/$\overline{W}$ pins connected

Each IC has A_7–A_0 connected to lowest 8 address lines (A_7–A_0)

3   D
C
B
A

A   B   C   D
7442
$\bar{0}\ \bar{1}\ \bar{2}\ \bar{3}\ \bar{4}\ \bar{5}\ \bar{6}\ \bar{7}$

$\bar{0}\ \bar{1}\ \bar{2}\ \bar{3}\ \bar{4}\ \bar{5}\ \bar{6}\ \bar{7}$

A   B   C   D
$\bar{0}\ \bar{1}\ \bar{2}\ \bar{3}\ \bar{4}\ \bar{5}\ \bar{6}\ \bar{7}$

$\bar{8}\ \bar{9}\ \overline{10}\ \overline{11}\ \overline{12}\ \overline{13}\ \overline{14}\ \overline{15}$

4

10 address lines

A_9 — 2102 RAM
$\overline{CE} \leftarrow$ — $\overline{ENABLE}$
A_0
READ/WRITE — R/$\overline{W}$
D   Q

data line

Problems on page 257

1 Still a D flip-flop, but falling edge triggered, with active high set and reset pins.

2 Put the NOT gate which is in the CLOCK line in front of the slave flip-flop instead of the master flip-flop.

3 Look at the $\overline{SR}$ flip-flop portion of figure 9.15.

Problems on page 261

1 Figure 9.20 with the flip-flop on the right missing; add an extra line to the timing diagram of figure 9.19.

2 $2^{10} = 1024$. So ten flip-flops can count from 0 to 1023. Or $1000_{10} = 11\,1110\,1000_2$ (a ten bit binary word).

3

A

B

C

counts down from 7 to 0 in binary

4 $+5$ — 0

5

D Q   A   B   C   D
$\overline{R}\,\overline{Q}$
0

RESET
0

A   B   C   D
A   B   C   D
7442
$\bar{0}\ \bar{1}\ \ldots\ldots\ \bar{9}$

$+5$

$\bar{0}\ \bar{1}\ \bar{2}$ $\bar{9}$

Problems on page 267

1 a)   A B C

b)   A B C D $+5$

c)   A B C D

d)   A B C D

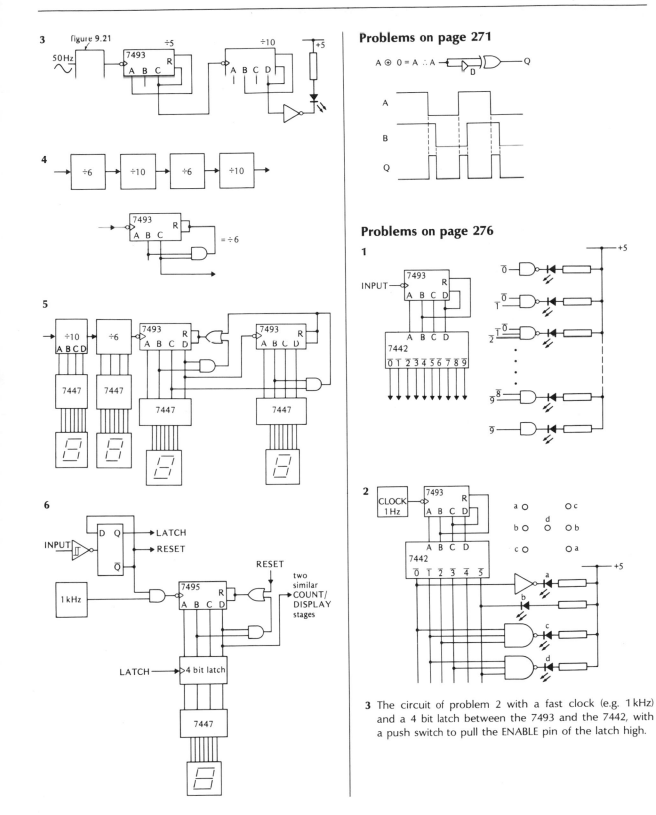

3

figure 9.21

50 Hz

7493 ÷5

A B C   R

÷10

A B C D

+5

4

÷6 → ÷10 → ÷6 → ÷10 →

7493

A B C   R

= ÷6

5

÷10
A B C D

÷6

7493
A B C   R D

7493
A B C   R D

7447

7447

7447

7447

6

INPUT

D Q → LATCH

→ RESET

Q̄

1 kHz

RESET

7495
A B C D   R

two
similar
COUNT/
DISPLAY
stages

LATCH → 4 bit latch

7447

Problems on page 271

A ⊕ 0 = A ∴ A

D

Q

A

B

Q

Problems on page 276

1

+5

INPUT

7493
A B C D   R

0̄
1̄ 0̄
1̄ 0̄
2
⋮
8̄ 0̄
9
9̄

7442
0̄ 1̄ 2̄ 3̄ 4̄ 5̄ 6̄ 7̄ 8̄ 9̄

A B C D

2

CLOCK
1 Hz

7493
A B C D   R

a ○   ○ c
        d
b ○   ○   ○ b
c ○   ○ a

7442
0̄ 1̄ 2̄ 3̄ 4̄ 5̄

A B C D

+5

a
b
c
d

3 The circuit of problem 2 with a fast clock (e.g. 1 kHz)
and a 4 bit latch between the 7493 and the 7442, with
a push switch to pull the ENABLE pin of the latch high.

4 A 0.2 Hz clock feeding a 7493 which resets on 12 (i.e. RESET = D . C) feeding its output into the one-of-sixteen selectors of figure 9.47. Then these four NAND gates:

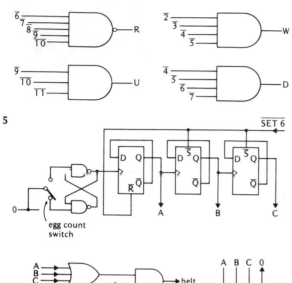

5

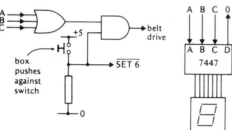

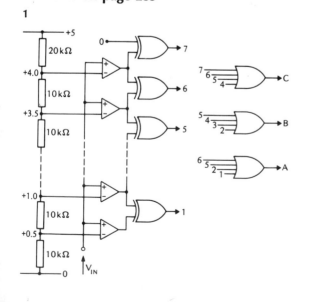

Problems on page 285

1

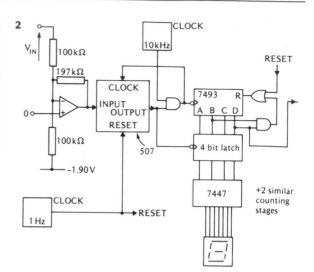

Problems on page 291

1 E goes high for one second every five seconds.

2

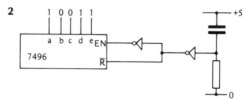

3

RIGHT/$\overline{\text{LEFT}}$

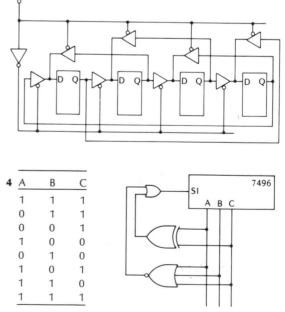

4	A	B	C
	1	1	1
	0	1	1
	0	0	1
	1	0	0
	0	1	0
	1	0	1
	1	1	0
	1	1	1

Problems on page 294

1 a) 0110, 0101   **b)** 0011, 0110   **c)** 0101, 0011

2 A = 0000, B = 1111, C = 0111

3 a) J = 1111, K = 0000, L = 0101
   b) J = 1111, K = 0000, L = 1111

4 Figure 10.12 without the quad NOR gate and each EOR gate replaced with:

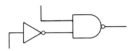

Problems on page 300

1

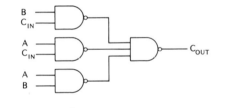

2

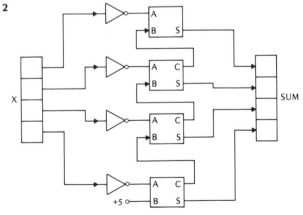

3 1000 0	4 1000 0
0100 1	1100 0
1010 0	1110 0
0101 0	1111 0

5 10111, 11000, 11001, 11010, 11011, 11100, 11101, 11110, 11111, 00000, 00001, 00010, 00011, 00100, 00101, 00110, 00111, 01000 and 01001

Problems on page 303

1

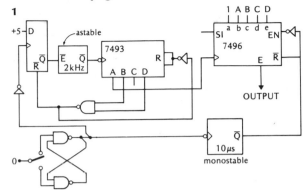

2

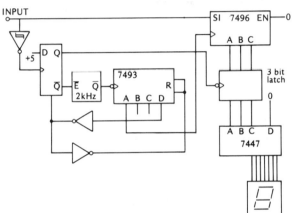

Problems on page 307

1 Counts up in binary from 0 to 7.

2

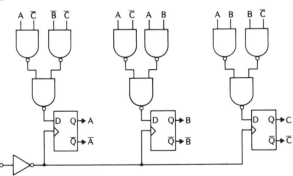

Problems on page 324

1 The top four LEDs appear to count up in binary, with a delay of 0.5 s between changes. The bottom four LEDs remain off all the time.

2 b) One lit LED chases from the lsb to the msb continually.

c) Modify the program of table 11.7 as follows; start off with 3E FE to set the initial value of A. Then have 07 followed by the listing of table 11.7.

3
0000	3E FF	0007	20 FC
0002	ED 79	0009	7B
0004	5F	000A	3D
0005	ED 78	000B	18 F5

Problems on page 331

1
0000	21 17 00
0003	16 0A
0005	7E
0006	ED 79
0008	0E 04 06 FF 10 FE 0D 20 F9
0011	23 15
0013	28 EB
0015	18 EE
0017	81 CF 92 86 CC A4 A0 8F 80 8C

2
0000	3E FF
0002	ED 79
0004	ED 78 E6 80 20 FA
000A	ED 78 E6 07
000E	47
000F	21 27 00
0012	04
0013	2B 10 FD
0016	7E ED 79
0019	EE 80 ED 79
001D	18 E5
001F	78 7C 7C 7E 7C 7E 7E 7F

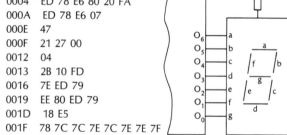

Problems on page 337

1
0000	3E 00 ED 79
0004	31 FF 03
0007	ED 56 FB
000A	76
000B	18 FD
0038	3C
0039	47
003A	E6 0F
003C	FE 0A
003E	20 04
0040	78
0041	C6 06 47
0044	FE A0 20 02
0048	06 00
004B	76 ED 79
004E	FB ED 4D

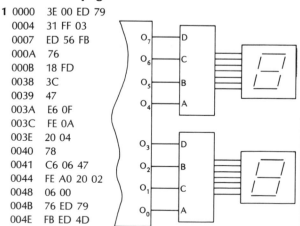

2
0000	31 FF 03
0003	1E 01
0005	ED 56 FB
0008	2F ED 79
000B	57 7B 47 10 FE

0010	7A 18 F5
0038	F5
0039	ED 78 E6 07 3C 5F
003F	F1 FB ED 4D

Problems on page 341

1 a) F5
ED 78 E6 80 20 FA
F1
C9

b) EE 80 ED 79
EE 80 ED 79
C9

c) F5
ED 78 E6 3C 5F
F1
C9

2 F5
3E 00 ED 79
ED 78 2F E6 03 28 F9
1E 00 3E FE
57 ED 79
ED 78 E6 02 28 0B
ED 78 E6 01 28 09
1C
7A 07
18 EC
7B C6 05 5F
3E 00 ED 79
ED 78 2F E6 03 20 F9
F1
C9

4 F5
3E 02 ED 79
3E 00 ED 79
1E FF
1C 3E 7F BB 28 10 3E 00
EE 01 ED 79
ED 78 E6 01 20 EE
F1
C9

5 06 08
7B E6 80
ED 79
00 00 00 00
7B 07 5F
10 F2
3E 80 ED 79
C9

3
0000	21 FF 02
0003	7E F6 70 ED 79
0008	2B 7E F6 B0 ED 79
000E	2B 7E F6 D0 ED 79
0014	2B 7E F6 E0 ED 79
001A	18 E4

Problems on page 357

1

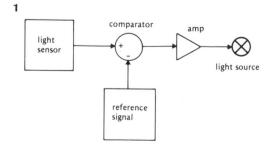

2 The on-off system will have a lot of overshoot, but the power transistor will run cool and the response time will be fast. The power transistor of the other system may get warmer than the heater! It will respond slower, with no overshoot.

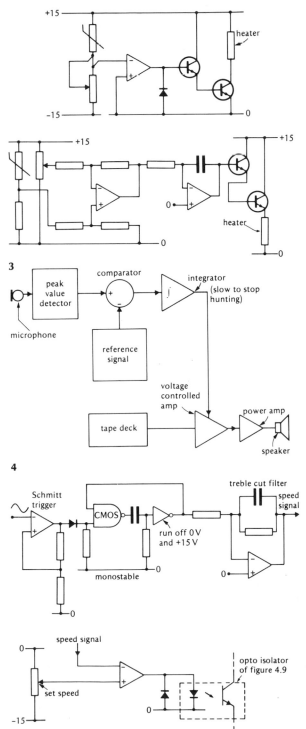

Problems on page 362

1 i)

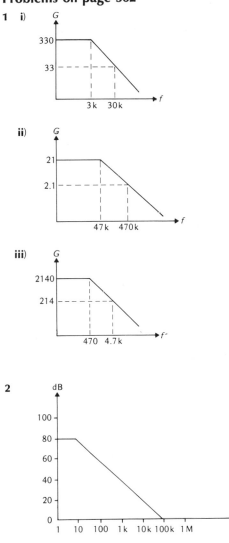

ii)

iii)

2

3 a) Referring to figure 12.20, a 10 kHz bandwidth implies a maximum closed-loop gain of 40 dB or 100.

b)

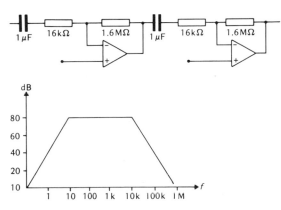

4 a)

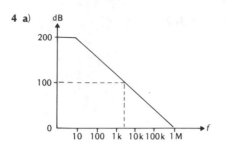

b) At the point where $AB = 1$, the phase change due to the op-amp is $180°$. So $G = A/1 - AB$ and therefore G is infinite.

c) 46^3 is roughly $100\,000$. A closed-loop gain of 46 gives a bandwidth of $22\,\text{kHz}$, whereas 316 (i.e. $\sqrt{100\,000}$) gives a bandwidth of $3\,\text{kHz}$. So you need three of these in series:

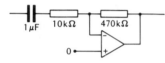

Problems on page 369

1 a) $16.3\,\text{V}$ **b) i)** $69\,\text{mV}$ **ii)** $690\,\text{mV}$ **iii)** $6.9\,\text{V}$

2 Maximum current is about $200\,\text{mA}$

3

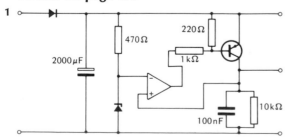

4 There is no voltage drop across R so it might as well not be there. Remove R and the system is exactly the same as figure 12.35. You might improve the ripple across the zener by putting a large capacitor in parallel with it.

5 a) $33\,\text{mA}$ **b)** $3300\,\mu\text{F}$ **c)** $36\,\Omega$, $40\,\text{mW}$

Problems on page 373

1

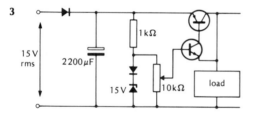

2 The signal transistor turns on when $V_{BE} = 0.7\,\text{V}$. This happens when the load current is $0.7/0.00068 = 1029\,\text{mA}$. The collector current of the signal transistor flows into the op-amp output, pulling it up; the op-amp refuses to sink more than $13\,\text{mA}$. As the op-amp output rises, the base current of the pass transistor decreases and so the load current falls, etc.

3

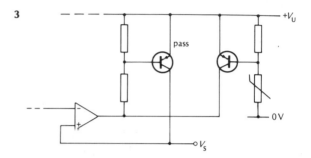

4

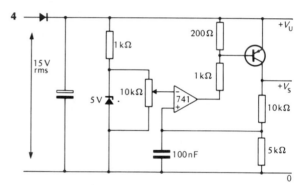

Problems on page 377

1 a) $16.3\,\text{V}$ **b)** 0 to $10\,\text{V}$ **c)** The op-amp driving the BC441 tries to keep the junction of the $10\,\text{k}\Omega$ resistors at $0\,\text{V}$. So $+V_S = -(-V_S)$. The other op-amp tries to keep the top end of the $10\,\text{k}\Omega$ chain at the same voltage as the wiper of the potentiometer.   **d)** $2200\,\mu\text{F}$.

2 a) $2.1\,\text{V}$ **b)** $2200\,\mu\text{F}$ **c)** $7.8\,\text{V}$
d) about $5\,\text{W}$ each
e) Makes system stable by acting as a treble cut filter.

Problem on page 380

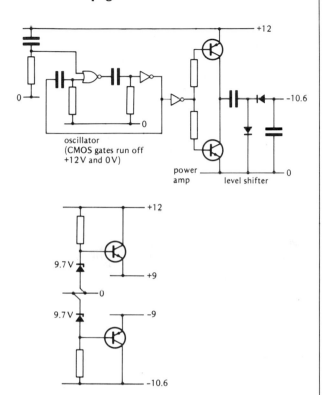

oscillator
(CMOS gates run off
+12V and 0V)

power
amp   level shifter

Problems on page 393

1

Input

Input of NOT gate

Output of NOT gate

Input of Schmitt trigger

Output of Schmitt trigger

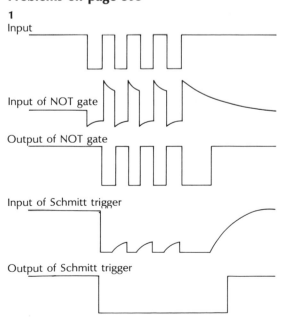

2

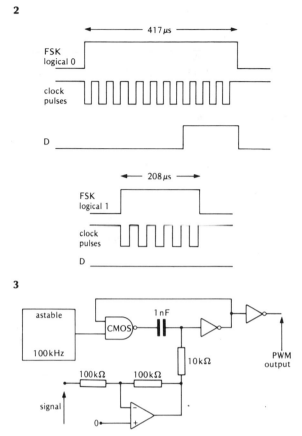

FSK
logical 0

clock
pulses

D

FSK
logical 1

clock
pulses

D

3

astable

CMOS

1 nF

100 kHz

PWM
output

10 kΩ

100 kΩ   100 kΩ

signal

0

4 Same as a PPM demodulator, i.e. a treble cut filter!

Problems on page 405

1   **i)** Pass filter 5 MHz
  ii) Pass filter 734 kHz
  iii) Stop filter 1.56 MHz
  iv) Stop filter 1.07 MHz

2 Use a 1.85 kHz carrier. Sidebands can spread out to 1.55 kHz on either side. Each bit needs at least 1 ms, so maximum byte rate is about 100 per second.

3 Use 3 kHz and 9 kHz as the two carriers, so that sidebands can stretch up to 3 kHz on either side. Each bit needs at least 0.5 ms so the maximum byte rate is about 200 per second.

4 a)

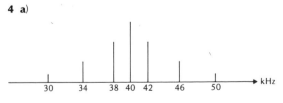

b) About 267 bytes per second. (Assume $6f_m = 8$ kHz.)
c) Use carriers at 38 kHz and 42 kHz. Each bit needs 0.75 ms. Maximum byte rate about 133 per second.

5 a) Bit rate is 800 per second (assuming carrier and four side bands, at least)

b) About 24 minutes   **c)** 0.3 mm

6 a) 13 MHz   **b)** 39 MHz

c) About 27 minutes if standard FSK is used.

d) 38 s   **e)** 16 s

Problems on page 417

1

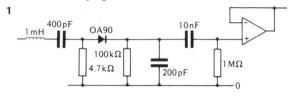

2 a) 14   **b)** 73   **c)** 12   (all approximate!)

3 Up to 15 kHz

4 200 kHz, about 100

5 About 100

6

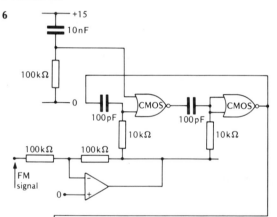

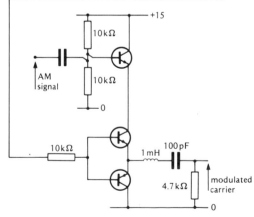

7

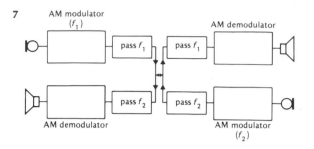

8 If the frequency is too large, the sidebands break into adjacent channels and give rise to crosstalk. The high frequencies will probably not get into the receiver at the other end.

If the amplitude is too large, the signal gets distorted. Here is an example:

✓OK   X over-modulated

Problems on page 419

1 2.3 MHz to 624 kHz

2

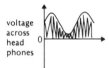

Problems on page 425

1

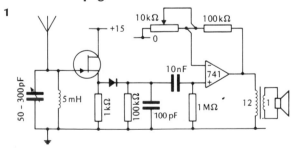

2 a) Modulated with a 7.8 kHz square wave, so bandwidth is 47 kHz.

b) 6.51 MHz   **c)** 6.51 MHz.

Index